BRIDGING DIVIDES

MONOGRAPHIAE BIOLOGICAE

VOLUME 83

Series Editor

H. J. Dumont

Aims and Scope

The Monographiae Biologicae provide a forum for top-level, rounded-off monographs dealing with the biogeography of continents or major parts of continents, and the ecology of well individualized ecosystems such as islands, island groups, mountains or mountain chains. Aquatic ecosystems may include marine environments such as coastal ecosystems (mangroves, coral reefs) but also pelagic, abyssal and benthic ecosystems, and freshwater environments such as major river basins, lakes, and groups of lakes. In-depth, state-of-the-art taxonomic treatments of major groups of animals (including protists), plants and fungi are also elegible for publication, as well as studies on the comparative ecology of major biomes. Volumes in the series may include single-author monographs, but also multi-author, edited volumes.

The titles published in this series are listed at the end of this volume.

Bridging Divides

Maritime Canals as Invasion Corridors

by

STEPHAN GOLLASCH
GoConsult, Hamburg, Germany

BELLA S. GALIL
National Institute of Oceanography,
Israel Oceanographic and Limnological Research, Haifa, Israel

and

ANDREW N. COHEN
San Francisco Estuary Institute, Oakland, CA , U.S.A.

Springer

A C.I.P. Catalogue record for this book is available from the Library of Congress.

ISBN-10 1-4020-5046-1 (HB)
ISBN-13 978-1-4020-5046-6 (HB)
ISBN-10 1-4020-5047-X (e-book)
ISBN-13 978-1-4020-5047-3 (e-book)

Published by Springer,
P.O. Box 17, 3300 AA Dordrecht, The Netherlands.

www.springer.com

Front cover: Miraflores Locks looking up toward Miraflores Lake, Pedro Miguel Locks and Centennial Bridge over the Culebra Cut. Courtesy of the Panama Canal Authority.

Printed on acid-free paper

dedicated to the scientists
whose studies on the role of canals in changing
the biota laid the foundation to our work.

Contents

Overall Introduction

Dan Minchin, Bella S. Galil, Matej David, Stephan Gollasch & Sergej Olenin

The Kiel Canal

The World's Busiest Man-made Waterway and Biological Invasions
Stephan Gollasch & Harald Rosenthal

The Panama Canal

Chapter I Cutting a Canal Through Central America
Andrew N. Cohen

Chapter II Shipping Patterns Associated with the Panama Canal: Effects on Biotic Exchange?
Gregory M. Ruiz, Julio Lorda, Ashley Arnwine & Kelly Lion

Chapter III Species Introductions and the Panama Canal
Andrew N. Cohen

The Suez Canal

The Marine Caravan – The Suez Canal and the Erythrean Invasion
Bella S. Galil

Epilogue

Canals, Invasion Corridors and Introductions
Chad Hewitt, Dan Minchin, Sergej Olenin & Stephan Gollasch

Contributors

Ashley Arnwine
Smithsonian Environmental Research Center, P.O.Box 28, Edgewater, Maryland 21037, USA

Andrew N. Cohen
San Francisco Estuary Institute, 7770 Pardee Lane, Oakland, CA 94621-1424, USA

Matej David
University of Ljubljana, Faculty of Maritime Studies and Transportation, Pot pomorscakov 4, 6320 Portoroz, Slovenia

Bella S. Galil
National Institute of Oceanography, Israel Oceanographic and Limnological Research, P.O.B. 8030, Haifa 31081, Israel

Stephan Gollasch
GoConsult, Bahrenfelder Str. 73 a, 22765 Hamburg, Germany

Chad Hewitt
National Centre for Marine and Coastal Conservation, Private Mail Bag 10, Rosebud, Victoria 3939, Australia

Kelly Lion
Smithsonian Environmental Research Center, P.O.Box 28, Edgewater, Maryland 21037, USA

Julio Lorda
Department of Ecology, Evolution, and Marine Biology, University of California, Santa Barbara, California 93106, USA

Dan Minchin
Marine Organism Investigations, 3 Marina Village, Ballina, Killaloe, Co Clare, Ireland

Sergej Olenin
Coastal Research and Planning Institute, Klaipeda University, H. Manto 84, Klaipeda, Lithuania

Harald Rosenthal
Schifferstr. 48, 21629 Neu Wulmstorf, Germany

Gregory M. Ruiz
Smithsonian Environmental Research Center, P.O.Box 28, Edgewater, Maryland 21037, USA

Preface

Navigable canals are as old as civilization – the first was constructed in the 6th century BCE and joined the Nile with the northern Red Sea. Another ancient canal, the Grand Canal in China, constructed in the 4th century BCE, connected Peking to Hangzhou, a distance of almost 1000 km. The technological innovations of the 18th century led to an expansion of the network of navigable inland waterways, followed in the 19th century and the early part of the 20th century by the excavation of two interoceanic canals: the Suez Canal opened a direct route from the Mediterranean Sea to the Indo-Pacific Ocean, and the Panama Canal afforded passage between the Atlantic and the Eastern Pacific.

Maritime canals dissolve natural barriers to the dispersal of marine organisms, thus providing them with many opportunities for natural dispersal, as well as for shipping-mediated transport. The introduction of alien species proved to be one of the most profound and damaging anthropogenic deeds – involving both ecological and economic costs. However, until recently marine bioinvasions were perceived as isolated mishaps. This book is the first to compare the impacts of the three principal maritime canals – Kiel, Panama, Suez – as invasion corridor for alien biota. The three differ in their geographic locations, hydrological regimes, and in their permeability to alien biota.

Globalization and climate change are projected to increase marine bioinvasions and reduce environmental resistance to invasion of thermophilic biota. Inter-oceanic canals offer a unique opportunity to study these processes in "*statu nascendi*". With ample evidence that some maritime canals serve as major invasion corridor, environmentally-considerate engineering may construct barriers to preclude future invasions. It is hoped that this book will stimulate further investigations in this field.

Neu Wulmstorf, Germany, March 2006
Harald Rosenthal

Haifa, Israel, March 2006
Bella S. Galil

Overall Introduction

DAN MINCHIN[1], BELLA S. GALIL[2], MATEJ DAVID[3], STEPHAN GOLLASCH[4] & SERGEJ OLENIN[5]

[1] *Marine Organism Investigations, 3 Marina Village, Ballina, Killaloe, Co Clare, Ireland*
[2] *National Institute of Oceanography, Israel Oceanographic and Limnological Research, P.O.B. 8030, Haifa 31081, Israel*
[3] *University of Ljubljana, Faculty of Maritime Studies and Transportation, Pot pomorscakov 4, 6320 Portoroz, Slovenia*
[4] *GoConsult, Bahrenfelder Str. 73 a, 22765 Hamburg, Germany*
[5] *Coastal Research and Planning Institute, Klaipeda University, H. Manto 84, Klaipeda, Lithuania*

Invasive alien species are considered as one of the key causes of biodiversity changes worldwide. The impacts of aquatic invasive alien species are immense, insidious, and usually irreversible. Some invaders are re-forming the structures, dynamics or functions of aquatic communities, or are imposing significant economic costs. The global rate of new aquatic invasions increased in recent years, driving efforts to evaluate their vectors and pathways (Fig. 1).

Shipping has been implicated in the dispersal of numerous aquatic organisms, from protists and macrophytes to fish. Yet, it is seldom possible to ascertain the precise means of transmission, as one species may be transported by a variety of vectors. The transport on the hulls of ships of boring, fouling, crevicolous or adherent species is certainly the most ancient vector of aquatic species introduction. Fouling generally concerns small-sized sedentary, burrow-dwelling or clinging species, though large species whose life history includes an appropriate life stage may be disseminated as well. Ballast (formerly solid, but for the past 130 years aqueous) is usually taken into dedicated ballast tanks or into empty cargo holds when offloading cargo, and discharged when loading cargo or bunkering (re-fuelling). Ballast water therefore consists mostly of port or near port waters. Water and sediment carried in ballast tanks, even after voyages of several weeks' duration, have been found to contain many viable organisms. Since the volume of ballast water may be as much as a third of the vessel's deadweight tonnage, it engenders considerable anxiety as a vector of introduction.

Fig. 1. Possible vectors (transfer mechanisms) of aquatic species. 1. shipping, 2. canals, 3. small craft, 4. intentional stocking, 5. release from aquaria, 6. release of organisms intended for human consumption, 7. release of bait species, 8. intentional and unintentional aquaculture introductions, 9. discharges of wastes following fish processing, 10. transport of fishing gear. Drawing Vitalija Gasiunaite, Vilnius, Lithuania.

Table 1. Comparative aspects of the world's three most important canals for ocean-going vessels. Shipping patterns in 2003 (Sources: Kiel Canal http://www.kiel-canal.org/english.htm, Panama Canal http://www.pancanal.com/eng/maritime/reports/table01.pdf, and Suez Canal Institute of Shipping Economics and Logistics (ISL) Bremen, ISL Shipping Statistics and Market Review (SSMR), Volume 48 (2004).

Canal	Opening	Length [km]	Canal features	Alien species movements	Number of ships in transit	Cargo in transit [mt]
Kiel Canal	21 Jun. 1895	98.6	Locks, marine-brackish	rare	39.797	72.296.794
Panama Canal	15 Aug. 1914	57.0	Locks, marine-freshwater-marine	medium	13.154	191.301.069
Suez Canal	17 Nov. 1869	162.3	No locks, marine-saline-marine	extensive	15.667	457.965.000

Canals serve as the world's greatest short cuts, nexus of major trade routes and the densest shipping lanes. The world's three principal navigable canals provide significant savings for sea borne trade: the Panama Canal eliminates travel through the Magellan Straits (saving 8,100 nautical miles (nm) on the route from Los Angeles to Philadelphia), the Suez Canal avoids the passage around Africa (a short-cut of up to 8,500 nm), and the Kiel Canal shortens the voyage between the Baltic and North Seas by up to 450 nm (Fig. 2, Tab. 1 & 2).

Table 2. Maritime route shortcuts. Ships´ speed put at 14 knots, assuming an average passage time through the Kiel Canal as 9 hours (www.ak190x.de/Bauwerke/Bau/Nord-Ostsee-Kanal.htm), average passage time through the Panama Canal as 24 hours (www.pancanal.com/eng/maritime/routes.html) and average passage time through Suez Canal as 14 hours (www.atlas.com.eg/scg.html).

Canal	Route	via	Distance [nm]	Distance via canal [nm]	Savings in distance [nm]	Savings in time
Kiel	Rotterdam - Klaipeda	Denmark	936	720	216	6,4 hours
Kiel	Hamburg - Rostock	Denmark	629	174	455	23,5 hours
Panama	Philadelphia - Tokyo	Cape Horn	16.298	9.684	6.614	18,7 days
Panama	Los Angeles - Philadelphia	Cape Horn	12.995	4.897	8.098	23,1 days
Suez	Mumbai - Koper	Cape of Good Hope	11.316	4.336	6.980	20,2 days
Suez	Mumbai - Haifa	Cape of Good Hope	11.672	3.215	8.457	24,6 days

About 6% and 3.4% of the global sea borne cargo passes through the Suez Canal and Panama Canal. Aquatic organisms progress through canals both as a result of "natural" dispersal, by autochthonous active or passive larval or adult movements, and are also transported by shipping. But in addition to serving as invasion corridors for autochthonous or shipping-transported invasion of alien species, canals facilitate aquatic invasions globally by increasing the overall volume of ship borne trade and changing the patterns of maritime transport. The accelerating globalization and greater economical interdependence between distant markets result in an increase in the volume of sea borne trade. World sea borne trade expanded in 2004 to 6.76 billion metric tons, driven by the economies of Asian countries and the USA (www.UNCTAD.org). The physical limitations of the intermodal (maritime & road/rail) cargo system have driven the success of the Kiel Canal, whereas the preference for "All-Water Routes" from Asia to Western Europe and the East Coast of the United States meant a surge in traffic through the Suez and Panama Canals. As growth outpaces capacity, authorities are under severe pressure to keep up with demand: the

Suez Canal Authority has been expanding the channel to accommodate ULCC with oil cargos of up to 350,000 dead-weight-tons by 2010, and the Panama Canal Authority plans to construct a third channel, and water-recycling new locks to accommodate ships twice as big as Panamax vessels.

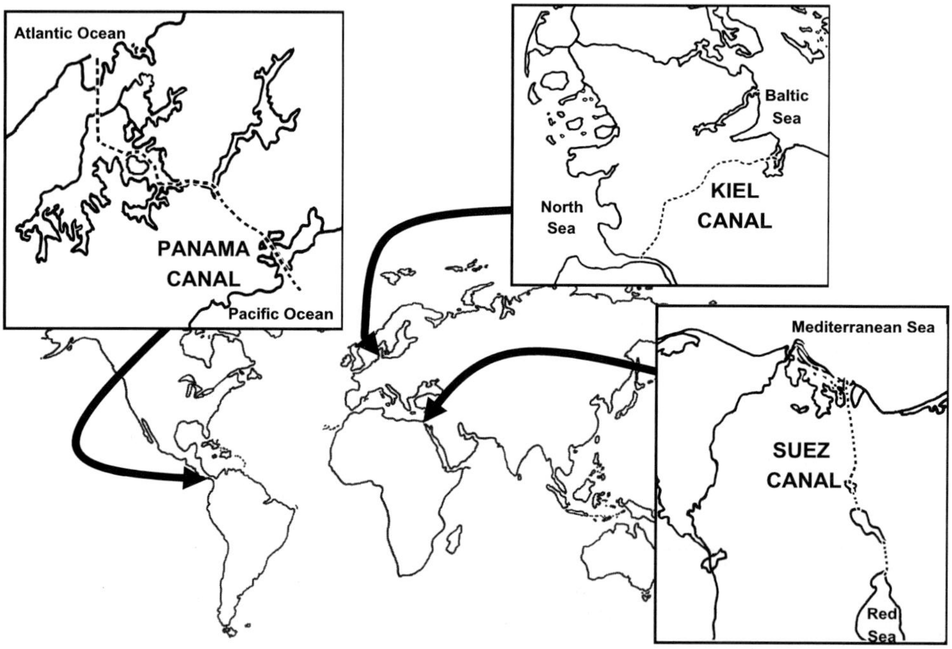

Fig. 2. Kiel, Panama and Suez Canals. Inserts show details of the canal route.

This book presents an account of the impact of the three principal maritime canals as invasion corridors for aquatic species. These canals differ in their 'permeability' to alien species. The Kiel Canal is characterized by a salinity gradient from seawater at one end to low halinity brackish water at the other, the seawater-fed Suez Canal had, for the first half of its existence, a hypersaline barrier in the form of the Great Bitter Lakes, whereas the Panama Canal is a triple-locked freshwater corridor between two oceans. The extent and distribution of the alien biota, together with their environmental impacts, and past as well as future trends, are discussed.

The Kiel Canal

The World's Busiest Man-made Waterway and Biological Invasions

STEPHAN GOLLASCH[1] & HARALD ROSENTHAL[2]

[1] *GoConsult, Bahrenfelder Str. 73 a, 22765 Hamburg, Germany*
[2] *Schifferstr. 48, 21629 Neu Wulmstorf, Germany*

1 Introduction

In all more than 25,000 kilometres (km) of canals exist in Europe. The longest inland waterway in Europe connects the southern North Sea at Rotterdam (the Netherlands) with the Caspian Sea, and consists of rivers linked by canal systems. This and other canal systems have been important corridors for the spread of species between previously separated regions.

The Kiel Canal in northern Germany, is Europe's longest man-made canal for ocean-going merchant vessels. It connects the North Sea (canal entrance at the mouth of the Elbe River estuary) to the Baltic Sea at the Kiel Fjord (Fig. 1), providing a more rapid and sheltered transit than the alternative passage though the Skagerrak, which is approximately 400 nautical miles (nm) longer.

The Kiel Canal is almost 100 km long and is the world's busiest artificial waterway: more than 40,000 merchant vessels and nearly 20,000 pleasure craft pass through it each year (Wasser- und Schifffahrtsdirektion Nord pers.com., www.ak190x.de/Bauwerke/Bau/Nord-Ostsee-Kanal.htm visited January 18th 2005).

Figure 2 shows the elevation of land along the route of the Canal. Figure 3 shows the major creeks and rivers draining into the Canal. The drainage area to the Canal covers ca 1,580 square kilometres (km^2), with an input of ca 630 million cubic metres (m^3) of freshwater per year. These inputs combined with the overall direction of water flow create a clear but seasonally and inter-annually varying decline in salinity from east to west, and a net outflow towards the North Sea.

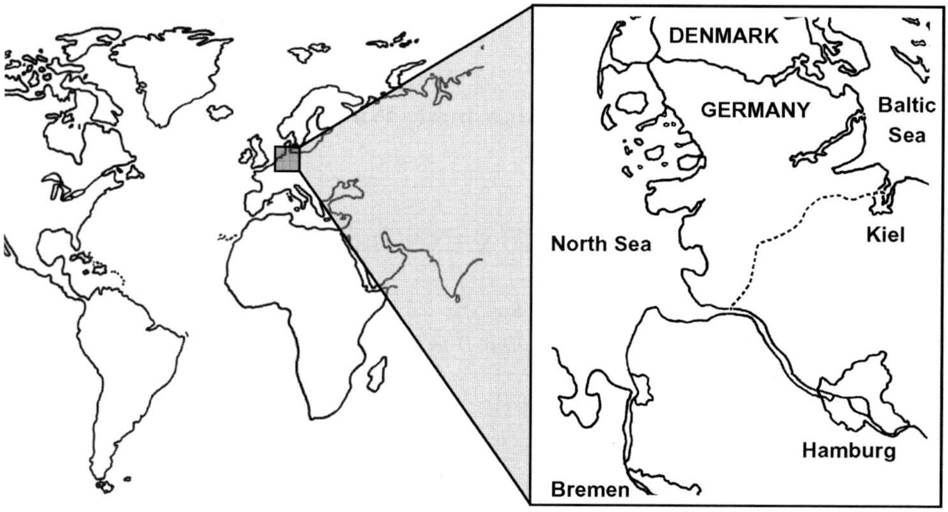

Fig. 1. Map. Insert = northern Germany with location of the Kiel Canal (dotted line).

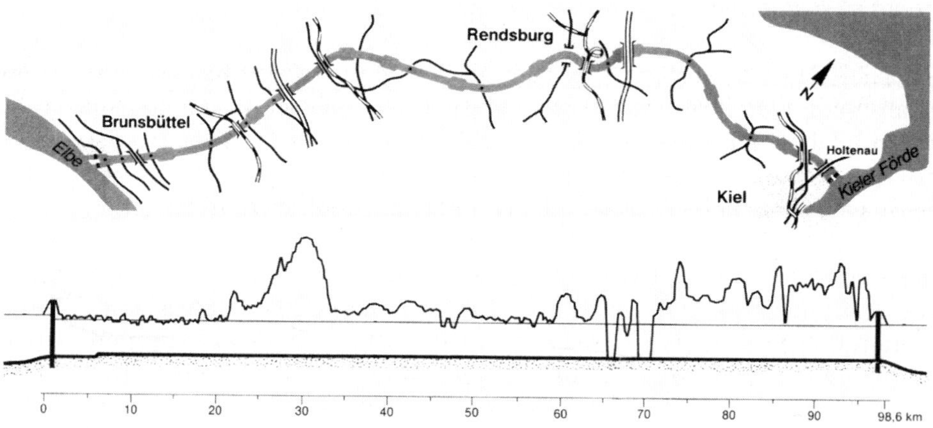

Fig. 2. Route of the Kiel Canal from Brunsbüttel to Kiel. Top: passing bays, bridges and tunnels. Bottom: upper line is the elevation of land along the route of the canal, the horizontal grey line is the canal water surface and the heavy black line the bottom of the canal bed, both canal entrances indicated with vertical black lines. (Modified after Hill in Wasser- und Schifffahrtsdirektion Nord, Kiel 2001).

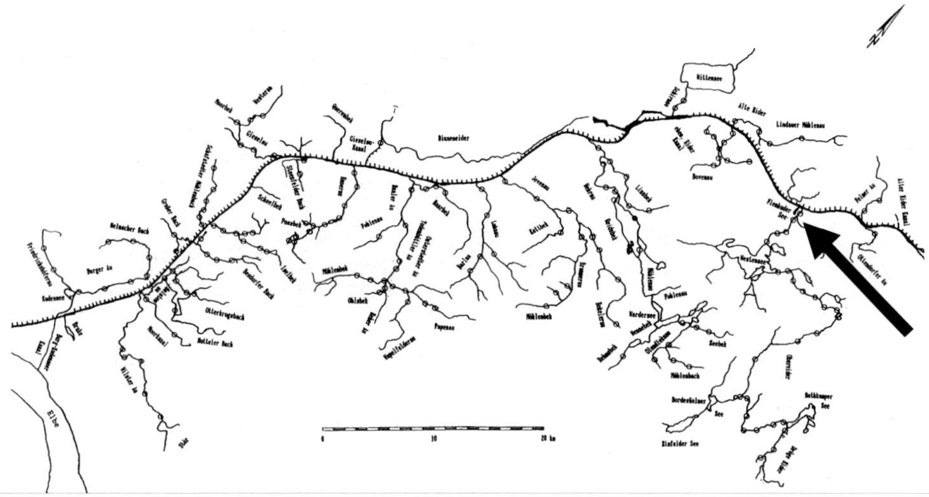

Fig. 3. Schematic overview of the principal watershed draining to the Kiel Canal, including ditches, creeks, small secondary canals and lakes. These are mainly located south of the canal, and represent a catchment area of ca 1,580 km². Scale = 20 km. Arrow gives location of Flemhuder Lake (see also Fig. 4). (Modified after Hartmann & Spratte 1995).

Figure 4 shows Flemhuder Lake, a small lake connected to the Canal at canal km 85 - 86. The volume of freshwater input into this lake results in seasonal changes in salinity. Flemhuder Lake is one example of many small water bodies connected to the Kiel Canal that provide refuges for fish and have rich benthic and planktonic communities.

The Canal is managed by the Wasser- und Schifffahrtsdirektion Nord. Other canal authorities located in Brunsbüttel and Kiel-Holtenau are responsible for traffic management, canal policing, building and maintenance as well as for running the canal facilities. Additional canal authorities in Rendsburg manage ship building and are responsible for mechanical and electrical engineering, communications, maintenance and improvement works (Wasser- und Schifffahrtsdirektion Nord 1993).

The official Kiel Canal homepage may be found at www.kiel-canal.org where two web cameras deliver real time images of the approaches to the canal locks. This homepage also provides an online canal fee calculator according to ship size.

The detailed traffic rules for the canal may be downloaded at www.kiel-canal.org/pages_english/vorschriften/regulations-KIEL-CANAL.pdf.

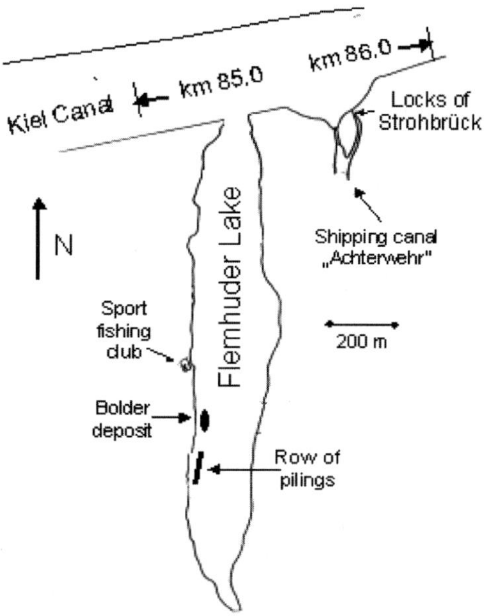

Fig. 4. The position, shape and overall topography of the Flemhuder Lake, one of the many small water bodies connected to the Kiel Canal. (Modified after Hartmann & Spratte 1995).

2 History and canal construction

Today's Kiel Canal was not the first connection between the North Sea and the Baltic Sea. There were repeated plans for a man-made shipping canal since Viking times (Fig. 5). However, most canal plans did not become reality.

2.1 The shipping route in Viking times

The first plans to build an inland waterway between the North and Baltic Seas date back to Viking times. In early medieval times, in approximately the 7th century, the Vikings started to search for an inland connection between the North and Baltic Seas.

By the end of the 9th Century Haithabu in northern Germany became the Viking's major trade hub. This sheltered settlement formed a strategic trade gateway between northern and southern as well as eastern and western regions of the Viking territory. At its prime, more than 1,000 inhabitants lived in Haithabu, making it one of the biggest settlements in the region.

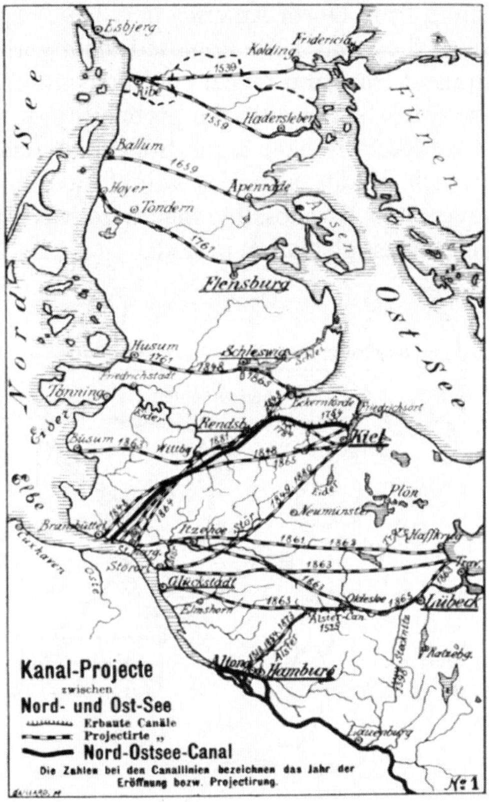

Fig. 5. Map of northern Germany in 1893 with an overview of planned canal projects (Beseke 1893 - with permission from Publishing House Lühr & Dircks, Hamburg, Germany).

Haithabu was located at the innermost part of the Schlei Fjord - approximately 40 km inland from the Baltic shores (Fig. 6). An overland distance of only 16 km was needed to reach the Treene River (at today's Hollingstedt), from which Viking ships sailed to the North Sea via the Eider River (Elsner 1994). However, at these early times, the building of a canal over this distance was not possible due to lack of construction knowledge, and the Viking vessels were transported overland either on carriages or by rolling them on wooden logs.

The settlement at Haithabu declined following its pillage and destruction in 1050 and 1066. Nearby settlements developed into the city of Schleswig, and in the early 13th Century the city of Lübeck subsequently became the most important trade hub in the region (Elsner 1994).

The Viking people were adventurers and may have been responsible for the introduction of the infaunal bivalve *Mya arenaria* to Europe (Petersen et al.

1992). Vikings returning from North America may have kept live *Mya arenaria* aboard as fresh food, or they may have imported them with the solid ballast on their ships. Excavations at Haithabu reveal enormous numbers of ballast stones at and near the landing pier, supporting the probability of an introduction with ballast. Viking ships are likely to have come from sheltered, muddy estuaries in North America, and such muddy estuaries would have had large numbers of *Mya arenaria*. However, it is also possible that there was a gradual reexpansion into Europe following the last glaciation period.

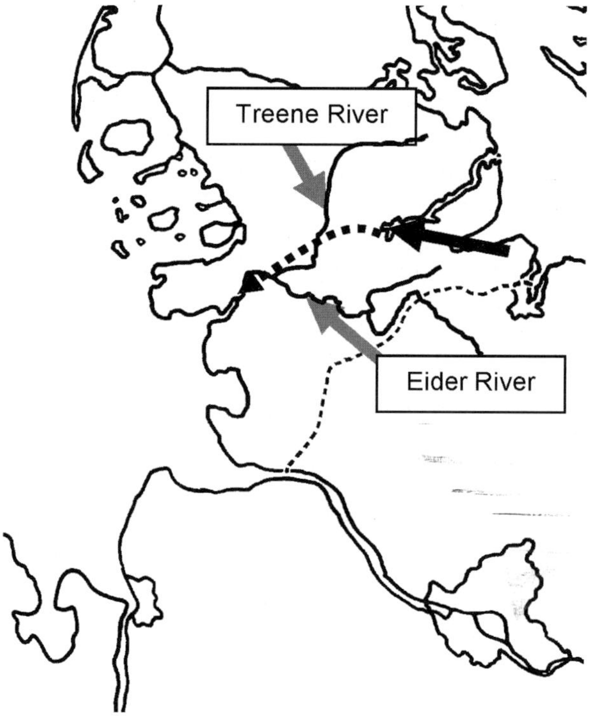

Fig. 6. Map of northern Germany with the Kiel Canal as a dotted line. Black arrow identifies the Viking settlement Haithabu on the innermost shore of the Schlei Fjord. Grey arrows point to the Treene and Eider Rivers. Dotted arrow indicate the possible Viking inland route connecting Baltic and the North Sea.

In contrast Wolff (2005) states that the transfer of *Mya* by the Vikings poses a problem. Except for an occasional vessel driven off course by gales, there was no direct transport between North America and Europe in Viking times (Marcus 1980). Greenlanders travelled to North America on a more or less frequent basis and also travelled between Greenland and Norway, but these trips were not undertaken by the same vessels. As a result *Mya* was probably first introduced from North America to Greenland and thereafter from Greenland to Europe (Ockelmann 1958, Höpner & Petersen 1978, 1999).

Legend or truth?

Mya arenaria, *was probably native to North America and introduced to Europe in Viking times which would have been the first species introduction into Europe. Viking ships used solid ballast, i.e. sand or gravel. It may well be that* Mya arenaria *was shovelled onboard together with ballast when the ships left America. Discharging the ballast in Europe may have introduced the clam to Europe. Alternatively, Vikings may have used the clam as a source of food during their voyages ... two possible scenarios, impossible to prove!*

Vikings may also have been the first to move a non-native species through today's German mainland with ships. Ships were moved between the Treene and Eider Rivers and Haithabu on logs or carriages to avoid the dangerous passage around Denmark. But it is likely that any solid ballast was removed to reduce a ship's weight before moving it over land.

However, one never knows!

Regular shipping trade demanded a safe passage from the North Sea to the Baltic Sea to avoid the dangerous passage around Denmark. As a result, canal planning projects were commissioned.

2.2 The Stecknitzkanal project

In the following centuries and at the end of the 14th Century, the cities of Lübeck and Hamburg developed as trade hubs with salt being the most valuable cargo. Salt was transported by road from Lüneburg (near Hamburg) to Lübeck. To facilitate cargo transportation the Stecknitzkanal was built between 1391 and 1398, enabling barges to be pulled by horses between Lauenburg (near Hamburg) and Lübeck. This canal was the first man-made waterway in northern Europe (Schütt 1991).

In the beginning, 13 locks were built along the water-way (total length 94 km); another four locks were added at the end of the 17th Century to ease transportation across the height difference of 18 m. The canal, 11.5 km long and 7.5 m wide, connected the Rivers Delvenau (a tributary of the Elbe River) and Stecknitz (connected to the Trave River near Lübeck). Its shallowest sections had a depth of 85 cm.

This canal was predominantly used by non-propelled barges to ship salt to Lübeck. The maximum allowable size of the barges was 12 m in length, 2.5 m in width with a maximum draught of 40 cm. Such vessels could carry 7.5 tonnes of salt. A canal passage took more than two weeks. The first vessels reached Lübeck in July 1398. Trade peaked in 1500 to 1550 when on average more than 12,000 tonnes of salt were shipped to Lübeck annually. The canal was in use for approximately 500 years and was eventually replaced by the Elbe-Lübeck-Kanal, which opened in June 1900 (www.geschichte.schleswig-holstein.de, www.schifffahrtslexikon.de/lexikon/lemma/def/stecknitzkanal_de.htm visited February 10th 2005).

2.3 The Alster-Trave-Kanal project

Another canal connecting Lübeck and Hamburg was first planned in 1448 – the Alster-Trave-Kanal (also known as Alster-Beste-Kanal). Construction started in the same year, but the excavation encountered technical difficulties, and the project was put on hold in 1452.

Construction began again in 1526 with support from the Danish King Friedrich I (1471 - 1533). Herzog Magnus II (1527 - 1603) tried to hinder this canal project as he believed the new canal would compete with and lower the toll income from the Stecknitzkanal (see above).

The Alster-Trave-Kanal opened in August 1529 providing an alternate route between Hamburg and Lübeck via the Alster and Beste Rivers and into the Trave River. Using the canal, the total distance between Hamburg and Lübeck was 91 km of which the canal itself was 8 km. The canal width was approximately 15 m and the depth nearly 2 m. It had 23 locks and was crossed by 26 bridges. Cargo was transported through the canal in approximately seven days using non-propelled lighters moved by men on the embankments.

Financially, this canal was unsuccessful. The building costs of approximately 43,000 Marks, more than the annual budget of the City of Hamburg at that time, could not be recovered and the canal was closed in 1550 (Schütt 1991, www.uni-kiel.de/ewf/geographie/forum/hintergr/sh1995/15_kanal.htm visited May 27th 2004, www.geschichte.schleswig-holstein.de/vonabisz/alstertrave kanal.htm visited February 10th 2005).

2.4 The Schleswig-Holsteinische Canal (Eiderkanal)

There were a number of plans to build a waterway across the Cimbrian Peninsula (northern Germany and Denmark) in the 16th century to handle the increasing trade demand and to enable a safe passage from the North Sea into the Baltic Sea for sea-going vessels.

Herzog Adolf I of Holstein (1526-1586), a brother of the Danish King Christian II (1481-1559), planned a canal between Kiel and the Eider River (Eiderkanal). He presented his plans to the German Emperor on the 10th of August 1571. However, no development took place or was considered until the end of the 18th Century (Schulz 1995).

In the 1770s, during the time of Danish rule over the Cambrian peninsula as far as the Elbe estuary, a Canal Commission was founded in Copenhagen under the control of Denmark's King Christian VII (1749-1808) and his brother Crown Prince Friedrich (1768-1839). In 1773, King Christian VII hired General Major Wegener to plan a canal between the Baltic and North Seas. This canal was to take advantage of existing navigable waters, using the Eider River from its mouth into the North Sea up to Rendsburg, and required the construction of a channel from there to the Baltic Sea at Holtenau, the "Schleswig-Holsteinische Canal".

The Canal Realization Commission was founded on May 11th 1774 (Schulz 1995), and the plan was changed to connect Glückstadt at the Elbe River with the Kiel Fjord using existing rivers and inland lakes, such as the Stör and Einbeck Rivers, the Einbeck and Bordesholm Lakes near Neumünster, the Eider River and Schulenberg Lake.

Some further changes resulted in the final plan (from east to west): Kiel Holtenau to Levensau River with locks to overcome the approximately 7 m height difference at Holtenau, Knoop and Rathmannsdorf; to Flemhuder and Western Lakes; to the Eider River with locks at Königsförde, Kluvensiek and Rendsburg (Schulz 1995).

Construction began in 1774 and the construction site became the largest non-military building site in continental Europe. The River Eider was deepened from

1776 onwards. Up to 4,600 construction workers were employed, more than half of which were killed by disease. Approximately 82 million m^3 of material were moved to build the canal and its six locks. The construction costs rose to approximately 2.5 million "Reichstaler".

The 43 km long Schleswig-Holsteinischer Canal was opened in October 1784 to become part of the 175 km long waterway from Kiel at the Baltic Sea to the mouth of the Eider River at Tönning near the North Sea. It was a narrow canal only 28.7 m wide at the water surface and 18 m wide at the bottom. The water depth was 3.45 m. The canal became an important shipping route for vessels up to 300 tons. Six identical locks were built to accommodate the height difference of the canal bed. Each lock was 35 m in length, 7.8 m in width and 4 m in depth. A canal passage took approximately 3 to 4 days. This canal was renamed the Eiderkanal in 1853. More than 5,000 vessels used it in the peak year of 1872 (Schulz pers. com., Schulz 1995, www.geschichte.schleswig-holstein.de visited February 10th 2005).

2.5 History of the Kiel Canal project

The Eiderkanal's modest size could not accomodate the emerging logistic, technical and shipping demands, particularly after 1864 when political rule passed from Denmark to Prussia and then to the German Reich. In addition, the German navy wanted a direct link between its bases in the Baltic and North Seas, so the German armada would not have to sail around Denmark. Several plans were made over some decades to built a wider canal. Construction of a new canal seemed likely when Chancellor Otto von Bismarck (1815-1898) gave the project support.

But because of political differences with field marshal Helmut Karl Bernhard von Moltke (1800-1891), the project did not proceed. Von Moltke summarized his objections to the project in a talk entitled "Rede gegen den Kanalbau" (=Speech against the canal construction). He argued that the canal would be of little value to the navy and had limited strategic importance and he recommended that financial resources should rather be used to strengthen land-based military forces. These arguments convinced the German government to not invest in building the canal, and by 1873, the new canal project seemed to have failed.

However, Bismarck returned to the idea of a new and larger canal connection. Moving the major Baltic navy port to Kiel, a strategic decision, strengthened the German province Schleswig-Holstein. Further, a commercial analysis by the

Hamburg shipowner and businessman Hermann Dahlström (1840-1922), supported by the waterworks inspector Mr. Boden, provided support for a new canal. Dahlström revised the original canal plan and financed an improvement study from his own funds.

Dahlström's plans, taking into account both commercial and naval considerations, were published in 1878 and 1879 ("*Ertragfähigkeit eines schleswig-holsteinischen Schiffahrtskanals*"). Bismarck eventually convinced the German Emperor Wilhelm I (1797-1888) to approve the project. In March 1880, Dahlström was contracted to develop a first working plan of the canal.

Safety aspects were a major motivation, as approximately 200 ships sank annually in the Skagerrak region on routes to and from the Baltic Sea – a canal would provide a safer and shorter route. The canal was originally planned to link the mouth of the Elbe river to Eckernförde on the Baltic shores. Dahlström planned the canal as a public-private partnership. However, the project remained in a stalled state and did not become reality!

Emperor Wilhelm I initiated a further plan in 1883, with the aim of building a canal of sufficient width and depth to enable the passage of German naval vessels. Two years later, Bismarck finally convinced Wilhelm I that work on the canal should begin.

Private Councillor Baench, a technical assistant for construction planning of the German State Department, also supported the project. Due to his and other support it was announced in 1886 that a shipping canal suitable for the German navy would be built.

The Canal Construction Act, passed by the German government, was signed by Wilhelm I in the same year[1]. Thus, the canal from the mouth of the Elbe river via Rendsburg to Kiel became a national project funded by the German government (www.ankieken.de/schleswigholstein/pages/nordostseekanal.htm visited May 27th 2004) and built by the Canal Construction Commission which was established.[2]

---

[1] Gesetz betreffend die Herstellung des Nord-Ostsee-Kanals (Reichsgesetzblatt Seite 58.). Wir Wilhelm, von Gottes Gnaden Deutscher Kaiser, König von Preußen verordnen im Namen des Reichs, nach erfolgter Zustimmung des Bundesraths und des Reichstags, was folgt: §.1. Es wird ein für die Nutzung durch die deutsche Kriegsflotte geeigneter Seeschiffahrtskanal von der Elbmündung über Rendsburg nach der Kieler Bucht unter der Voraussetzung hergestellt, daß Preußen zu den auf 156.000.000 Mark veranschlagten Gesammtherstellungskosten desselben den Betrag von 50.000.000 Mark im Voraus gewährt. [...] §.3. Von den nicht zur Kaiserlichen Marine und zur Bauverwaltung gehörenden Schiffen, welche den Kanal Benutzen, ist eine entsprechende Abgabe zu entrichten. [...] Gegeben Berlin, den 16. März 1886.

[2] Allerhöchste Verordnung betreffen die Errichtung einer besonderen Commission für die Herstellung des Nord-Ostsee-Kanals (Reichsgesetzblatt Seite 233). Wir Wilhelm, von Gottes Gnaden Deutscher Kaiser, König von Preußen verordnen im Namen des Reichs, mit Zustimmung des Bundesraths, was folgt: Für die Herstellung des Nord-Ostsee-Kanals wird eine dem Reichsamt des Inneren unmittelbar untergeordnete besondere Commission unter der Bezeichnung "Kaiserliche Kanal-Commission" errichtet, welche innerhalb

Fig. 7. Ceremony for laying the Kiel Canal foundation stone in Kiel-Holtenau on June 3rd 1887. Photo courtesy of Walter Schulz, Brunsbüttel, Germany (Schulz 1995).

The Canal Construction Commission was launched on October 1st 1886 in Kiel. Subcommittees were established to build canal locks and approaches. The canal construction itself was supervised by eight construction departments, each of them responsible for approximately eleven kilometres (Schulz 1995, www.holtenau-info.de/history/kanal2.htm visited January 18 th2005).

2.6 History of the Kiel Canal construction works

On June 3rd 1887, Emperor Wilhelm I (then over 90 years of age) laid the foundation stone for the canal in Holtenau near Kiel (Fig. 7). He died soon afterwards, in March 1888. The document embedded in the foundation stone refers to the national and international importance of the canal.[3]

---

des ihr zugewiesenen Geschäftskreises für die Dauer ihres Bestehens alle Rechte und Pflichten einer Reichsbehörde haben soll. [...] Gegeben Schloß Mainau, den 17. Juli 1886.

[3] Wir Wilhelm, von Gottes Gnaden Deutscher Kaiser, König von Preußen [...] thun kund und fügen hiermit zu wissen: Die Herstellung einer unmittelbaren Verbindung der beiden deutschen Meere durch eine für den Verkehr der Kriegs- und Handelsflotte ausreichende Wasserstraße ist seit langer Zeit Ziel patriotischer Wünsche gewesen. [...] Durch das Reichsgesetz vom 16. März 1886 ist die Verbindung beider Meere nunmehr sichergestellt worden. [...] Wir beschlossen, dass ... der Grundstein zum Bau des Nord-Ostsee-Canals, und zwar an der Stelle gelegt werde, an welcher sich in Zukunft die Eingangsschleuse bei Holtenau erheben wird. [...] Möge der Bau dem Deutschen Vaterlande, möge er den Elbherzogtümern zu Heil und

Fig. 8. Johann Fülscher. Photo Courtesy of Walter Schulz, Brunsbüttel, Germany (Schulz 1995).

Johann Fülscher (Fig. 8), vice-chairman of the Canal Commission, managed the construction works. Because of his success, he was later appointed as an advisor to the Panama Canal Project (Schulz 1995).

Building started in 1888 and took eight years to completion in 1895. During this time the Eiderkanal was in operation and every effort was made not to interfere with its traffic. Where the two canals met, a movable bridge was built so as to simultaneously allow (a) construction work of the Kiel Canal and (b) passage of vessels in the Eider Canal. Figure 9 shows the canal construction site with a tug boat crossing at the movable train bridge. The movable bridge allowed a navigational water surface in the Eider Canal of 18 m in width. Here, both canals were separated by a dam (Schulz 1995). For comparison today's Kiel Canal and the Eider river are shown in Figure 10.

Up to 8.900 workers from Germany, Spain, Poland, Italy, Denmark, Austria and Russia were employed during the building phase. The minimum age for employment was 17, although younger workers could accompany their fathers.

Segen gereichen! Möge durch ihn das Gedeihen der deutschen Schiffahrt und des deutschen Handels, die friedliche Entfaltung des Weltverkehrs, die Stärkung vaterländischen Seemacht und der Schutz Unserer Küsten kräftig gefördert werden! [...] Gegenwärtige Urkunde haben Wir in zwei Ausfertigungen mit Unserer Allerhöchsteigenhändigen Unterschrift vollzogen [...] Wir befehlen, die eine Ausfertigung mit den dazu bestimmten Schriften und Münzen in den Grundstein der Schleuse Holtenau nieder-zulegen, die andere in Unserem Archiv aufzubewahren. Gegeben Holtenau, den 3. Juni 1887.

Fig. 9. Kiel Canal construction site (in the background) in 1892 with the Eider Canal in the front. The train tracks crossing the Eider Canal were needed to allow the construction works of the Kiel Canal. Photo courtesy of Walter Schulz, Brunsbüttel, Germany (Schulz 1995).

The treatment of the workers was exceptionally up-to-date and included housing in camps organised by commissioned military officers in several locations near the canal. Food, health care, sanitary facilities and religious services were also supplied. All workers had daily meals – prepared according to strict regulations allowing an optimum nutrition. This health regime prevented disease outbreaks and almost certainly maintained the work-force in good condition, especially when in the nearby City of Hamburg, a severe cholera outbreak in 1892 resulted in approximately 8,600 deaths (Schütt 1991, Grahl & Kelm (eds.) 1992). The disease also spread among the canal workers, but due to proper treatment only three died from cholera.

The canal was completed on time and within the calculated budget of 156 million Gold-Marks.

It was built at mean sea level of the North and Baltic Seas – with a light house at each end. The tidal amplitudes are different, with the greatest tidal range in the North Sea. These level differences required the construction of locks at both canal entrances (Fig. 11 & 12). Without locks, the tides at the North Sea canal entrance would have flooded it at high tide and drained it at low tide.

Fig. 10. Areal view of the Kiel Canal near Rendsburg, with the Eider River to the right of the canal. Photo courtesy Joachim Eicke, Rendsburg, Germany.

Presently, the level is maintained at approximately 1.4 m above mean low tide and 1.4 m below mean high tide in the Elbe estuary (North Sea canal entrance). At Holtenau the water level approximately equals the mean water level of the Kiel Fjord (Baltic Sea canal entrance).

Fig. 11. Construction work on the canal locks in Kiel-Holtenau in September 1893. Photo courtesy of Walter Schulz, Brunsbüttel, Germany (Schulz 1995).

Fig.12. Construction work on the canal locks in Brunsbüttel in May 1893. Photo courtesy of Walter Schulz, Brunsbüttel, Germany (Schulz 1995).

The building of the canal was a challenge because of intersecting rivers, ground water influence and a topography ranging from 25 m above sea level, to 3 m below sea level en route (Fig. 13). Due to careful planning, locks were only required at both ends of the canal.

The canal construction began simultaneously at both ends and more than 80 million cubic meters of material were excavated. Locks, bridges and ferry terminals were built. The last section to be dug, finally connecting both seas was located near Reitmoor (41 km position) and was cut in February 1895 (Arndt 1931/32).

The work involved a great deal of equipment, including 65 excavators and dredges for wet and dry material, 94 railway locomotives with 2,750 carriages, 270 vessels (such as tug boats and lighters), 10 steam powered rams, 20 cranes and other technical equipment including lorries and wheelbarrows.

Concrete at that time was only in marginal use, mainly in the canal locks. Most of the canal bed was built with bricks – a common procedure for construction works at that time (Schulz 1995).

To ease canal construction groundwater levels were lowered using wells and an extensive drainage system. Especially challenging were the boggy and marshy sections of the canal. Embankments in these parts needed to be renewed and repaired several times before the construction was completed (Schulz 1995).

Fig. 13. Building phase of the Kiel Canal. Inner section of the canal at Grünental in 1889. Photo courtesy of Walter Schulz, Brunsbüttel, Germany (Schulz 1995).

Fig. 14. German navy ship Pelikan used prior to opening of the canal as a training vessel for canal pilots. Photo courtesy of Walter Schulz, Brunsbüttel, Germany (Schulz 1995).

Before the canal was opened officially, the German navy vessel *Pelikan* was used to train the canal pilots to ensure the operation of the locks and to provide safe navigation (Fig. 14 & 15). A canal lighting system was installed soon after the opening of the canal. The locks were illuminated with headlights. The almost 100 km long construction was equipped with a chain of light signals enabling navigation at night (Schulz 1995).

2.7 The official opening

The official opening ceremony took place in Hamburg on June 19th 1895 following two months of final preparation to ensure that it was fully operational.

The German head of Government, German "Royalties", ministers and representatives from foreign countries together with a cheering crowd of thousands of Germans attended the ceremony. After the ceremony Emperor Wilhelm II (1859-1941) and his "guests" boarded navy vessels waiting in the Port of Hamburg and left the port for the Brunsbüttel locks of the Kiel Canal – where they arrived in the early hours of June 20th (Schulz 1995) (Fig. 16).

Fig. 15. The first large merchant vessel, the Palatia, passes the bridge at Grünental on June 5th 1895 during the testing and training phase preceding the canal opening. The vessel is seen releasing ballast water in the canal. The old Grünental Bridge was replaced by a new bridge in 1986. Photo courtesy of Walter Schulz, Brunsbüttel, Germany (Schulz 1995).

The canal locks and embankments were decorated with flags, wreaths and garlands. In the early morning of June 20th 1895 the German imperial yacht *Hohenzollern* with Emperor Wilhelm II aboard entered the lock at Brunsbüttel (North Sea) (Fig. 17). Music was played and several thousands of observers welcomed the vessel. A ribbon crossing the entrance of the canal was cut as an indication of the pre-opening of the canal.

The *Hohenzollern* was followed by a convoy of 24, predominantly naval ships, fourteen of which represented other seafaring nations. To name a few:

- the British *Osborne* and *Enchantress,*
- the Italian *Savoia* and *Aretusa,*
- the Austrian-Hungarian *Trabant,*
- the French *Surcouf,*
- the Russian *Grosjaschtschi,*
- the Spanish *Marques de la Ensenada,*
- the Swedish-Norwegian *Edda* and *Viking,*
- the US *Marblehead,*
- the Romanian *Mircea,*
- the Danish *Hecla*, and as last vessel in the convoy,
- the Dutch *Alkmaar*.

Fig. 16. Formation of the convoy through the Kiel Canal during the opening ceremony. Photo courtesy of Walter Schulz, Brunsbüttel, Germany (Schulz 1995).

Approximately eight hours later, many thousands welcomed the convoy in Kiel-Holtenau at the eastern end of the canal and a salute was fired from the convoy vessels. The following day, June 21st, a ceremony was held in Holtenau where Emperor Wilhelm II officially opened the canal and named it Kaiser-Wilhelm-Kanal (=Emperor-Wilhelm-Canal).

Fig. 17. The German imperial yacht Hohenzollern arriving in the Kiel-Holtenau locks after its canal passage on June 20th 1895. Photo courtesy of Walter Schulz, Brunsbüttel, Germany (Schulz 1995).

The celebration ceremonies attracted an enormous number of people, similar to a ticker-tape-parade in New York. Platforms were built in Kiel Holtenau to accommodate invited participants.

The opening ceremony came to an end with a speech by Reichskanzler Fürst Hohenlohe (1819-1901).[4]

[4] Wir Wilhelm, von Gottes Gnaden deutscher Kaiser, König von Preußen thun kund und fügen hiermit zu wissen: Das Werk, zu welchem Unseres in Gott ruhenden Herrn Großvaters Kaiser Wilhelm's I Majestät am 3. Juni des Jahres 1887 im Namen des Reichs den Grundstein gelegt hat – die unmittelbare Verbindung der deutschen Meere – steht vollendet vor unseren Augen. [...] Und wenn wir heute mit hoher Befriedigung die Erwartungen der Erfüllung näher geführt sehen, welche das Reich an die Herstellung einer für die Zwecke der Kriegs- und Handelsflotte ausreichenden Wasserstraße zwischen Nord- und Ostsee geknüpft hat, so gereicht es Uns zu besonderer Freude, daß Wir, umgeben von dem erlauchten Kreise Unserer hohen Verbündeten, der Vertreter des Volkes und unter der dankenswerten Betheiligung der Abgesandten befreundeter Mächte, deren Geschwader wir in unserem ersten, ihnen gastlich geöffneten Kriegshafen willkommen heißen, diese Straße dem Verkehr übergeben können. [...] Aber nicht nur dem Vaterlande und seinem Handel, ferner Schiffahrt und seiner Wehrkraft soll der Kanal förderlich sein. Zudem Wir ihn in den Dienst des Weltfriedens stellen,

Fig. 18. Navy training vessel Moltke using the canal locks in Brunsbüttel in 1896. Photo courtesy of Walter Schulz, Brunsbüttel, Germany (Schulz 1995).

The official canal opening diploma was imbedded in the canal keystone together with other documents and medals. The keystone was laid when the German National Hymn was played and the flags lowered. The day was concluded with an official dinner at which commemoration medals, six centimetres in size and made of gold, silver or bronze, were given to selected guests (Schulz 1995).

A highlight of the ceremony was the laying of the keystone at the Kaiser Wilhelm I memorial monument (Fig. 19). Beside this an encryption of the official canal opening was later fixed to the wall inside the canal opening memorial hall in the canal lighthouse Kiel-Holtenau.[5] A further memorial plaque was fixed outside of the canal lighthouse at Holtenau.[6]

eröffnen Wir neidlos allen seefahrttreibenden Völkern die Theilnahme an den Vorteilen, welche seine Benutzung gewährt. Möge er, ein Friedenswerk, allezeit nur dem Wettkampfe der Nationen um die Güter des Friedens dienstbar sein! Indem Wir befehlen, daß der Kanal für die Schiffahrt aller Völker geöffnet werde, wollen Wir zugleich, daß an der Stelle, an welcher derselbe in Unseren Kriegshafen mündet, ein Denkmal errichtet werde, welches der Nachwelt Kunde giebt von der durch Uns in Gegenwart Unserer hohen Verbündeten vollzogenen denkwürdigen Eröffnung der neuen Wasserstraße. Mit diesem Denkmal wünschen wir zugleich einen Theil des Dankes abzutragen, den das deutsche Volk dem großen Kaiser schuldet, welcher vor nunmehr 25 Jahren die heutigen Stämme zu einem ewigen Bunde geeint und in weiser Voraussicht das jetzt vollendete Werk begonnen hat. Der reiche Segen, welcher das Walten des unvergesslichen Kaisers begleitet hat, möge auch auf diesem Werke ruhen! Gegenwärtige Urkunde haben Wir in zwei Ausfertigungen mit unserer Allerhöchsteigenhändigen Namensunterschrift vollzogen und mit Unserem größeren Kaiserlichen Insiegel versehen lassen. Wir befehlen, die eine Ausfertigung mit den dazu bestimmten Schriften und Münzen in den Grundstein des Denkmals niederzulegen, die andere in Unserem Archiv aufzubewahren. Gegeben Holtenau, den 21. Juni 1895.

[5] Seine Majestät Kaiser Wilhelm II legte bei der feierlichen Eroeffnung des Kanals am 21. Juni 1895 den Grundstein fuer das auf diesem Huegel errichtete Standbild seiner Majestaet Kaiser Wilhelms des Grossen

Fig. 19. Keystone laying ceremony of the Kiel Canal, June 21st 1895 at Kiel-Holtenau. Photo courtesy of Walter Schulz, Brunsbüttel, Germany (Schulz 1995).

The canal was managed at the outset by almost 300 civil servants taking care of maintenance planning and construction works; more than 500 workers were employed to maintain and run the locks, bridges, ferries and to regularly check the canal bed condition (Schulz 1995).

2.8 First canal enlargement phase

As a result of the building of ever larger naval vessels, the canal was considered as too small by 1903.

To meet the increase in traffic and an expanding German navy with larger ships, the canal cross-section was widened to almost twice its original size. This enlargement took place between 1907 and 1914, costing 242 million Marks. This operation was more expensive than the construction of the original canal but by this stage the canal was accepted as being an asset. Almost 100 million cubic metres of material were moved at a rate of approximately 40,000 cubic

und begleitete seine Hammerschlaege mit den Worten "Zum Gedächtniss Kaiser Wilhelms I der Grossen taufe ich dich: Kaiser Wilhelm-Kanal" Zum Ruhme des Deutschen Reichs – zur Wohlfahrt aller Nationen.

[6] Kaiser Wilhelm II vollzog die Weihe des Nord-Ostsee-Kanals und übergab ihn dem Weltverkehr am 21. Juni 1895.

metres per day and a laying of up to 2,000 cubic metres per day. The enlargement project was completed with the construction of two additional and larger locks at each end, in Brunsbüttel and Holtenau.

To span the canal steel bridges were built. The Rendsburg steel bridge had a length of 2.5 km and was the largest steel construction in Europe at that time. The improvement works were completed in 1914 and another opening ceremony was held on June 14th. The newspapers reported the canal could now accommodate the largest naval vessels.

Shortly thereafter, in early August, World War I broke out. One may argue whether or not the completion of the Kiel Canal, a strategic link, "prompted" the beginning of World War I.

2.9 Kiel Canal and World War I & II

The canal was a political issue. Restrictions in canal use applied during both World Wars. After World War I (1914-1918) the Treaty of Versailles decreed in 1919 that the waterway was of international importance and that it was an international waterway, open to all.[7]

The canal administration was left in German hands, but in effect it was put under international surveillance. Adolf Hitler revoked this policy in 1936, and restrictions to canal use applied again during World War II (1939-1945). Following this war the Treaty of Versailles was re-implemented and ever since the canal has been open to all traffic.

In 1948 the canal was renamed as Nord-Ostsee-Kanal (=North-Baltic-Seas-Canal), better known as the Kiel Canal.

2.10 Second enlargement phase

Naval traffic progressively increased and in the 1960s a second enlargement project was launched to allow for the transit of more ships. The project began in the 1960s and to date has cost approximately 600 million Euros. Of the 98.6 km of the canal 88 km were widened and reinforced to prevent bank erosion which is of particular importance in the western section of the canal. The remaining narrows at the bridges in Kiel-Holtenau were widened in 2001 (n.n. 2003).

---

[7] The Kiel Canal is to remain free and open to war and merchant ships of all nations at peace with Germany. Goods and ships of all states are to be treated on terms of absolute equality, and no taxes to be imposed beyond those necessary for upkeep and improvement for which Germany is responsible.

2.11 Future improvement plans

An Automatic Identification System is to be put in place to facilitate computer based canal navigation and improve the efficiency of ship passage through the canal. With this system about twice the number of ships may be handled. The project envisages completion in 2007 with additional improvement of tunnels, ferries and canal locks. The allocated budget is 115 million Euro (n.n. 2003).

2.12 Celebrating the 100th year anniversary

On June 20th 1995 the 100th year anniversary of the Kiel Canal was celebrated. The highlight of the ceremony was again a convoy of ships travelling from Brunsbüttel to Holtenau led by the German naval training sailing vessel *Gorch Fock* followed by the British Royal yacht *Britannia* and more than fifty ships representing 20 nations (Schulz 1995).

3 Topography and canal details

The Kiel Canal has a length of 98.6 km (=53.3 nautical miles). At the waterline it is presently 162 m wide, 102,5 m in narrow sections. The floor width is 90 m, 44 m where the canal narrows. Its depth is 11 m (Tab. 1, Fig. 20). The banks are built with bricks and gravel; some sections have concrete (Arndt 1931/32).

3.1 Topography

The canal was cut through a terrain ranging from 25 m above to 3 m below sea level. At the western Elbe entrance an extensive region of estuarine marshland, coastal moorlands, grass land, forest and boggy soils occurs. Bog, lakes, creeks and small rivers intersect the canal for up to 50 km of its length. A dam separates the canal from the Eider River (see Eiderkanal above). The easternmost stretch is made up of glacial moraines (Arndt 1931/32, Ax 1955).

Table 1. Canal dimensions at its opening and after two improvement projects (www.nok-wsa.de/nok/nok.html May 27th 2004).

Year	Width at water line	Width at bottom	Water depth
1895 (opening)	66.7 m	22.0 m	9.0 m
1914 (completion date of first improvement project)	102.5 m	44.0 m	11.0 m
1965 (second improvement project, almost completed)	162.0 m	90.0 m	11.0 m

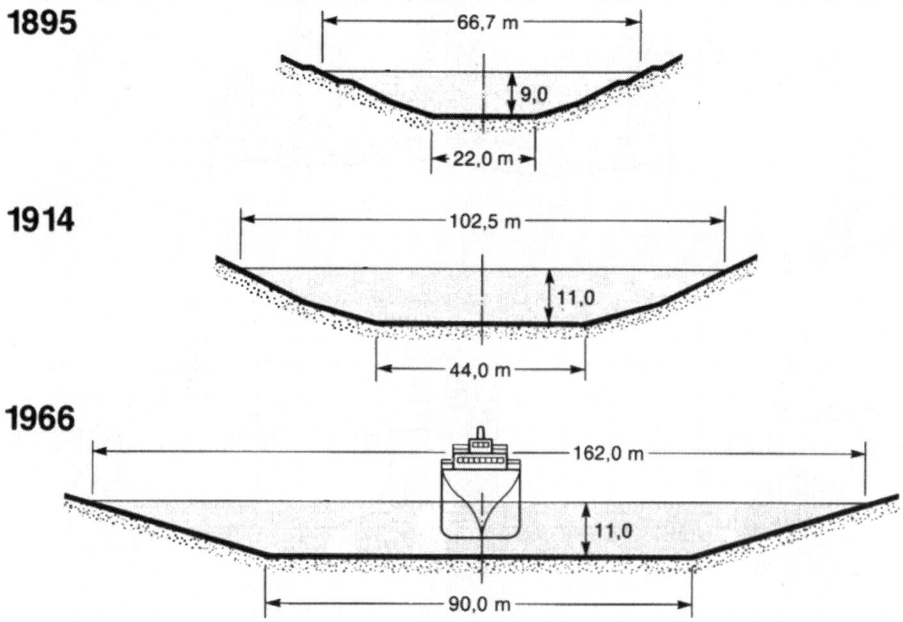

Fig. 20. Canal dimensions at opening and after two improvement projects. (Modified after Hill in Wasser- und Schifffahrtsdirektion Nord, Kiel 2001).

3.2 Locks

Each canal entrance has four locks, two old and two new (Fig. 21). The old locks were the largest locks constructed at that time and were operated by hydraulic systems. Up to 30,000 cubic metres of concrete were used to build them and almost 100 million bricks were used for the construction of the locks and jetties (Schulz 1995).

The usable length of the old locks is 125 m, the usable width 22 m. The lock depth at Brunsbüttel is 10.20 m, and at Kiel-Holtenau 9.80 m. These locks have gates with two opening and two closing doors per chamber and are filled with water through two side channels each with 12 branch channels. The time required for passage is approximately 45 minutes.

The more recent locks have a usable length of 310 m, a usable width of 42 m and a depth of 14.0 m (Fig. 21 & 22). Each lock chamber has three gates. In Brunsbüttel the locks have revolving gates, whereas in Holtenau two side channels exist each with 29 branch channels. The transit time is 45 minutes.

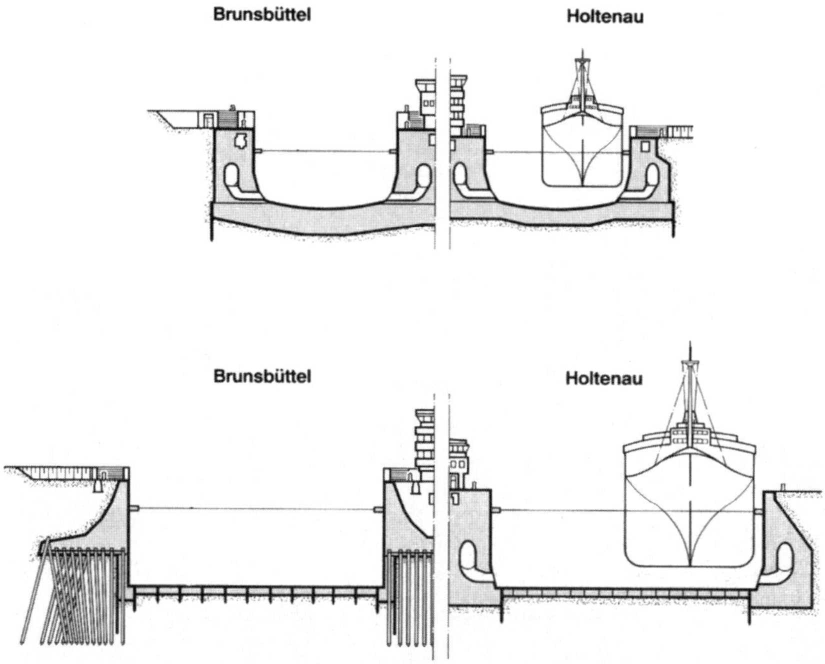

Fig. 21. Sectional drawing of the old (top) and new locks (bottom). Brunsbüttel locks left, Kiel-Holtenau locks right. (Modified after Hill in Wasser- und Schifffahrtsdirektion Nord, Kiel 2001).

3.3 Passing bays

Twelve passing bays (sidings), widened parts of the canal, of 586 to 5,656 m in length are located alongside the canal. They allow larger ships to pass unhindered and to organise priorities among ships. One of these sidings is automated, i.e. is supervised by remote control from the control station Holtenau (Wasser- und Schifffahrtsdirektion Nord 1993).

3.4 Bridges, tunnels and ferries

Important transport routes cross the Kiel Canal; presently ten train and road bridges cross the canal, providing an airdraft of 40 m (Tab. 2). Most spectacular is the railway bridge at Rendsburg, built 1911 - 1913 (Fig. 23). It was made of 19,700 tonnes of steel connected with 1,3 million rivets to form a 2,486 m bridge, still in use today. Also at Rendsburg a pedestrian tunnel and a road tunnel pass under the canal (Fig. 24). There are thirteen ferry crossings for pedestrians and/or vehicles, and beneath the Rendsburg Bridge a unique suspension ferry is in operation.

Fig. 22. Aerial view of the canal bridge and locks (background) in Kiel-Holtenau. Photo courtesy Klaas Hinderk Rosenboom, Hamburg, Germany.

Fig. 23. The railway bridge at Rendsburg built 1911-1913. Courtesy Wasser- und Schifffahrts-verwaltung (WSV) Germany.

3.5 Canal navigation

The waterway is a two-way traffic system. The ship passages and canal navigation are controlled by a team of 24 traffic control officers in control centres in Holtenau and Brunsbüttel. In management of the canal, ship classes are defined according to their size and the nature of their cargo. Ships transporting dangerous cargo and large vessels are only permitted to pass in one of the twelve passing bays. Such vessels are not permitted to overtake slower vessels.

Fig. 24. Street tunnel in Rendsburg. Measurements in metres. Courtesy Mr. Witt, Foto Focus, Brunsbüttel, Germany.

The traffic control centres of the canal provide a traffic bulletin every half hour in order to allow vessels in passage to plan their voyage.

A Pilot-Order is in place. Piloting is mandatory for ships with a draught larger than 3.1 m. However, some exemptions from obligatory pilot service exist. Mandatory piloting is needed, because of the unusual hydrodynamic forces in some sections of the canal and the passage through the locks. Two categories of pilots are available: Firstly, lock pilots who guide vessels through the entrance locks, and secondly canal pilots who navigate vessels through the canal. In the mid-section of the canal a pilot exchange takes place with the first canal pilot leaving and a second canal pilot taking over until the end of the canal. Two additional canal navigation officers are employed on larger vessels (www.ak190x.de/Bauwerke/Bau/Nord-Ostsee-Kanal.htm visited January 18th 2005).

Table 2. Bridges, tunnels and ferries of the Kiel Canal from west (0 km) to east (98 km) (Wasser- und Schifffahrtsdirektion Nord 1993).

Bridge/tunnel/ferry	**Canal km** (west to east)	**Built**
Ferry Brunsbüttel	2.1	
Ferry Ostermoor	4.4	
Road bridge Brunsbüttel	6.1	1979-1983
Ferry Kudensee	7.4	
Ferry Burg	14.8	
Railway bridge Hochdonn	18.8	1915-1920
Ferry Hochdonn	19.1	
Ferry Hohenhörn	24.0	
Highway bridge Hohenhörn	24.9	1985-1989
Railway and street bridge Grünental	31.1	1983-1986
Ferry Fischerhütte	35.6	
Ferry Oldenbüttel	41.2	
Ferry Breiholz	49.9	
Road tunnel Rendsburg	60.9	1957-1961
Pedestrian tunnel Rendsburg	61.4	1962-1965
Railway bridge Rendsburg with Suspension Ferry Rendsburg	62.7	1911-1913
Ferry Nobiskrug	65.3	
Highway bridge Rade	68.1	1969-1972
Ferry Sehestedt	75.3	
Ferry Landwehr	86.8	
Railway and street bridge Levensau	93.5	1893-1894
Road bridge Levensau	93.6	1980-1983
Road bridge Holtenau I	96.6	1992-1995
Road bridge Holtenau II	96.6	1969-1972
Ferry Kiel-Holtenau	97.3	

3.6 Running costs

Except for the period prior to World War I, the canal has been operating at a loss. The management and maintenance exceeds 50 million Euros annually and canal dues generate only 40% of this. However, the canal employs 2,500 workers and supports the industrial sectors and the leisure industry in the region. At Rendsburg an “inland” dockyard provides an important local industry.

3.7 Canal "cleaning"

The Elbe Estuary carries a high sediment load in comparison to the Kiel Fjord (Fig. 25). As a result the Brunsbüttel locks accumulate sediments which need to be removed. Keeping the Holtenau locks constantly open and selective opening of the Brunsbüttel locks during low tide in the Elbe Estuary causes a controlled water flow in the canal from the Baltic towards the North Sea. This flow lasts for approximately three hours, at which time the tide in the Elbe Estuary changes (Arndt 1931/32). Besecke (1893) calculated that each three hour water flow carries 4 million cubic meters of water into the Elbe Estuary. As a result the docks at Brunsbüttel are cleaned and sediment is washed back into Elbe River.

Cleaning locks using water flow was regularly carried out twice a day soon after the canal opening. This considerably reduced the sediment accumulation in Brunsbüttel. Brandt (1897) calculated that water from the Baltic reached the Elbe Estuary within two weeks as a result of the water turnover in the canal following a twice daily flushing. However, this activity added some further risks to navigation and to erosion of the canal embankments. To reduce the negative impact the canal cleaning was later reduced to once per day.

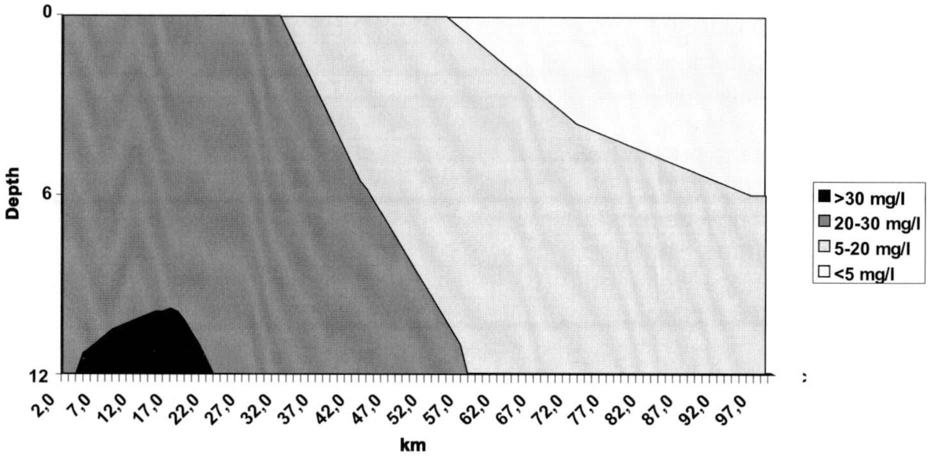

Fig. 25. Sediment content (mg/l) of water in the Kiel Canal. (Modified after Krumm & Rheinheimer 1966).

After the canal enlargement works completed in 1914, the purging of sediments by means of water flow was discontinued and was replaced by a dredging programme. Currently, 6.5 million tonnes of mud are dredged from the Brunsbüttel locks and the nearby canal section each year. Dredging activities along the other canal sections result in 0.1 million tonnes of mud removed annually, reflecting the different sedimentation rate across the canal (Wasser- und Schifffahrtsdirektion Nord 1993).

To manage the natural inflow to the canal by river runoff, draw wells were installed (Arndt 1931/32). Today, major flows only occur following exceptional periods of run off through rivers and drains that enter the canal.

3.8 Canal fee

Canal fees are proportional to ship size. Fees are required for using the Elbe approach and Kiel Fjord as well as for the canal itself. The high fees applied may have reduced the shipping traffic until the 1990s. As a result fees were dropped by 10% in 1993 and a discount for vessels regularly using this passage was introduced to attract more business. An Internet based calculation scheme for canal fees is available at www.kiel-canal.org. On July 16th 2004 a canal passage for a ship of a 10.000 dead weight tonnage (DWT) cost approximately 4,000 Euros. In addition to pilot fees and transit costs (applicable for all merchant vessels) further costs apply for pilotage and helmsman services. The canal fees for a 40,000 DWT vessel were calculated as 7,041 Euros plus 2,802 Euros for the Elbe approach and 1,316 Euros for the Kiel Fjord approach (total 11,159 Euros).

3.9 Recreational aspects

The canal is also used for recreational purposes. However, apart from those leisure crafts that have permanent berths within the canal, such craft are only permitted to use the canal during daylight and only for passage. Sailing is not permitted.

Fishing is also of importance but is only permitted with permission from the waterway and shipping offices at either Brunsbüttel or Kiel-Holtenau. Target species for fishing include Eel, Carp, Herring, Smelt, Pikeperch, Cod, Wittling, Flounder, Plaice and Rainbow Trout (see below).

The canal also draws visitors. Each canal entrance has exhibition centres documenting historical aspects, the building phases and shipping pattern in the canal. Ships passing through the canal may be of general interest. Especially when larger cruise liners or famous yachts pass through the canal, thousands of interested viewers visit the embankments. In addition tourists on cruise liners appreciate passing through the canal; the passage is one of the highlights of a cruise (see below).

4 Importance of the canal for the shipping industry

In ancient times the shipping passage around Denmark was a dangerous challenge. Many ships have been wrecked and an alternative route was sought resulting in various projects to build canals across northern Germany (see above).

Historically naval strategies were important; today the Kiel Canal is an essential trade link for Baltic countries and an important shipping route in Europe, reducing cargo transport by trucks, resulting in more environmentally friendly transportation.

The Kiel Canal is the shortest link between the North and Baltic Seas. Compared with other shipping canals the voyage duration gain using the Canal is small: about 400 nautical miles are saved by using the canal instead of the shipping route around Denmark. On a voyage from Hamburg to Rostock approximately 1.5 days of travel time are saved (446 nautical miles) by using the canal. It is estimated that the shortcut using the canal is at least worth 20,000 Euro.

The canal is open every day. However, in winter 1928/29 a severe period of frost occurred with temperatures dropping to –21°C. The water surface was frozen and the canal was closed for three weeks. Where usually ferries crossed the canal the traffic was now "guided" directly over the frozen canal water surface (Schulz 1995).

The canal can be used by ships with a maximum length of 235 m and maximum width of 32.5 m. The maximum allowable draught is 9.5 m and the maximum ship height is 40 m – limited by the numerous bridges crossing the canal (www.nok-wsa.de/nok/nok.html May 27th 2004). Acceptance of greater draughts are given in individual cases and under special conditions.

Most large cruise liners cannot pass through the canal because of the bridges. However, one of the larger cruise liners, the *Norwegian Dream* (50,764 gross tonnage, length 229.8 m, width 28.6 m, draught 6.4 m), has specially designed funnels and masts which can be folded to the side to enable its passage through the canal. Transits of this ship are a major tourist attraction.

The maximum allowable cruising speed of ships in the canal varies according to ship size, type and cargo. Ships with a draught of more than 8.5 m are permitted to travel at a maximum speed of 6.5 knots (kn) (=12 km/h). For all other ships the maximum speed is 8.1 kn (=15 km/h). The passage through the canal usually takes 6.5 to 8.5 hours excluding the lock passage (approximately 45 minutes) resulting in a total time of 8.0 to 10 hours (www.ak190x.de/Bauwerke/Bau/Nord-Ostsee-Kanal.htm visited January 18th 2005, Lloyds Register-Fairplay 2002).

4.1 Shipping statistics

The number of merchant ships transiting the canal and the cargo transported varies greatly (Fig. 26). On average 47,026 ships passed through the canal (eastbound and westbound) each year, with a lowest number of 16,215 in 1915 (excluding the opening year) and the highest number, 85,919, in 1965. The number of ships using the canal dropped significantly during both World Wars. However, there has been an increase in number of ships passing through the canal since the turn of the last century with eastbound and westbound traffic being fairly balanced.

Although the canal fees were reduced in 1993 as a re-attraction measure the number of transit ships has only been increasing slightly since then. However, the Kiel Canal is still the world's busiest artificial waterway. In 2004 41,682 ship movements were recorded, on average 114 vessels per day, transporting more than 70 million tonnes of cargo between the North and Baltic Seas, the highest amount of cargo transported through the canal ever (www.kiel-canal.org July 16th 2004). Leisure craft are documented in a separate statistic: more than 18,000 craft use the canal annually (www.kueste.de/kanal/kanal2.htm May 27th 2004).

The biggest ship ever passed through the canal was the German naval vessel *Bismarck.* She transited on March 9th 1941. The 56,551 DWT vessel was 251 m in length, 36 m in width with a draught of 9.9 m – exceeding today's vessel size limits for passage (see above).

The majority of cargo transported through the canal consists of general cargo followed by oil, bulk cargo, chemical products, iron and steel, wood and wood chips, fertilizers and grain (www.kiel-canal.org July 16th 2004).

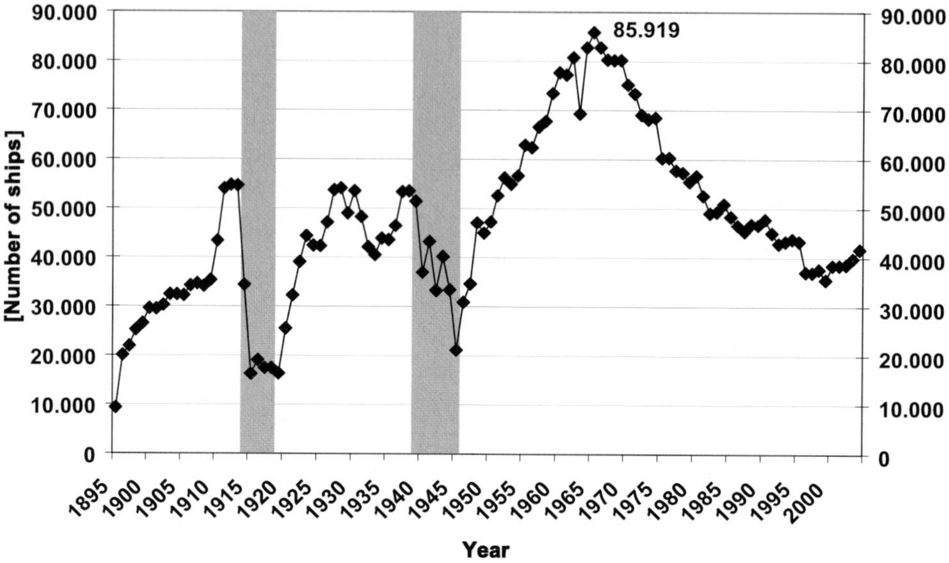

Fig. 26. Number of all non-pleasure boats, i.e. merchant ships, navy vessels and others, passing annually through the Kiel Canal since 1895 (canal opening) to 2004. The maximum number of passings is indicated. Grey bars indicate World Wars One and Two. Data source: Wasser- und Schifffahrtsdirektion Nord, Kiel, Germany.

4.2 Traffic accidents

Although the most recent technologies and safety practices are used to guarantee a safe passage some accidents occur. Most result from technical failures, i.e. rudder system failure leading to ship, canal lock or embankment damage. In 2003, 89 ship accidents were reported.

Most accidents are of minor consequence. However, deaths resulting from accidents have occurred. In rare situations the canal was closed to enable salvage measures (German Federal Bureau of Maritime Casualty Investigation 2004, www.ism-center.de/d2600.htm visited February 10th 2005).

5 Canal water characteristics

5.1 Temperature

Northern Germany has a cold-temperate climate and the Kiel Canal shows water temperatures comparable to adjacent waters. In severe winters ice cover occurs (see above). During the exceptionally cold winter of 1928/29 ice breakers were employed for 70 days to allow navigation (Arndt 1931/32).

5.2 Salinity

While other major canals connect different waters and are mainly isolated from rivers and lakes, the Kiel Canal is unique because of its direct connection with many smaller rivers and creeks that drain directly to it. Further, several lakes are also linked with draining creeks and ditches to the canal, allowing movement of fish between them while also permitting migratory fish to enter.

The river systems that the canal crosses are used as spawning and nursery grounds for fish that forage in the canal. This makes the Kiel Canal unique, affecting not only natural freshwater bodies and their biota, but also being affected by these freshwater systems. Perhaps these influences are as important as those caused by biological invaders which enter the canal via the locks. In fact, canal immigrants (native or non-native to the region) may particularly benefit from this setting by finding seasonal refugia in adjacent waters the canal itself cannot offer.

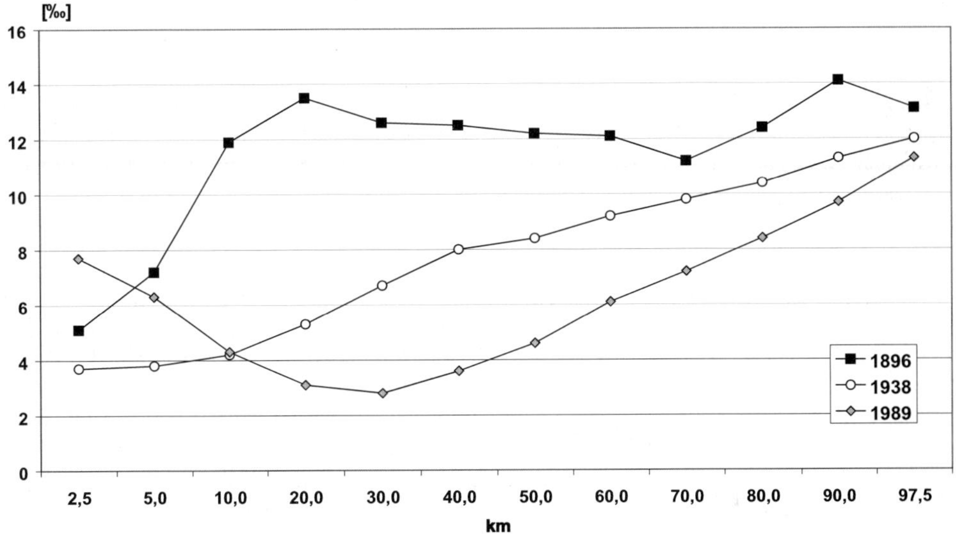

Fig. 27. Mean salinity in the Kiel Canal according to Canal kilometres in 1896, 1938 and 1989. (Modified after Czerny 1996).

The salinity in the Kiel Canal varies according to the run-off from adjacent lakes, creeks, small rivers entering the canal and also by marine influence from the Elbe River estuary and the Baltic Sea. In the Elbe Estuary salinity fluctuates from approximately 3 - 20 PSU at the bottom water layer. In Friedrichsort, close to the canal entrance in Holtenau, the bottom water salinity ranges from approximately 10 - 29 PSU (Arndt 1931/32).

The Baltic Sea locks at Holtenau were open during the first few years of the canal and were only closed when the water level in the Kiel Fjord was more than 0.5 m lower or higher than the canal's level. This normally happens for 25 days per year. After 1898, the locks were more frequently closed and in 1901 they were only open for 7 days (Arndt 1931/32). When the locks are open at Holtenau there is an increase in salinity throughout the canal and also its western part becomes brackish. When the locks are closed, significant river run off occurs, and salinity declines (Fig. 27) (Czerny 1996).

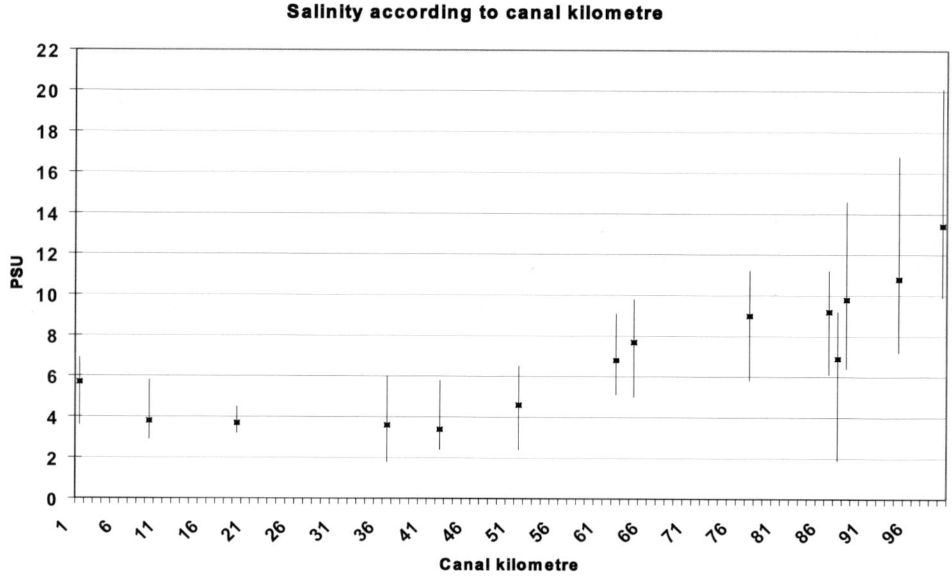

Fig. 28. Surface water salinity gradient in the Kiel Canal in PSU according to canal kilometres based on data for Summer 1953. (Modified after Schütz & Kinne 1955). Kilometre measured from west to east, i.e. Kilometre 1 = Brunsbüttel, kilometre 97.5 = Holtenau. Boxes = salinity mean, upper end of column = maximum, lower end of column = minimum salinity.

Today, the canal water is brackish with a salinity gradient along its length. In summer, salinity is highest near the Holtenau entrance because of the influence of bottom water inflow of the Baltic. The lowest salinity occurs in the middle canal section (canal km 20 - 40). Progressing towards Brunsbüttel, salinity increases slightly due to the brackish water entering the canal via the Brunsbüttel locks. The salinity inflow in Holtenau has a longer (wider) reach until the middle section of the canal. The influence of the Brunsbüttel inflow at the canal bottom extends for a maximum of 18 km inland (Arndt 1931/32, Schütz & Kinne 1955, Czerny 1996).

Near the Brunsbüttel end the salinity of surface and bottom water fluctuates between 4 and 7 PSU. In the middle section of the canal at km 50, surface and bottom water have a salinity of 4 and 6 PSU. The salinity is highest in Holtenau. Here, the surface water salinity lies between 10 and 20 PSU (Fig. 28), the bottom water 15 to 19 PSU (Arndt 1931/32, Schütz & Kinne 1955).

The inflowing, saline Baltic water sinks to the bottom while the surface water is influenced by the discharge of freshwater run off to the canal. The total catchment area is 1,580 square kilometres (Wasser- und Schifffahrtsdirektion Nord 1993). Fresh water inflow is highest in spring. A stable stratification does not occur, due to ship movements.

5.3 Canal zonation

Based on the biological and environmental features of the canal, Czerny (1996) in his literature review divided the Kiel Canal in four sections:

(a) the stretch from Brunsbüttel harbour to the ferry of Hochdonn (canal km 1.7 – 19.1),
(b) from ferry Hochdonn to ferry Breiholz (km 19.1 – 49.9),
(c) from ferry Breiholz to ferry Sehestedt including the Obereidersee (km 49.9 – 75.3), and
(d) the section from Sehestedt to inner harbour of Kiel – Holtenau including the Flemhuder See (km 75.3 – 98.8).

6 History of biological studies in the Kiel Canal

The first faunal study of the canal was carried out by Brandt in November 1895 approximately four months after the official opening of the canal (Brandt 1896 a, b). Brandt published two reports: one on the fauna of the Kiel Bight and the Kiel Canal (Brandt 1986 a) and a further one on the migration of marine fauna into the canal (Brandt 1986 b), the first publication analysing colonization of the canal.

Additional studies on the canal fauna were published by Brandt in 1897 while Hinkelmann (1897) published a series of papers on fisheries and fish occurrences in the canal, showing a high variability in the number and quantity of species invading and recording the development of the fish fauna and the fishery for a full decade (Hinkelmann 1897, 1898, 1899, 1900, 1903, 1905, 1906, 1907). In particular, he reports early "invaders" of such species as the Herring (1897) with the first recorded spawning of the species in the canal in 1902, the Bream (*Abramis brama*) in 1898, the River Lamprey (*Lampetra fluviatilis*) in 1902.

Eel, Flounder, Herring and Cod were found in 1898 during fishing activities near Rendsburg in the Meckel- and Flemhuder Lakes (Hinkelmann 1903, Barfod 1904). In 1903 a dense population of Flounder was observed at canal km 71-85 (Barfod 1904).

In 1904 Barfod published a study on Baltic species migration into the canal.

The first study on algal species in the canal was carried out by Reinke and Darbishire in 1896, published in 1898. However, the research on canal biota was dominated by fishery related studies (Andresen 1899, 1900, Barfod 1903, 1904).

The first PhD thesis based on sampling canal biota was by Dechow (1920), with a focus on the bottom-living fauna.

Brandt (1897) calculated that Baltic organisms may have reached the Elbe Estuary via the surface water of the canal within a week, aided by the flushing activities often performed (see above). This is consistent with the initial canal colonisation by Baltic species. However, after 1913 no "cleaning" (flushing) occurred. Species also moved into the canal by ship movements. Hinkelmann (1896) observed that the bow wave of ships transported species for several kilometres into the canal. However, this observation is anecdotal and one may suspect that if at all effective, this transportation means could operate in either direction.

Arndt (1931/32) lists 115 zooplankton and zoobenthos taxa from various sampling stations in the canal and states that the known taxa from the canal are underestimated, especially since several groups of taxa had not been studied to species level. One third of the taxa he identified were freshwater species; two third were brackish or marine taxa. He presumed that freshwater species were washed into the canal by river run off and were unlikely to survive the brackish conditions in the canal.

As may be expected, marine species are more frequent in the eastern part of the canal where salinity is highest. There is a division that separates the majority of brackish from freshwater taxa at Rendsburg (approximately at canal km 60). Dechow (1920) concluded that brackish zoobenthos numbers in the eastern part of the canal were similar to those in the western Baltic. Whereas the number of zoobenthos taxa in the western part of the canal were similar to those of the Bothnian Gulf (eastern Baltic). Dechow also observed the well-known phenomenon that mussels in the western, least saline canal section had small and more fragile shells than individuals in the eastern, more saline section.

In 1951, the fouling of wooden pilings of a mooring pier near the canal entrance in Holtenau was investigated by Ax (1952). He sampled at depths of 2 to 9 m. The species composition was dominated by hard bottom biota including some species unknown from the adjacent Kiel Fjord: *Victorella pavida*, *Pseudomonocelis cetinae*, *Promesostoma bilineatum*, *Pentacoelum fucoideum* and a species unknown to science, was found: *Jensenia luetjohanni,* named after Mr. Luetjohann who took the samples via diving (Ax 1952).

The study was repeated in 1955 by Schütz & Kinne (1955): wooden pilings were qualitatively sampled from September to November 1953 for mobile organisms at depths of 0.5 to 3 m. Sampling stations were located throughout the canal, starting in Brunsbüttel (km 0), followed by kms 8, 18, 34.5, 41, 50, 61.5, 62.5, 75.5, 85.5, 86, 87, 89.5, 92.5 and 97.5 (Holtenau). For comparison, samples were taken outside the canal at three sites in the Kiel Fjord. Sessile organisms were not included in the analysis. The principal species were Nematoda, Harpacticoida, Polychaeta, Rotatoria, Oligochaeta, Turbellaria and Halacarida.

From the 1950s to the 1970s, studies on the Kiel Canal biota were conducted by Kinne (1956), Schütz (1961, 1963, 1969) and Kothe (1973) and more recently by Bothmann (1998), who studied the fouling communities on wooden pilings, Spratte & Hartmann (1998) studied fish abundances and Kafemann (2000) made the first time ever investigation of trophic interactions of fish. Kafemann (2000) also provides the first study on the feeding and food composition of fish in the canal through extensive gut analysis, reporting on food composition and biomass in relation to fish size.

Biological invasions were not a primary concern, although it was speculated that the Kiel Canal acted as a corridor for the spread of alien species in both directions (e.g. Jensen & Knudsen, submitted).

More investigations on the fish fauna were conducted during the 1980s and 1990s. These were undertaken in support of the state policy and regulations in Schleswig-Holstein and to monitor and control fishing activities in the canal and the riverine and lake habitats draining into it.

7 Colonization of the canal by native and non-native species

Aquatic species colonised the canal either by active or passive immigration from both the Baltic and North Seas as well as from freshwater run off from adjacent streams and lakes (Fig. 29).

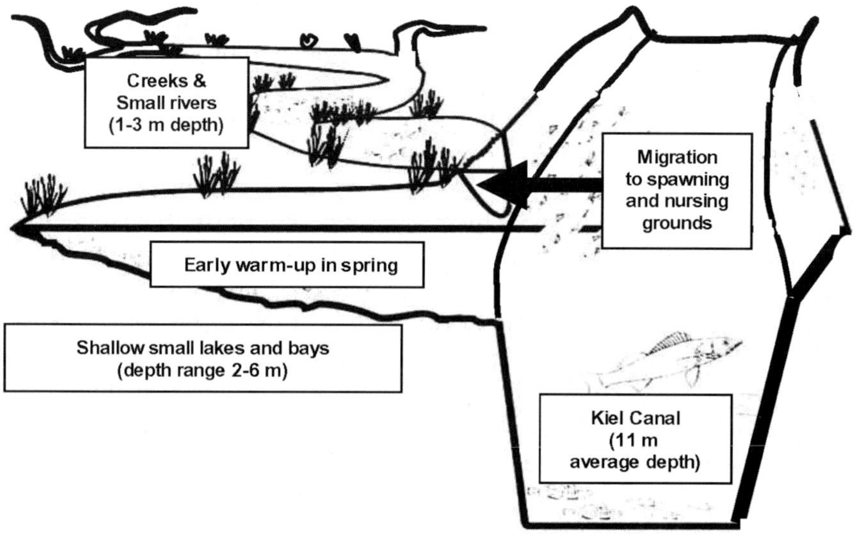

Fig. 29. Schematic cross-section of the Kiel Canal and the draining lakes and creeks, depicting their depth and habitat differences. (Modified after Czerny 1995).

It has been calculated that more than 90% of these taxa originated from the Kiel Fjord and only 2% from waters near the western end of the canal. The remaining 8% could have colonised the canal from waters adjacent to the canal (Dechow 1920, Arndt 1931/32).

7.1 Overall distribution of the benthic fauna in relation to salinity

Kothé and Sindern (1973) examined the macro-zoobenthos. Apparently, the biocoenotic characteristics of the canal have been influenced by the initial flushing of the canal with Baltic Sea water while the limnic benthos relates to a spectrum of freshwater species known from those freshwater systems now connected to the canal. The authors summarize their findings on the zoobenthic species by categorizing them into four groups:

(a) euryhaline marine species,
(b) genuine brackish water species,
(c) brackish-limnic adapted species, and
(d) limnic species with some euryhaline tolerance.

Kothé and Sinder (1973) plotted the distribution of benthic species according to their "halinity" in a similar manner as the "Saprobien-Index" to describe the benthic community's response to pollution (Fig. 30). Figure 31 shows the distribution of the four species groups revealing a predominance of euryhaline marine species in the eastern part of the canal while strictly limnic species are restricted to the western section.

7.2 Bacteria

Rheinheimer (1997) studied bacterial abundance along salinity gradients in German estuaries and the Kiel Canal. Overall, this study, along a transect, indicated a slight decline in abundance from west (North Sea) to east (Baltic Sea) for bacteria growing in freshwater medium whilst a clear relationship between heterotrophic bacteria (saprophytes) and the salinity gradient was observed. Although the counts are less than 10% of the total, these seem to be a very active group of microorganisms, somewhat depressed at the lowest salinity but quickly recovering once exposed to seawater. Apparently their importance in self-purification of water bodies through nutrient conversion is important also in the Kiel Canal and needs to be considered also in similar canals with estuarine conditions. Coliforms and bacteria growing on nutrient agar seem to be at low level throughout the canal compared to the growth potential of marine heterotrophs.

When comparing with microbial studies in general it was concluded that many bacteria have a narrow salinity optimum, leading to significant changes in their activity in a salinity gradient of a few PSU. Especially during times with heavy rainfall and little ship traffic (small inflow of high salinity water through the locks) these bacteria may quickly loose their competitiveness. Different from other estuaries with quickly changing salinity regimes, the Baltic Sea Canal has a relatively stable salinity gradient across seasons and therefore hosts a unique potential for adaptations to certain salinity ranges. A distinction between several ecological canal sections therefore seems as justified as that for fishes and benthic invertebrate fauna. This, however, requires further bacteriological studies.

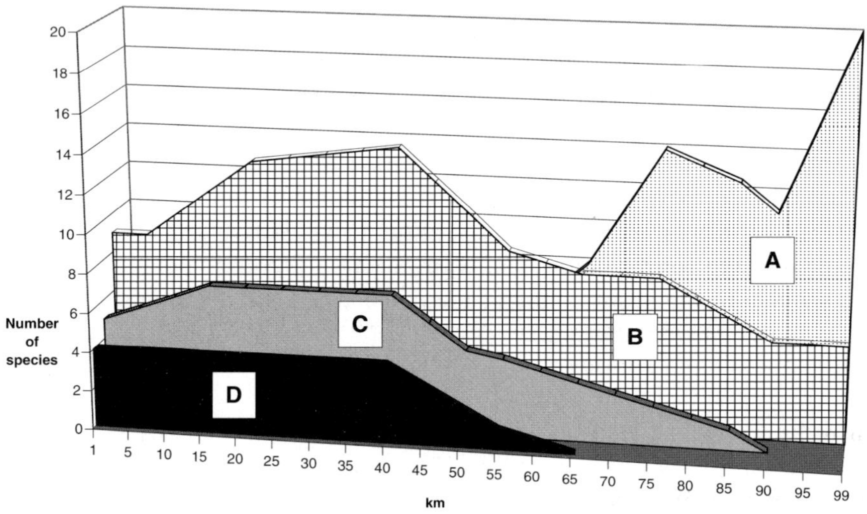

Fig. 30. Distribution and frequency of occurrence of benthic macrofaunal species in the Kiel Canal in accordance with their "halinity" (salinity preferences): a) euryhaline-marine, b) genuine brackish, c) brackish-limnic, d) mainly limnic. (Modified after Kothé & Sindern 1973).

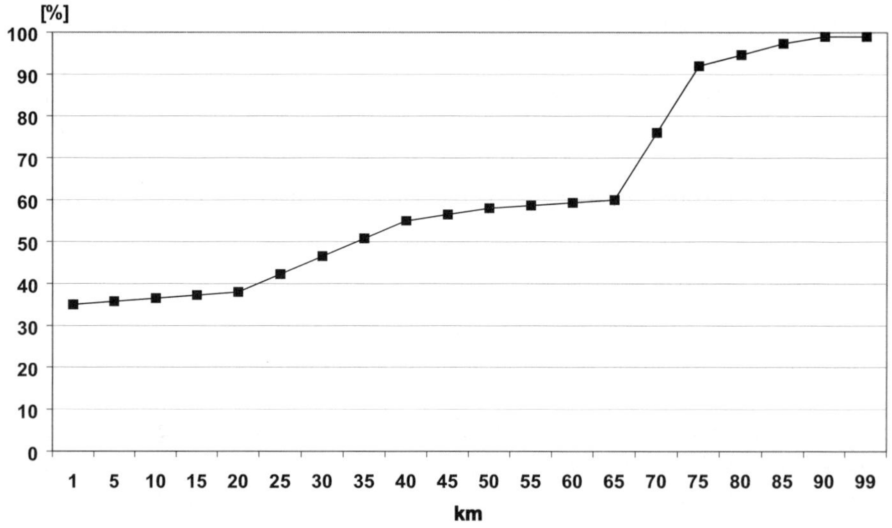

Fig. 31. Percentage of marine and brackish water species compared to all taxa across the canal. (Modified after Kothé & Sindern 1973).

7.3 Algae

Baltic macroalgae and sea grasses were first recorded during the spring of 1896 (Brandt 1986 b, Arndt 1931/32, Aleem & Schulz 1952). Dense sea grass beds extended into several parts of the canal in 1898, and *Fucus* spp. formed dense populations from 1899 onwards, gradually expanding westward. Brandt (1895, 1896) also noticed a similar westwardly directed colonisation of marine phytoplankton (e.g. *Chaetoceros* spp.).

Aleem & Schulz (1952) observed a zonation of benthic algal communities along the supra- and sublitoral of the canal and distinguished between four communities which they characterized during a September study:

(a) Cyanophycean dominated community,
(b) *Urospora-Ulothrix-Capsosiphon* assemblage,
(c) *Enteromorpha* community, and
(d) Assemblage dominated by diatoms.

Aleem & Schulz (1952) also described in the supra-littoral zone an extensive lichen community, on wood and stone pilings, principally composed of *Verrucaria maura* (Wahlbaum).

The Cyanophycea-community is well established in the **first zone** in the splash water level along the beaches of the canal, regularly wetted by the wash of passing ships. This zone is dominated by *Calothrix pulvinata* meadows, usually with *Gloeocapsa crepidinum*. These species are resistant to desiccation and well adapted to splash conditions.

The **second zone** (*Urospora-Ulothrix-Capsosiphon* assemblage) is best developed close to the water level and up to 10 cm above, and is dominated by *Urospora penicilliformis, Capsosiphon aureolis* and *Ulothrix subflaccida* which also support several epiphytes. Aleem & Schulz (1952) found this community predominantly near the locks of Holtenau and the canal stretch towards the Schülp bridge.

The **third zone** overlaps with the previous community and is mainly formed by *Enteromorpha intestinalis* and closely related species (*E. ahlneriana* and *E. linza)*. These form meadows and their density depends on the beach profile and substrate type, while their vertical distribution is mainly light-limited. The *Enteromorpha* belt extends to a depth of 50 cm and does not exceed 2.5 m. Several other species co-occur in this zone, e.g. *Cladophora glomerata.*

The **fourth zone** is dominated by diatoms and expands to a depth of about 2.5 m (Aleem & Schulz 1952) with *Melosira juergensii* as the dominant species, often accompanied by the genera *Achnanthes*, *Navicula*, *Pleurosigma*, *Cocconeis* and others. Within this community, the cyanophyceaen *Spirulina subsalsa* is also frequent, as well as *Ectocarpus* and *Phyalla* spp.

Soon after the canal opening *Fucus vesiculosus* was recorded but it was not found by Aleem and Schulz (1952). These authors indicated similarities in species composition with the Schlei "Fjord" in the north of the Baltic coast of Germany, although some species are completely absent from the canal. Brandt (1895, 1896) observed a westward colonisation of many marine plankton algae (e.g. *Chaetoceros* spp.), in contrast with the phytoplankton species *Coscinodiscus wailesii* that colonized the canal from the Elbe Estuary.

7.4 Invertebrates

Brackish water invertebrates colonised the canal soon after its opening. Two months after the opening, at a time when the Holtenau locks were constantly open, *Mytilus edulis* was found up to canal km 90. Other common taxa in the Kiel Fjord, such as *Aurelia aurita*, *Cyanea capillata* and *Mysis vulgaris*, were also frequently reported at the same time (Arndt 1931/32). Shortly after its first record in the canal, *Aurelia aurita* was found in Brunsbüttel, making it the first "immigrant" to appear along the entire canal length (Dechow 1920). In November 1895, *Mysis vulgaris* was collected by Brandt (1896 b) from various sites almost as far as Brunsbüttel. In winter 1895/96 and spring 1896 the first brackish water crabs, fish (including pipe fish) and seastars were noted. However, upon terminating the use of the water pulse system for removing sediment from the canal (see above), the salinity decreased and many species again disappeared. Whilst *Amphibalanus (=Balanus) improvisus* successfully spread and became more common, *Mytilus edulis* numbers declined (Arndt 1931/32). Freshwater sponges, such as *Spongilla lacustris* have not been reported from the canal since the salinity increased at the end of the 19th Century (Hinkelmann 1898).

Amphipods can be found throughout the canal. Seven species invaded it, four of which are hemi-sessile. Two are distributed according to their salinity preference. *Corophium insidiosum* has not yet been found to reproduce in the canal and its population represents an annual immigration that dies off in winter. *C. insidiosum* occurs only in the easterly, brackish water sections while *Corophium multisetosum* is restricted to the western part at low salinities. *Leptocheirus pilosus* is the most common species and is often accompanied by *Corophium lacustre*. *L. pilosus* is often found between filaments of *Cladophora glomerata* and is rare on sandy substrates or areas low in oxygen. The species shows two annual reproductive maxima (June to early August and end of September to October) (Arndt 1931/32, Ax 1952).

Until 1979, of all isopods known in German brackish waters, only one species of *Sphaeroma* (*S. hookeri*, Leach 1814) became established in the Kiel Canal. Betz (1979), re-analysed previous finds and compared these with his own investigation, confirming that his record of *S. rugicauda* represents the first record in the canal. Before that, Betz believed it was already present but was overlooked. The incomplete knowledge of the ecological requirements of the two species permits nothing more than speculation why S. *hookeri* is more abundant compared to *S. rugicauda*. Frier (1976) provided some hints as to the different osmoregulatory capabilities of the species, favouring *S. hookeri* at lower salinities. Beside these two species of isopods, *Jaera albifrons* is also common on pilings along the entire canal. However, its abundance declines with decreasing salinity. Two additional isopods are rare: *Idothea viridis* (a euryhaline species) and *Cyathura carinata* (a strictly brackish water species).

During summer, dense patches of *Neomysis vulgaris* can be observed (Schütz 1969). At temperatures below 5°C, it disappears from surface water. A comprehensive description of its life cycle is given by Kinne (1955).

Boje (1965) reported growth variability in *Mytilus edulis* in the Kiel fjord and Kiel Canal. He transplanted mussels from one source to various test sites in both habitats and found that the best growth was in the inner Kiel Fjord. This may be because food was more abundant here during most of the year. Growth in the outer Fjord and the Kiel Canal growth was similar only during summer. However, during fall mussels in the canal have low tissue content and lower shell weight than at Baltic test sites. Also during spring growth was retarded, probably due to abundant organic detritus in the water column. The seston load follows a gradient with increasing intensity from east to west along the canal (Krumm & Rheinheimer 1963). Boje (1965) who studied the eastern part of the canal near the locks of Holtenau (Kiel) suggested bacteria as an important food source for mussels, confirming work by Rheinheimer. *Skeletonema costatum* – an alternative food source – is largely missing from the canal.

7.5 Fish records over time and by species

It is to be expected that fish entered the canal soon after its construction. While some have already been mentioned, we here focus on fish not only during the early years of the canal but on the total fish species ever observed, whether native or non-native, and penetrating this new habitat by invasion from adjacent sea, brackish or freshwater areas.

To understand the richness of the fish fauna of the Kiel Canal we refer again to the numerous small creeks and small lakes draining into the canal or connected to it via small canals (see above). These water bodies provide a wealth of habitats suitable as refuge or spawning/nursery areas, permitting species to survive and immigrate or retreat from the canal, depending on season, salinity and other environmental conditions. Although it is impossible to completely address all these linkages and their influence on the stationary and semi-migratory fish fauna, several well documented cases demonstrate the uniqueness of the fish fauna invading and inhabiting the canal. Without these linkages, many species and their population sizes could not be maintained.

7.5.1 Fisheries in the Kiel Canal

Catch statistics of the total fishery have been sporadic. It is only for the past few years that the sport fishery is obliged to fill in catch protocols (Czerny 1994). The commercial fishermen have recorded their catches since 1982 for the section between km 65.4 and km 93.1. For Herring catches, a fairly reliable statistic exist since 1936 for the western part of the canal and since 1939 for the eastern part. Since 1987 the catches are recorded separately for three canal sections in order to account for the different salinity regimes while trying to reflect the specific fish fauna per section. Since 1990, it was found advisable to re-arrange the areas and divide the canal into 4 sections.

7.5.2 Early records

Hinkelmann (1898) reported Pike, Perch, Bream, Pikeperch, Roach, Bleak in high numbers and in all size classes. Pipefish, commonly recorded in 1895 and 1896, vanished after salinity depletion in the canal (see above) (Arndt 1931/32).

In 1899, thirty individuals of pikeperch were found by fishermen in the Kiel Fjord and Arndt (1931/32) assumed that these individuals must have been migrants from the canal.

Barfod (1904) reported that “suddenly” several smelts were caught by commercial fishermen in the Kiel Fjord in October 1903 and he suggested that these were immigrants from the Kiel Canal. A similar occurrence of smelt was documented by Neubaur (1926).

In 1926, a young Sturgeon was caught in the canal (Mohr & Dunker pers. com. in Arndt 1931/32).

Since 1906, the Baltic Herring has migrated through the canal into the Elbe Estuary, as documented by investigations at the Brunsbüttel locks (Hinkelmann 1907).

7.5.3 Overall occurrence of fish species

Occurrences of fish larvae in the Kiel Canal have been repeatedly recorded. This signals either active spawning, or passive drift of larvae with the water masses through the locks, accompanied by a good survival potential.

Figure 32 depicts the relative abundance of fish species in the Canal in 1995, but this is certainly nothing but a "snapshot" description for this particular year (Kafemann et al. 1998). Comparing earlier studies with the one performed later by Kafemann (2000) demonstrated that both the frequency and biomass of a species per unit area fluctuates seasonally and between years.

River Lamprey (*Lampetra fluviatilis*)
Early records of individual specimens have been made by Hinkelmann (1902 1904); it now regularly occurs. It had been seen spawning in one of the tributaries of the canal, the Haaler Au (Bley 1989) and Czerny (1995) considers its frequency of occurrence in the entire canal in the range of 11,100 specimens since 1984. However, near the locks of Holtenau, Kardel (1995) reported several specimens caught as by-catches of the silver eel night fishery in the mid 1980s. Up to seven specimens were captured per night. Nellen and Dehus (1985) present a map on which records are given for the Todenbüttler Au (which drains into the canal) and from the canal itself near the mouths of the rivers Wehrau and Besdorfer Bach.

Sturgeon (*Acipenser sturio*)
Spratte & Hartmann (1992) list the lower Eider River ("Untereider"), a tributary of the Kiel Canal, as the last spawning ground of native sturgeon in Germany until 1936. Dallmer (1890) optimistically predicted that the sturgeon would adopt the Kiel Canal as suitable habitat for feeding although not as a suitable path to migrate into the smaller creeks where spawning may possibly occur. Hartmann & Spratte (1995) refer to a personal communication by Czerny who searched historical records of sturgeon catches in the Kiel Canal and its

tributaries, and found a total of four records: two juveniles (about 1 - 1.25 kg) in the Schirnauer Lake near Rade in 1926 (Duncker 1964) and 1927 (OFM 1927), and two juveniles weighing between 1 - 2 kg at Canal km 21 (near Dückerswisch) in 1936. In July 1990, an individual of 1 meter length was reported by a sport angler (weight about 4.5 kg) from a place near the ferry Hochdonn. However, it is unclear whether this was a native sturgeon or a non-native species.

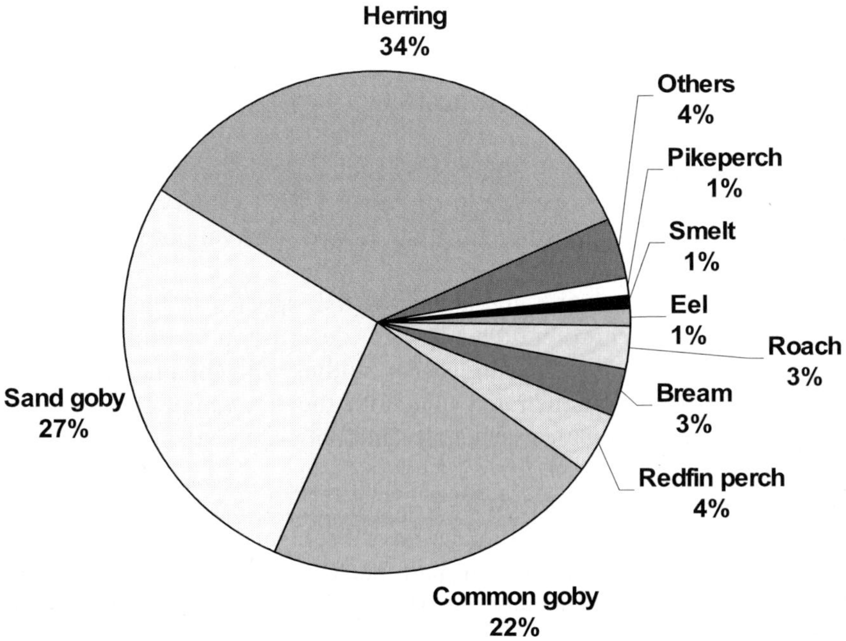

Fig. 32. Composition of the fish community in the Kiel Canal in 1995 in terms of relative frequency of occurrence of key species. (Modified after Kafemann et al. 1998).

Atlantic Salmon (*Salmo salar*)

Individuals of salmon (*Salmo salar*) were occasionally captured by both commercial fishermen and anglers. It is assumed that these – while on their feeding migration - were following migratory herring schools. A total of 8 specimens were recorded in the period 1984-1994. Wiese (no year) reported that during the early 1970s a salmon, tagged in Sweden, was caught in the Kiel Canal but no further records have become known since then.

Sea Trout (*Salmo trutta* forma *trutta*)

This species is stocked to support a recreational fishery in rivers of the State of Schleswig-Holstein. The recreational fishery in the Kiel Canal benefits from the releases of this hatchery-produced fish. On average the yield is about 200 kg per year.

Coregonus albula
A small number of specimens have been captured as by-catch in other fisheries in the canal and these may have originated from the Wittensee via the Schirnauer Au. This species is also hatchery-produced and is stocked in the Wittensee for recreational fishing.

Smelt (*Osmerus eperlanus*)
Smelt is an abundant species, occurring along the entire canal but apparently not migrating into the small tributaries draining to the canal.

Pike (*Esox lucius*)
This species occurs naturally in most rivers and creeks draining into the Kiel Canal. However, there are also some regional stock enhancement measures. Pike is relatively abundant in the canal and migrates upstream into rivers and creeks draining into it in order to spawn.

The canal itself does not posses suitable spawning habitat for pike. During the years 1990 and 1991, juvenile pike were stocked in the canal. Presently, the total landings are made up of about 100 specimens of about 2 kg average weight. The largest size recorded is approximately 10 kg. Regulations limit the catch to minimum total length of 40 cm (Hartmann & Spratte 1995).

Bream (*Abramis brama*)
This species was first reported by Hinkelmann (1898) while Henning (1939) considered bream to be frequent, particularly between Rendsburg and Brunsbüttelkoog. It is now the most abundant fish in the canal (Hartmann & Spratte 1995, Czerny 1995). The larger rivers and creeks draining directly into the canal are used for spawning.

Bleak (*Alburnus alburnus*)
This small fish schools near the surface of shallow stretches in clear rivers and lakes, thriving in or near dense vegetation and overwintering in deeper water. It occurs in the western and central part of the Kiel Canal (Hartmann & Spratte 1995) but is not abundant and is considered endangered in the state of Schleswig-Holstein. One fisherman reported catches of this species between 1927 and 1988, but declining over time. During the past 12 years, no records from commercial fishermen are available. Since the late 1980s, anglers have reported no more than five specimens (Hartmann & Spratte 1995).

Eel (*Anguilla anguilla*)

The immigration of elvers into the Kiel Canal takes place from the western entrance at Brunsbüttel (North Sea) as had been reported by Eichelbaum (1924) and Neubaur (1933). Although glass eels and elvers are commonly attracted by a freshwater outflow into a marine environment, the situation in the Kiel Canal is different. At Brunsbüttel, the Elbe estuary is already low in salinity and except for the attracting current from the locks, there is no salinity gradient to guide the migratory juveniles. Eichelbaum (1924) notes this peculiarity as being the reason why relatively few elvers enter the Kiel Canal compared to the numbers entering the Elbe estuary (Fig. 33).

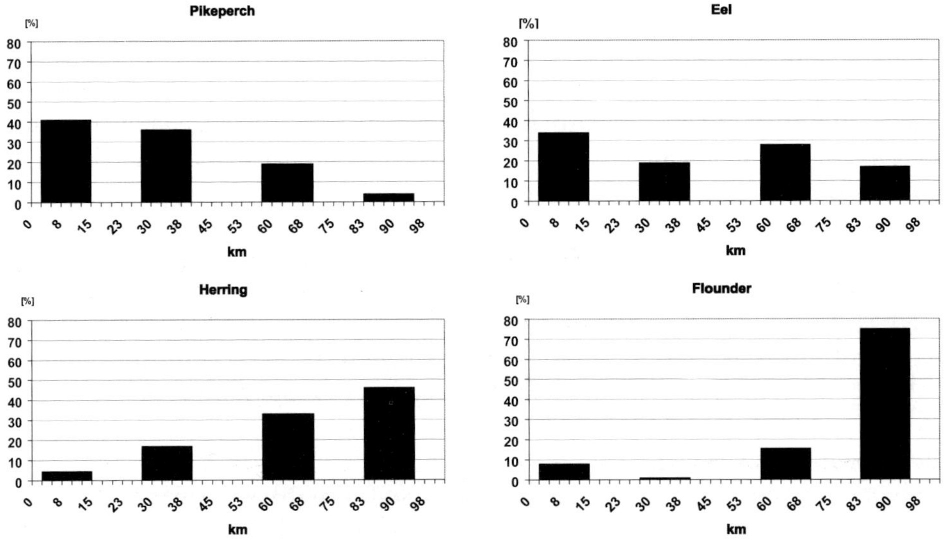

Fig. 33. Occurrence of Pikeperch, Eel, Herring and Flounder in percent of all fishes caught according to canal kilometres. (Modified after Kafemann et al. 1998).

Herring (*Clupea harengus*)

Herring entered the Kiel Canal from the very beginning (Hinkelmann 1896, 1898, 1899). A detailed account of the spring spawning Herring population appearing in the Kiel Canal and entering through the Locks at Kiel Holtenau, was given by Brandhorst (1955).

7.5.4 The Kiel Canal as a spawning ground for Herring

A striking phenomenon is the importance of the Kiel Canal as a herring spawning ground. The first study on the spawning of Baltic Herring near the locks of Kiel-Holtenau was by Brandhorst (1956). Kils (1992) impressively showed the behavioural aspects of mature fish schools, ready to spawn, utilizing the water currents and micro-turbulences along a minor salinity and temperature gradient at the time when the lock gates open. They actively swim into the locks, not chased away by moving ships. Fiedler and Kils (1990) also caught larvae of the following fish species: *Liparis liparis* (km 90), *Taurulus bubalis* (km 70, 80, 90, 97), *Agonus cataphractus* (km 50 - 97) and *Pholis gunellus* (km 50 - 97).

In addition to the importance of the Kiel Canal as a migration pathway of species it also is an important nursing area for certain fish (Arndt 1931/32). Arndt assumed that the canal is more important as a recruitment habitat than for species migrations.

The key recruitment species is the herring (Hinkelmann 1899). In 1896 Hinkelmann observed large swarms of herring and herring larvae. In 1899 he located a Herring spawning ground of approx. 1 km length near canal km 74 with up to 5.500 eggs per 10 cm^2. Later he documented other herring spawning grounds in the canal and the migration of young herring into the Baltic. Herrings became so numerous that certain canal regions were leased to commercial fishermen from nearby Eckernförde. Sprat ready to spawn were also found. However, after the salinity depletion in the 1910s, sprat ready to spawn did not re-occur.

Weber (1971) listed the spawning grounds of spring-spawning herring in the western Baltic, including the Kiel Bight and adjacent sea areas. His study showed that the Kiel Canal is one of the major spawning grounds. While other man-made activities along the Baltic coast (in particular in harbours) have reduced the availability of coastal spawning habitat, here a positive man-made effect is obvious, making the Kiel Canal currently the second major spawning area along the German Baltic coast next to the Schlei estuary. The ripe and running fish caught between 1960 and 1969 ranged from 34 - 290 tonnes. However, these data reflect the catches of one fisherman only using the same gear at all times.

Certainly, the traditional recreational fishery on Herring during the short spawning season (which varies over the years between March and May) adds additional unaccounted tonnes to the overall yield. As can be seen from Fig. 34, Weber (1971) found that spawning of spring-spawners took place throughout a long stretch of the Kiel Canal. While the northern beaches are preferred spawning habitats of the eastern sections, the westerly shores are preferred in the western sections of the Canal (Hinkelmann 1902, Neb 1952).

Baltic spring-spawning herring spawn in shallow waters, seldom below the 3 to 4 m depth line. The boulders and rocks along the shores of the Kiel Canal offer ideal habitat, however, since it has no or a weak current only. One may wonder how dense egg depositions on and between the boulders and coarse gravel material survive, because a high water exchange in the interstitial spaces is required to bring oxygen-rich water to the egg surfaces.

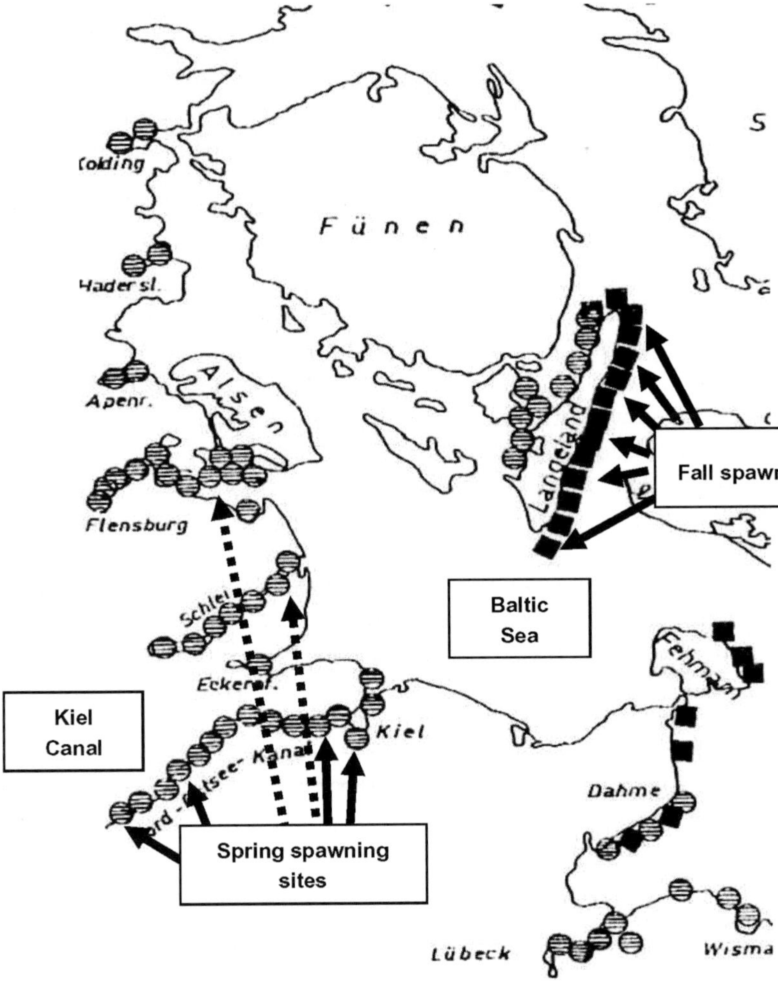

Fig. 34. Distribution of the major spawning grounds of Baltic Herring as identified by Weber (1971), indicating the Kiel Canal (solid arrows) next to the Schlei and Flensburg fjords (dotted arrows) as the major spawning habitats for spring spawners (circles) while those for autumn spawners are located mainly along the shores of islands (squares). (Modified after Weber 1971).

Kils (1992) was able to demonstrate that ship traffic substantially contributes to the survival of Herring spawn. The movement of ships creates a substantial current along the beaches, leading to effective water exchange between the rocky crevices in which Herring eggs are deposited. These artificially created but fast short-term currents provide oxygenated water also to the innermost interface between the egg masses, allowing the embryos to develop. Without this regular and frequent exchange, mass mortality would occur. This study was done during the early 1990s when ship traffic peaked during the first years after the reunification of Germany, occasionally making the canal the busiest inter-oceanic waterway. No follow-up study has been done since, while traffic through the canal has declined but recreational fishermen still observe larvae at high densities every year during and shortly after the peak of the spring spawning period.

The beginning of the Herring spawning period along the German Baltic coast is always temperature-dependent, and always starts in the somewhat warmer inner parts of the Kiel fjord. Because of earlier warming of the Kiel Canal each spring, the spawning starts earlier here than in the Kiel Bight (Neb 1952). This author also indicates that multiple spawners (older fish) seem to prefer the Kiel Canal as spawning ground, leading to a higher proportion of younger year classes spawning in the inner Kiel Fjord.

Neb (1952), however, followed the condition of the larvae hatched in the Kiel Canal before migrating out through the locks to the Baltic Sea and assumes limited feeding conditions for larger juveniles (about 7 cm total length) in the canal. However, it is not known whether these conditions prevail or how they vary between years.

While the regular annual immigration of large quantities of mature and ripe herring schools (up to 1,800 tonnes per year) is a spectacular event, the question still arises why the spawning along the beaches of the Kiel Canal results in large-scale successful hatching and high survival of embryos.

7.5.5 Non-native fish

A table of the known number, places and years of stocking fish species at various sizes has been compiled by Hartman & Spratte (1995) using historic records, published and unpublished sources, covering the years after 1954 till present. Since the early 1980s the canal is regularly stocked with various species and has become an important recreational fishing area for people of economic centres such as Kiel and Hamburg. Therefore, stocking of fish has been organised and the sport fishery had to be regulated by licensing. The utility of the canal to recreational fishermen as well as their landings have gone up slightly in the 1990s, accounting for about 50 - 70 licences per canal km of which an increasing number (656 in 1984 and 2225 in 1993) report their catches (Czerny 1994, cited in Hartmann & Spratte 1995).

Statistics from a ten year period (1984 - 1993) indicate that herring was the most important species (with an annual catch of more than 50 tonnes), followed by eel (approximately 45 tonnes), pikeperch (approximately 40 tonnes), bream (approximately 35 tonnes) and introduced carp and rainbow trout, in all about 19 tonnes and 1.5 tonnes, respectively (Czerny 1995, Hartmann & Spratte 1995).

Rainbow Trout (*Oncorhynchus mykiss*)
Hatchery produced rainbow trout are released to many water bodies in northern Germany and some also escape from aquaculture sites. In the Kiel Canal, relatively large rainbow trout are caught by recreational fishermen, in the order of about 150 kg annually. They migrate between January and May from the Kiel Canal upstream into small creeks and tributaries once they reach a total length of 45 to 55 cm. (Czerny & Spratte 1988). Fish occur because of stocking or escape from aquaculture; natural reproduction has not yet been reported from the State of Schleswig-Holstein.

American Mud-minnow (*Umbra pygmea*)
A species deliberately introduced into some water bodies of northern Germany from the USA around 1898. A few localized self-sustaining populations evolved from this introduction. The fish resists low oxygen levels and water pollution. Few individuals have been found in the Kiel Canal but most have been recorded from the small creeks draining into the canal. Occasionally, fish enter the canal as part of a natural recruitment from a small ditch from which they migrate via the Quistenhofbek creek through the Hanerau into the canal and further upstream into other waters of the drainage system, e.g. the Mühlenbek creek (Holm & Neumann 1990).

Prussian Carp (*Carassius auratus gibelio*)
This species has been extensively dispersed by human activity and still continues its range extension. It forms stable self-sustaining populations in Schleswig-Holstein (Dehus 1990). There are no direct reports on its occurrence in the Kiel Canal, although there is one from the Moorkanal and a few from nearby creeks and ditches. It is expected that the species will continue to expand its range and it is likely to become more common.

Asian Carp (*Cyprinus carpio*)
Brought to northern Germany during the Middle Ages for cultivation, the Asian carp has since been selectively bred and adapted to local conditions. Its occurrence and distribution in Schleswig-Holstein is described by Schütt (1927) and Duncker & Ladiges (1960). Some stocked fish gain access to the Kiel Canal. Hartmann & Spratte (1995) report on the regular stocking of the species in tributaries of the Kiel Canal and it has also been directly released into the canal by sport fishing clubs who report annual landings of 1,500 and 3,000 kg (1968-1994). Tagging experiments show that carp migrate during spring from the deeper (warmer) Kiel Canal to the lower reaches of some creeks. However, they do not spawn there, as the water temperature never reaches the minimum required. All carp occurring in the Kiel Canal and its drainage system are from artificial stocking.

Pseudorasbora parva
This species has been introduced accidentally. It was first recorded from a few specimens in the early 1990s (Neumann 1993 a) and also by Neumann (1994 b) from the Papenau River draining into the Kiel Canal. These occurrences are either due to escapees from pond culture, from releases of aquarium specimens, or from their use as baitfish in sport fishing.

American Sunfish or Pumpkin Seed (*Lepomis gibbosus*)
The sunfish was introduced to Germany from North America between 1880 and 1890. The fish is recorded in the tributaries of the Kiel Canal or in the canal itself and it is claimed to be due to accidental releases from a nearby pond fish farm escaping via creeks into the Papenau River that drains to the Kiel Canal (Holm & Neumann 1990).

7.5.6 Overall summary of the fish fauna of the Kiel Canal

Shortly after opening the canal, annual fishing surveys were conducted by Hinkelmann between 1897 and 1915, using a variety of fishing gear to assure a full coverage of habitats and species (Tab. 3).

Not all species reported from the Kiel Canal have inhabited the canal at all times. Some are permanent, some are temporary visitors on a regular basis, some are rare or of occasional occurrence, and some are maintained by regular stocking. In total, during the more than 100 years of its existence, 63 fish species and two lampreys have been recorded.

There have also been some deliberate releases of freshwater fish native to the region. These attempted to enhance the local commercial and recreational fishery.

For example, Bley (1989) reports on stocking of *Thymallus thymallus* during the years 1961 - 1963 and of *Salvelinus fontinalis* in 1963. Both species occurred in the fishery during subsequent years but disappeared few years after releases ceased. Czerny (1996) analyzed the available information in published and unpublished reports and concluded that the succession in the fish fauna assemblages can be divided into four periods.

Table 3. Number of species calculated from results of annual fishing surveys in the Kiel Canal.

Period (years)	Lampreys	Freshwater Fish	Marine Fish	Total
1896-1915	1	21	14	**36**
1920-1939	2	21	13	**36**
1953-1981	2	23	14	**39**
1984-1993	2	27	27	**56**
Total since 1896	**2**	**30**	**33**	**65**

Table 3 summarizes the information for each of these periods, indicating that most of the freshwater species were recorded already during the first period while 100 years after the opening of the canal only additional 33% of freshwater species had to be added as "regularly occurring". The list of marine species remained short for over 80 years (13 - 14 species) but increased drastically between 1986 and 1993 (plus 13 species). This increase in species number reflects most likely the intensified research on the fish fauna in the canal rather than a true increase in species diversity.

7.6 Mammals

Arndt (1931/32) noted that a bottle nosed dolphin (*Tursiops tursio*) was caught in the canal in 1929 (Mohr & Dunker pers. com. in Arndt 1931/32). This "invader" did likely penetrate the canal from the Brunsbüttel entrance at the North Sea. A ringed seal (*Phoca hispida*) was also caught – likely having migrated into the canal from the Baltic Sea (Arndt 1931/32). He assumes that these species erroneously migrated into the Canal.

7.7 Impact of canal improvement projects

The canal improvement works negatively impacted canal biota, especially the benthos and some species became extinct (Arndt 1931/32).

8 Introduction of non-indigenous species

Several introduced species are known from the Kiel Canal itself and from water adjacent to the canal mouths. First records of invaders in the canal and their subsequent spread suggest that the canal may have been a migration pathway, especially for brackish water species (both canal openings are brackish).

Two considerations argue against this assumption:

1. species may have been independently introduced into both seas with e.g. ballast water release. But ballast water release is unlikely as ballast water operations in the canal are kept to an essential minimum, and
2. migrating species tolerant of marine salinity conditions may have been able to colonise both seas by spreading along the Danish coasts, not using the canal as a migration route. However, truly brackish water species are not able to migrate by natural means along the Danish peninsula as marine conditions occur along the west coasts of Germany and Denmark.

Although the Kiel Canal shows different salinity regimes, Arndt (1931/32) calls the canal an important invasion corridor between the Baltic and North Seas. He suggests that the locks in Brunsbüttel do not limit the westward spread of the species that colonized the canal, and many species meet their "relatives" in the Elbe Estuary after migrating through the canal (Arndt 1931/32).

8.1 Non-native species in German coastal waters

Coastal waters are defined here as navigable for ocean-going ships. A wealth of publications document the occurrence and spread of non-indigenous species along the German North Sea and Baltic Sea coasts (see references at end of Table 4). Surprisingly, the canal was not studied as a migration pathway for biological invasions of non-native species. Selected publications mention first records of non-native species in the canal, but no comprehensive overview was undertaken in the past. Arndt (1931/32) noted that the canal is a migration pathway in both directions.

In total 34 non-native species are known to occur in the Kiel Canal or adjacent waters, in close proximity of which 20 species occur in self-sustaining populations. Most non-native species being recorded from the Kiel Canal region have been found in other German waters previously (Tab. 4).

Table 4. First records of non-native species in German coastal waters (defined here as navigational waters for ocean-going ships with adjacent waters) and the Kiel Canal region – * = in adjacent waters, (x) = invasion status unknown, ((x)) = not established. ➔ = first record in the North Sea with subsequent findings in the Baltic Sea, 🡐 = first record in the Baltic Sea with subsequent findings in the North Sea. For references see end of table.

Group and taxon	First record in German North Sea & adjacent waters	Ref	First record in Kiel Canal region	Ref	First record in German Baltic Sea & adjacent waters	Ref
Bacteriae						
Vibrio vulnificus	2004	89				
Phytoplankton						
Alexandrium minutum 🡐	<1998	83			<1993	15
Chattonella cf. *verruculosa*	1998	37				
Corethron criophilum	<1998	83				
Coscinodiscus wailesii	1977	81			1977	81
Fibrocapsa japonica	<1999	82				
Gymnodinium aureolum	<1999	82				
Gymnodinium catenatum					1993	22
Nodularia spumigena					<1963	91
Odontella (=Biddulphia) sinensis ➔	1903	29			1904	29
Prorocentrum minimum					1980s	48
Prorocentrum redfieldii	<1999	82				
Rhizosolenia indica	<1998	83				
Thalassiosira hendeyi	<1998	83				
Thalassiosira punctigera	<1983	87			<1983	87
Thecadinium mucosum	(2002)	96				
Macroalgae						
Aglaothamnion halliae	1960	101				
Ascophyllum nodosum	1990s	48				
Alaria esculenta	(<1999)	85				
Antithamnionella ternifolia	(<1999)	85				
Bonnemaisonia hamifera	<1890s	38				
Chara connivens					1858	101
Codium fragile ssp. *tomentosoides*	1930s	48				
Dasya baillouviana					2002	101
Devaleraea ramentacea	(<1999)	85				

Table 4. continued

Group	First record in German North Sea & adjacent waters	Ref	First record in Kiel Canal region	Ref	First record in German Baltic Sea & adjacent waters	Ref
Fucus evanescens					<1999	85
Gracilaria vermiculophylla	<2002	94				
Laminaria ochotensis	<1999	85				
Mastocarpus stellatus	1970s	48				
Neosiphonia (=*Polysiphonia*) *harveyi*	<1978	85				
Porphyra cf. *insolita*	(<2000)	84				
Sargassum muticum	1980s	45				
Magnoliophyta						
Elodea canadensis			1899	2		
Anthophyta						
Spartina anglica	1927	30				
Porifera						
Eunapius carteri	1993*	86				
Cnidaria						
Bimeria franciscana			<1952	55		
Bougainvillia macloviana	((1895))	69				
Cereus peduculatus	(1921)	67				
Clavopsella navis			(<1994)	99		
Cordylophora caspia →	1858*	56	1899	90	1870	49
Craspedacusta sowerbyi	(1953*)	64				
Diadumene cincta	1928	65				
Diadumene lineata (=*Haliplanella luciae*)	(1920)	68				
Gonionemus vertens (=*murbachi*)	(1947)	70				
Muggiaea atlantica	((1989))	98				
Nemopsis bachei	1942	57				
Bryozoa						
Pectinatella magnifica	1883*	62				
Victorella pavida			(<1952)	7	(1880)	35
Turbellaria						
Pseudomonocelis cetinae			(<1943)	8		
Nematoda						
Anguillicola crassus ←	1982*	48			1970s	93
Gastropoda						
Crepidula fornicata	1934	17				
Potamopyrgus antipodarum ←	1893*	62	<1900	2	1887	21

Table 4. continued

Group	First record in German North Sea & adjacent waters	Ref	First record in Kiel Canal region	Ref	First record in German Baltic Sea & adjacent waters	Ref
Lithoglyphus naticoides	1887*	62				
Physella acuta →	1992*	64	1990	64	<1996	88
Bivalvia						
Corbicula fluminalis	1984*	61				
Crassostrea angulata	((1911))	23				
Crassostrea gigas	1991	16				
Crassostrea virginica	((1911))	24			((<1887))	25
Dreissena polymorpha ←	<1835*	42	<1896	3	1824	62
Ensis americanus (=*directus*)	1978	19				
Mya arenaria	<1200	34	<1931	2	<1200	34
Mytilopsis (=*Congeria*) *leucophaeta*	<1994*	59	<1928	6		
Petricola pholadiformis →	1896	17			1927	13
Teredo navalis →	<1808	40	1951	10	<1993	41
Polychaeta						
Aphelochaeta marioni	(1938)	73				
Ficopomatus (=*Mercierella*) *enigmaticus* →	1975*	20	<1980	21	<1980*	21
Marenzelleria cf. *viridis* ←	1996*	63	1996	63	1985	27
Marenzelleria cf. *wireni* [8]	1983*	28				
Microphthalmus similes	(1962)	74				
Nereis (=*Neanthes*) *virens*	1923	75			1920s	76
Polydora (=*Boccardiella*) *ligerica*			<1932	77	<1962	54
Polydora redeki			1960s	46	1960s	46
Tharyx killariensis (=*marioni*)	1972	31				
Xiphosura						
Limulus polyphemus	((1866))	72				
Cladocera						
Cercopagis pengoi					<2002	95
Copepoda						
Acartia tonsa →	1931	36			<1981	14
Cirripedia						
Amphibalanus (= *Balanus*) *improvisus* →	1858	56	<1899	1	<1873	52
Elminius modestus	1953	18				

[8] According to Sikorski & Bick (2004) *Marenzelleria* cf. *wireni* is now considered to be *M. neglecta*. However, we refer to *M. wireni* here to allow for comparison with older literature.

Table 4. continued

Group	First record in German North Sea & adjacent waters	Ref	First record in Kiel Canal region	Ref	First record in German Baltic Sea & adjacent waters	Ref
Lepas anatifera	((1830))	26				
Lepas fascicularis	((1865))	26				
Amphipoda						
Chaetogammarus ischnus					<2002	100
Chelicorophium (=*Corophium*) *curvispinum* ➔	1920s	58			1932	50
Corophium multisetosum			<1989	33	<1989	33
Corophium sextonae	(1997)	66				
Gammarus tigrinus ➔	<1957*	11	1978	5	1975*	12
Pontogammarus robustoides					1994	51
Orchestia cavimana	(1920*)	44				
Isopoda						
Idotea metallica	1994	78				
Proasellus coxalis	<1987*	59				
Decapoda						
Atyaephyra desmarestii	(1975*)	64				
Callinectes sapidus	((1964))	32				
Eriocheir sinensis ➔	1912	60	<1926	4	1932	53
Orconectes limosus			(<1990)	71		
Palaemon longirostris	1920s*	79				
Portumnus latipes	1936	47				
Rhithropanopeus harrisii 🡐	<1977	20	1936	9	1936	21
Chronomidae						
Telmatogeton japonicus 🡐	<2002	96			1962	96
Tunicata						
Aplidium nordmanni	<1994	48				
Styela clava	1997	48				
Pisces						
Acipenser sp. (non-native, hybrids?)	(1980s*)	92	(1990)	92	(1996*)	92
Carrassius auratus gibelo			<1980*	92		
Ctenopharyngodon idella			(1992)	92		
Cyprinus carpio			((<1500))	92		
Hypophthalmichthys molitrix			((1988))	92		
Hypophthalmichthys nobilis			((1995*))	92		
Ictalurus nebulosus			(1990s)	92		

Table 4. continued

Group	First record in German North Sea & adjacent waters	Ref	First record in Kiel Canal region	Ref	First record in German Baltic Sea & adjacent waters	Ref
Lepomis gibbosus			(1991*)	92		
Neogobius melanostomus					1996	39
Oncorhynchus mykiss			(1960s*)	92	(1960s)	
Pseudorasbora parva			(1996*)	92		
Salvelinus fontinalis			((1980s*))	92		
Umbra pygmea	1924	102	<1910*	92		
Subtotal established taxa	**60**		**20**		**35**	
Total	**83**		**34**		**39**	

References: 1 Dechow 1920, 2 Arndt 1931/32, 3 Brandt 1896a, 4 Neubaur 1926, 5 Bulnheim 1980, 6 Boettger 1933a, 7 Ax 1952, 8 Meixner 1943, 9 Neubaur 1936, 10 Schütz 1961, 11 Schmitz 1960, 12 Bulnheim 1976, 13 Knudsen 1989, Jensen & Knudsen submitted, 14 Arndt & Schnese 1986, 15 Nehring 1994, 16 Reise 1998a&b, 17 Kuckuck 1957, 18 Kühl 1954, 19 Essink 1985, 20 Kühl 1977, 21 Cole 1982, 22 Nehring 1995, 23 Meyer-Waarden 1964, 24 Rady 1913, 25 Möbius 1887, 26 Luther 1987, 27 Bick & Burckhardt 1989, 28 Essink & Kleef 1986, 29 Ostenfeld 1908, 30 Dijkema 1983, 31 Hauser 1973, 32 Kühl 1965, 33 Köhn & Gosselck 1989, 34 Petersen et al. 1992, 35 Kraeplin 1887, 36 Klie 1933, 37 Backe-Hansen et al. 1998, 38 Kylin 1930, 39 Szaniawska & Dobrzycka-Krahel 2004, 40 Hahn 1956, 41 Sordyl et al. 1998, 42 Reinhardt et al. 2003, 43 Grabow et al. 1998, 44 Schlienz 1922, 45 Wallentinus 1992, 46 Eliason & Haahtela 1969, 47 Müllegger 1937, 48 Bartsch & Kuhlenkamp 2000, 49 Schulze 1981, 50 Neuhaus 1933, 51 Rudolph 1997, 52 Möbius 1873, 53 Peters 1933, 54 Jaeckel 1962, 55 Schütz 1963a&b, 56 Kirchenpauer 1862, 57 Kühl 1962, 58 Schlienz 1923, 59 Post & Landmann 1994, 60 Marquard 1926, 61 Kinzelbach 1991, 62 Thienemann 1950, 63 Bick & Zettler 1997, 64 Tittizer et al. 2000 and references therein, 65 Pax 1936, 66 Reise pers. com., 67 Müllegger 1921, 68 Pax 1920, 69 Hartlaub 1897, 70 Werner 1950, 71 Dehus 1990, 72 Wolff 1977, 73 Caspers 1950, 74 Hartmann-Schröder & Stripp 1968, 75 Hagmeier & Kändler 1927, 76 Reibisch 1926, 77 Augener 1940, 78 Reise 1998, 79 Schnakenbeck 1933, 80 Dörjes 1987, 81 Wiltshire & Dürselen 2004 and references therein, 82 Elbrächter 1999, 83 Nehring 1998, 84 Bartsch & Kuhlenkamp 2000, 85 Wallentinus 1999, 86 Nehring 2002, 87 Hasle 1983, 1990, 88 Jungbluth 1996, 89 Jark pers. com., 90 Hinkelmann 1899, 91 Pankow 1971, 92 Spratte & Hartmann 1997 and references therein, 93 Minchin & Rosenthal 2002, 94 Nehls 2004, 95 Gruzka pers. com., 96 Kerckhof 2005, 97 Hoppenrath et al. 2004, 98 Greve 1994, 99 Barnes 1994, 100 Jazdzewski & Konopacka 2002, 101 Wallentinus pers. comm. and 102 Duncker 1939.

8.2 Non-native species known from Germany along the Baltic and North Sea prior to the opening of the Kiel Canal

The following species were recorded from German coastal waters in the North and Baltic Seas prior to the opening of the Kiel Canal. These species may have migrated through the canal after its opening (for references see Table 4):

- *Mya arenaria* is one of the earliest invaders in Europe, with records dated prior to 1200. The first record in the canal was in 1931.

- *Teredo navalis*. The shipworm was likely introduced to Europe several centuries ago. Its key distribution area in Germany is the North Sea region. It was also found in wooden pilings of bays in the Kiel Canal. Although salt water inflow from the North Sea carried populations into the Baltic the species may have colonised the Baltic from the North Sea via the Kiel Canal, or with ballast water of ships. However, the Baltic population did not reproduce until the 1990s. Today, the ship-worm is well established in the Baltic, with self-sustaining populations.
- *Dreissena polymorpha* was first recorded in the German Baltic in 1824, in the German North Sea estuaries in 1835 and is one of the first invaders reported from the Kiel Canal, introduced prior to 1896.
- *Amphibalanus (=Balanus) improvisus* has occurred in Europe for more than 125 years. It was first found in German coastal waters of the North Sea in 1858 and <1873 in the Baltic. It was also found during the first studies of Kiel Canal biota published in 1899.
- *Cordylophora caspia* was first recorded in the North Sea in 1858 and subsequently in the Baltic in 1870. It has been recorded in the Kiel Canal since 1899.
- *Potamopyrgus antipodarum* (=*jenkinsi*). Steusloff (1909) documented the occurrence of *Paludestrina* (=*Potamopyrgus*) *jenkinsi* in the Baltic Sea and published a note in 1927 on the spread of this invader, native to New Zealand and first recorded from Europe in the Thames Estuary in 1859. The species was first recorded in the Baltic in 1887 (Leppäkoski 1984) and from the Elbe River in 1893. Both records occurred before the opening of the canal. Dechow's (1920) first record of the snail in the Kiel Canal was in 1900 when 3 individuals were found at canal km 12.3. Its first record in the canal near Brunsbüttel indicates that the snail may have colonized the canal from the Elbe Estuary (Arndt 1931/32). Until 1907, the species spread throughout the western part of the canal (Brunsbüttel to Nobiskrug) and was caught in high numbers. Dechow (1920) concluded that the introduction occurred between 1897 and 1900. Arndt (1931/32) suggested that the snail may have been introduced by ships or birds. In contrast, Dechow (1920) assumes introduction by canal flushing from the Baltic. Nehring (2000) assumes that the species migrated through the Kiel Canal to colonize Western Europe.

8.3 Species migration to/through the Kiel Canal

Three migration pathways are theoretically possible:

- species introduction with freshwater run off,
- eastward species migration from the Baltic towards the North Sea, and

- westward species migration from the North Sea via the Elbe River to the Kiel Fjord at the Baltic sea shore.

When selecting species for closer consideration in the following sections emphasis was on species with an unwanted ecological or economic impact.

8.3.1 Migration to the canal with freshwater run-off

In 1899, *Elodea canadensis*, native to North America, was found in the canal (Hinkelmann 1903, Arndt 1931/32). The species had previously been recorded from German waters and it is therefore likely that it colonised the canal from nearby localities. *Elodea canadensis* is a freshwater species, and its introduction to the canal may have occurred with river run-off. As this species is popular in ornamental aquaria it may also have been accidentally introduced to German waters. During mass occurrences the species may clog fishing nets and can also reduce water currents.

Non-native freshwater fish found in the canal or adjacent waters are *Acipenser* sp. (hybrids of non-native species), *Ictalurus nebulosus*, *Lepomis gibbosus*, and *Umbra pygmea*. Stratte and Hartmann (1997) assume that ornamental species' release is the introduction vector for these species.

8.3.2 Westward migration (from the Baltic Sea to the North Sea)

Species first recorded in the Baltic with subsequent occurrence in the North Sea are described here. For references see also Table 4.

Anguillicola crassus
This eel parasite was likely introduced to Europe with live imports of eels. Although *A. crassus* was first found in the Baltic in the 1970s and subsequently in the North Sea a canal migration is unlikely as no records are known from the inner Kiel Canal (Rosenthal pers. com.).

Rhithropanopeus harrisii
The crab *Rhithropanopeus* is native to the northwest Atlantic, from New Brunswick to Florida and the Caribbean Sea. It was introduced to Europe, presumably with ship fouling, around 1874, and it was also collected in the Panama Canal in 1969. It was first recorded along the German Baltic coast in 1936 and thereafter in the North Sea (<1977). Nehring (2000) assumes that a

migration through the Kiel Canal is likely. It was first recorded in the canal in 1936 (Neubauer 1936).

8.3.3 Eastward migration (from the North Sea to the Baltic Sea)

For species not known to occur in the Kiel Canal, a canal migration is unlikely. These invaders may have migrated around Denmark or were introduced with shipping:

- *Odontella sinensis* appeared in the North Sea in 1903, in the Baltic in 1904. It is assumed that this phytoplankter reached the Baltic with salt water influx from the North Sea.
- *Chelicorophium curvispinum*: in the North Sea in the 1920s, in the Baltic in 1932.
- *Acartia tonsa:* first recorded in the North Sea in 1931, in the Baltic <1981.

Species first recorded in the North Sea with subsequent occurrence in the Baltic are described below, assuming that they used the canal as a migration pathway. For references, see Table 4.

Eriocheir sinensis

The Chinese mitten crab was first recorded in the Aller River, a tributary of the Weser River draining into the North Sea, in 1912 and subsequently spread to the Weser and Elbe Rivers. Its native range is temperate and tropical regions between Vladivostok (Russia) and South-China (Peters 1933, Panning 1938), including Japan and Taiwan. Its centre of occurrence is the Yellow Sea (Panning 1952).

Its lifecycle is characterised by migrations to waters with different salinities. Larvae develop in marine waters, and perform a spring upstream migration, aided by estuarine currents. Young crabs and young adults actively migrate upstream. In their native region, living crabs have been found more than 1,000 km upstream in the river Jangtsekiang.

Crabs feed on a wide variety of plants, invertebrates, fishes and detritus. Plants, snails and clams are the main food component. Two-year old adults migrate down to the marine environment in summer, during which time they become reproductively mature (Peters 1933, Panning 1938, Panning & Peters 1932).

In 1926, Neubaur published the first finding of a single specimen of the Chinese mitten crab in Lake Wittensee, three kilometres from the Kiel Canal, near

Eckernförde. Lake Wittensee is connected with the Kiel Canal by a small creek, the Habyer Au. It is likely that the Chinese mitten crab migrated into the canal from the Elbe River and was first recorded on the German Baltic coast in 1926 (Boettger 1933b, Panning 1938).

Today, crabs are reported in the Baltic up to Estonia, Finland and Russia. However, it is unlikely that the species is able to attain self-sustaining populations in the Baltic Sea. Due to the low salinity, the reproduction cycle cannot be completed here.

It is assumed that specimens captured in the Baltic immigrated from estuaries adjacent to the North Sea. In 2002 and 2003, an annual catch of approx. 100 individuals was documented near Saaremaa Island (Gulf of Riga). The migratory distance from the nearest reproductive area (Elbe river estuary) via the Kiel Canal to the most distant location in Estonia (Narva Bay in the eastern Gulf of Finland) exceeds 1,500 km. This distance almost doubles the recorded migration maximum upstream in the Elbe River (Ojaveer at al., in prep.)

Ficopomatus enigmaticus

The polychaete *F. enigmaticus* is native to tropical and warm-temperate brackish coastal habitats of India, the Indo-Pacific, and Australia (Walford & Wicklund 1973).

During mass occurrences, reef like aggregates of densely packed tubes form, perpendicularly to the substrate. Their density ranges from 70,000 to 180,000 per square meter. Small tubes of newly settled individuals encrust interweaving tubes of adult specimens. As a result, the reef-like aggregates become highly consolidated (Aliani et al. 1995).

F. enigmaticus was first collected from German North Sea estuaries in 1975 – the first record from the German Baltic coast was before 1980 and it was also found in the Kiel Canal prior to 1980. This fouling organism frequently lives on ship hulls (e.g. Gollasch 1996). From the North Sea it may have reached the Baltic via the Kiel Canal or with the fouling of vessels.

Gammarus tigrinus

Native to North America, *Gammarus tigrinus* was first recorded in Europe on the west coast of England in 1931 (Sexton 1939). The species tolerates salinities of 1 – 25 PSU, and would not have been able to migrate into the Baltic around Denmark as local waters are too saline.

The first record in Germany occurred in 1957 (Schmitz 1960), in the Werra River. Schmitz intentionally released about 1,000 individuals to re-establish an amphipod in the highly polluted river where native amphipods had become depleted. In 1967, the species was found in the Weser Estuary near Bremerhaven (Klein 1969).

The first record in the Baltic region was in 1975 in the Schlei Fjord (Bulnheim 1976). During an analysis of Kiel Canal amphipods in 1978 and 1979, *G. tigrinus* was recorded in the lower saline middle part of the canal in a stretch of about 40 km (Bulnheim 1979).

Although it was found in western Germany decades before its record in the Kiel Canal, it is doubtful whether it colonized the canal from the Elbe Estuary. During an investigation of Elbe Estuary amphipods in 1978 *G. tigrinus* was not found. It was also absent from the Kiel Fjord in 1978 (Bulnheim 1979). Bulnheim (1980) suggests that it was introduced to the canal with ballast water. However, this is unlikely as ballast water operations in the canal are strictly limited. Therefore, and in contrast to Bulnheim, we assume that this amphipod colonised the Kiel Canal from the North Sea.

Petricola pholadiformis

P. pholadiformis, from the north-western Atlantic (Gulf of St. Lawrence to Uruguay), was first recorded in 1890 along the southeastern coast of England. It may have been introduced with live American oyster shipments. The first record along the German North Sea coast was in 1896. The first record in Denmark was in 1905, in the Wadden Sea. Thereafter, the species spread and was found in 1912 in Esbjerg harbour (Wadden Sea), in 1915 in Lønstrup (Skagerrak coast), and in 1926 in the Limfjord. In the northern Kattegat it was found in 1931 and it in Storebælt in 1943 (Knudsen 1989, Jensen & Knudsen in prep.).

There is a record from the western Baltic in 1927, to where the species may have migrated through the Kiel Canal. It became established and today can be collected in the Wadden Sea, along the Skagerrak coast, the Limfjord and in northern Kattegat (Knudsen 1989, Jensen & Knudsen in prep.). No record in the canal may also indicate a migration of this species around Denmark.

8.4 Unintentional species introductions during canal improvements works

During the first improvements to the Kiel Canal from 1910 to 1914, some construction equipment was transported from the Netherlands to Germany. This equipment had previously been used in the Netherlands for water works in canals near the Dutch Noordzeekanaal which drains into the Zuiderzee (Redeke 1937). Nehring & Leuchs (2000) assume that several species were moved to the Kiel Canal with this construction equipment:

- *Bimeria fransiciana*,
- *Mytilopsis leucophaeta*,
- *Rhithropanopeus harrisii*.

8.5 Future invaders which may take advantage of the Kiel Canal to spread across Europe

Several species recently recorded in Europe are currently spreading. Due to their known salinity tolerance, some have the ability to migrate through the Kiel Canal and in using this shortcut rapidly colonize the North and Baltic Seas.

The following chapters document that the Kiel Canal is of continuous concern as a migratory pathway, now even more than 100 years after its opening. It should be noted that the examples given are not exhaustive, but should give an overview.

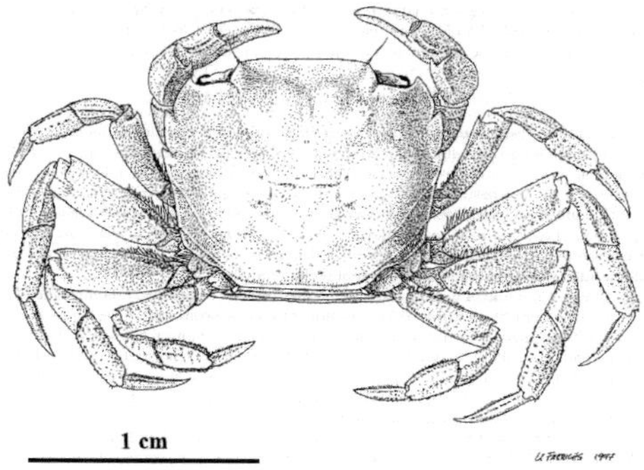

Fig. 35. Hemigrapsus penicillatus. Drawing U. Frerichs, Hamburg, Germany. Look out for this invader along the German coasts!

8.5.1 Species known to spread in the North Sea region, but not (yet) known from the German coast along the Baltic

Hemigrapsus penicillatus

The crab *H. penicillatus* (Fig. 35), with native range from northern Japan to China, was first recorded in Europe in 1994 by a record of a one-year old specimen in France (Noël et al. 1997). Since then, it has been spreading and is now found along the coasts stretching from northern Spain towards Belgium (Dumoulin 2004).

The crab occupies the same niche as the indigenous (young) green crab *Carcinus maenas* and is supposed to be a competitor and predator of various native species. Wide salinity and temperature tolerance enable it to migrate through the Kiel Canal and to colonise a wide range of habitats along the Baltic shores (Gollasch 1999). So far, it has not been found in German waters, however.

8.5.2 Species known to spread in the Baltic region, but not (yet) known from the German coast along the North Sea

Cercopagis pengoi

The brackish water cladoceran *Cercopagis pengoi* (fishhook water flea) (Fig. 36) was first recorded in the Baltic Sea in 1992. The upper salinity tolerance of 15-18 PSU does not permit it to spread around Denmark to the North Sea and its adjacent water bodies and estuaries. However, by using the Kiel Canal it may colonize the Elbe River estuary. Recently, it was found in the Pomeranian Lagoon, the first record from German coastal waters (Gruzka pers. com.).

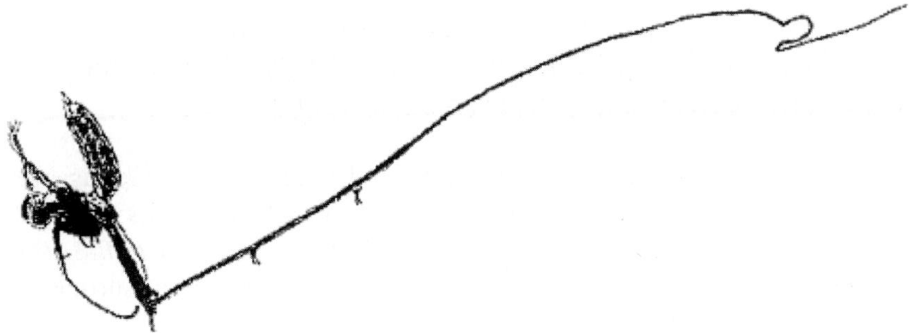

Fig. 36. The cladoceran *Cercopagis pengoi* from the Gulf of Finland – instar III parthenogenetic female with embryos in brood poach. Drawing courtesy of Vadim Panov, St.Petersburg, Russian Federation, see also www.zin.ru/rbic.

Neogobius melanostomus

Another Baltic invader may have used the Kiel Canal: *Neogobius melanostomus* (Fig. 37). This fish of Ponto-Caspian origin tolerates salinities up to 45 PSU in laboratory experiments and temperatures from –1 to 35°C. Spawning occurs above 8°C. Its scattered distribution along the southern Baltic shores indicates multiple introductions or a spreading potential of this invader (Sapota pers. com.).

In contrast to *Cercopagis pengoi*, this invader tolerates the higher salinity of the western Baltic. However, it requires hard bottom habitats, rare along the south-western Baltic coasts. It is assumed that the artificial hard bottom substrate of the Kiel Canal bed may attract *Neogobius*, especially in the absence of alternative hard substrates in the region.

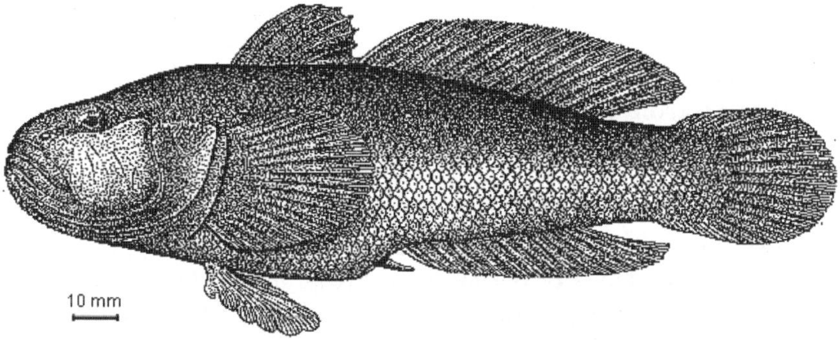

Fig. 37. Neogobius melanostomus. Drawing courtesy of Kryzstof Skora, Hel, Poland.

It was recently recorded from the Dutch North Sea coast (Beek pers. comm.) but it remains unclear whether it reached the North Sea from its Baltic population or whether it was introduced from its natural range.

It is also interesting to note that two additional species of the Genus *Neogobius* have recently been found in the Baltic Sea: *N. gymnotrachelus* (first record in 1995) and *N. fluviatilis* (first record in 1997). As *N. melanostomus*, they have the potential to colonize a wide range of habitats in northern Europe and their spread may be facilitated by the Kiel Canal.

9 Impacts of non-native species

A comprehensive impact assessment of non-native species in German coastal waters cannot be presented due to a lack of relevant information. Data are not available in a consistent and comprehensive format. In addition the limited information available is geographically scattered.

The three most impacting species in German coastal waters are the Zebra mussel *Dreissena polymorpha,* the ship-worm *Teredo navalis*, and the Chinese mitten crab *Eriocheir sinensis*. For the latter two, a tentative economic impact is calculated below.

9.1 Ecological impact

Reise et al. (1999) concluded that the North Sea is not impacted severely by invaders and that established non-native species in the region are rather additive than destructive, and without major consequences. However, as several examples have shown worldwide, each invader poses a potential risk to the environment and economy of a region. Today, the Pacific Oyster *Crassostrea gigas* is spreading in the Wadden Sea; in the future, it may cause unwanted impacts as a competitor of the native blue mussel *Mytilus edulis* (Reise pers. com).

9.2 Economic impact

An overall impact calculation for all known introduced species in German coastal waters remains a challenge, because of a lack of relevant information. During an unpublished study carried out by S. Gollasch for the German Federal Agency for Shipping and Hydrography in 2004, a questionnaire was sent to relevant stakeholders. Indications on the monetary impact of two species, the Chinese mitten crab and the ship-worm were received. The following sections summarise the impact of these two invaders. However, this attempt at a monetary calculation is preliminary, pending additional information.

9.2.1 *Dreissena polymorpha*

The Ponto-Caspian Zebra mussel *D. polymorpha* was first recorded in Germany in 1824, in the Baltic. This mussel causes severe damage to hard structures by dense fouling, resulting in the clogging of industrial water intakes, fishing gear and boats (Olenin et al. 1999).

It is also known from the inner estuaries of the Elbe and Weser rivers. Water dependent industries are threatened here by it, and modifications of water intakes had to be implemented. Selected industrial plants were moved to marine conditions to avoid it (Rosenthal pers. com.). However, its impact to German waters cannot be quantified.

9.2.2 *Teredo navalis*

The worm-shaped bivalve *T. navalis* has been known from German waters (North Sea) for more than 200 years (Schütz 1961). First recorded in the Netherlands in 1731 (Sellius 1733), the species destroyed wooden flood protection installations. It is likely that this ship-worm also colonised wooden sailing vessels of the Spanish Armada, while they were waiting in French and Portuguese ports for the invasion of England in 1588, resulting in structural damage to the ships and loss of the war (Gollasch & Riedel-Lorje 2000, Hoppe 2002). Möbius (1872) noted its presence in the Eiderkanal (see above). It was also known to occur in the Kiel Fjord (Schütz 1961).

Sporadic mass invasions lasted for two or three years in the 1930s and 1950s - also along the Baltic shores and the species was also found in the Kiel Canal in 1951. By 1961 it was found until canal kilometre 93, i.e. in brackish water, almost 5 km from the Kiel locks (Schütz 1961).

In the Baltic, however, the bivalve disappeared again after a few years, presumably because reproduction was impossible at the local low salinities. The latest invasion in the Baltic took place in the early 1990s, possibly due to much salt water inflow from the North Sea. Today the population appears to be established at a self-sustaining level. The easternmost limit of shipworm distribution in Germany is along the Island of Rügen, where low salinities negatively impact larval survival. Wooden pilings used in marinas, harbours and sea bridges are still attacked, with large pieces of wood destroyed within just two years (Hoppe 2002).

Tentative cost calculation

The damage caused in the Baltic alone is calculated as 25 Mio € since 1993 (confirmed). The total damage along all German coastal waters is estimated as 50 Mio € since 1993 (Hoppe, pers. comm.).

9.2.3 *Eriocheir sinensis*

The Chinese mitten crab was introduced to Germany by shipping (see above). The first sighting of an adult crab was from the Aller river in 1912 (Peters 1933, Panning 1938, Panning & Peters 1932). The impact of this invader became especially clear during the mass occurrences of the 1930s, 1940s, 1950s, 1980s and 1990s. In total, mass developments were reported for approximately 30 years (Fladung pers. comm.).

Table 5. Tentative calculation of cost since the first findings of the Chinese mitten crab in German waters. (Modified after Gollasch & Fladung, unpublished).

Cost item (data adjusted from Fladung pers. comm.) costs from 1930s and 1940s calculated to current value	**Estimated sub-total** [in €] conservative calculation	**Estimated sub-total** [in €] maximum calculation
• Costs of catchment gear installation During **1935-1945** in total 35 catchment installations, i.e. barriers, ramps, collection buckets were in use. The average cost per installation was 750 € During **1996-1998** four catchment systems were in use, capital cost total to	26,250 10,000	26,250 10,000
• Labour to clean and maintain catchment gear During the **1935-1945** the catchment season lasted for 8 to 10 weeks with 1 or 2 employees (estimated salary per week 300 €) During **1996-1998** labour cost totalled to	24,000 40,000	60,000 40,000
• The impact on bank erosion and feeding on native species are difficult to quantify. The assumption results in several 10,000s €	20,000	50,000
• Loss in commercial fisheries (estuaries and rivers) Assuming that 250 fishermen were affected during **1930-1950** costs are estimated as 70,000,000 € (for 20 years annually ca. 14,000 €per fisherman) including repair of nets as crabs tend to cut net ropes. 60 fishermen were affected during the period **1994-2004** costs are estimated as 8,400,000 € (annually ca. 14,000 € per fisherman).	65,000,000 8,000,000	75,000,000 9,000,000
• Loss in commercial fisheries (pond fisheries), estimated for **1994-2004**. Impacts include predation of fish food and cultured pond fish	75,000	100,000
• Loss in commercial fisheries due to the predatory impact of the crabs on macrozoobenthos (fish food) resulting in e.g. poor growth of fish is calculated as 10,000 to 20,000 € annually during the 30 year duration of mass occurrences.	300,000	600,000
Estimated total	**73,495,250**	**84,886,250**

During four severe mass developments in the last century up to 140 t of juvenile crabs were caught annually. A single fishing net collected 50 - 60 kg of crabs per day (Fladung pers. comm.).

In the 1930s, 1940s and 1990s, attempts were undertaken to catch and destroy as many crabs as possible. This implied labour costs and some catchment gear production at the German Rivers Elbe and Havel.

It was calculated that the monetary impact caused to German waters has totalled approximately 80 million Euro since 1912 (Tab. 5).

Other cost implications

Additional negative impacts exist, but cannot be quantified:

- impacts on biodiversity,
- impacts on recruitment of commercial species, and
- increased erosion rate due to crab burrowing activities in river banks.

It should also be noted that one positive effect was documented. During mass occurrences crabs were and continue to be sold for 1 to 3 € /kg to the industry e.g. for industrial use and for direct human consumption (Asian markets).

During 1994 - 2004 crabs were sold to a value of approximately 3,000,000 to 4,500,000 €. This amount needs to be deducted from the impact cost figures to take account of "beneficial" effects.

10 Summary of species migrations through the Kiel Canal

The Kiel Canal has not been studied as a migratory pathway for biological invasions of non-native species. Selected publications mention first records of non-native species in the canal, but a comprehensive overview is lacking.

Arndt (1931/32) highlights the canal as an important invasion corridor between the Baltic and North Seas. He also suggests that the canal locks would not limit the spread of species.

Most of the established invaders nowadays occurring in both the North Sea and the Baltic were previously first recorded in the North Sea. As a result species migrations in eastwards direction, i.e. from the North Sea to the Baltic, were more frequent.

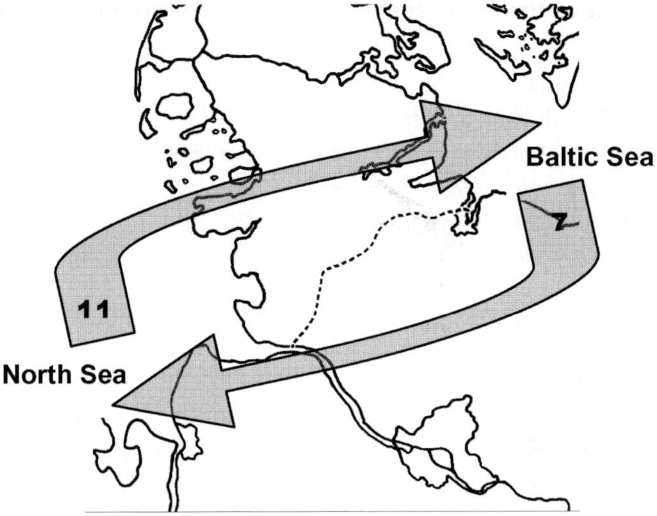

Fig. 38. Spread of established non-native species between the North and Baltic Seas – possibly via the Kiel Canal. For species details see Tab. 4.

In all, 83 non-indigenous taxa have been recorded from the German coasts of the North Sea (60 established species) and 39 from the Baltic Sea (35 established). In total 18 species are known to be established on both, along the German coast of the North and Baltic Seas (Tab. 4).

When comparing the year of first record, 11 of these common species have been found first in the North Sea and later in the Baltic (Fig. 38) indicating the likely eastern route of secondary spread (species known to occur in the Kiel Canal in bold):

1. *Acartia tonsa*
2. ***Amphibalanus (=Balanus) improvisus***
3. *Chelicorophium curvispinum*
4. ***Cordylophora caspia***
5. ***Eriocheir sinensis***
6. ***Ficopomatus enigmaticus***
7. ***Gammarus tigrinus***
8. *Odontella sinensis*
9. *Petricola pholadiformis*
10. ***Physella acuta***
11. ***Teredo navalis***

The average time lag between the first record along the North Sea coast and the subsequent finding of the organism in the Baltic is approximately 19 years.

Nearly half (7 species) have first been recorded in the Baltic and were found thereafter in the North Sea or adjacent waters (Fig. 38), indicating a westward spread (species know to occur in the Kiel Canal in bold):

1. *Alexandrium minutum*
2. ***Anguillicola crassus***
3. ***Dreissena polymorpha***
4. ***Marenzelleria* cf. *viridis***
5. ***Potamopyrgus antipodarum***
6. ***Rhithropanopeus harrisii***
7. *Telmatogeton japonicus*

The average time lag between the first findings along the Baltic Sea coast and the subsequent finding of the organism in the North Sea is approximately 12 years.

For two species it remains unclear where they were recorded first (species known to occur in the Kiel Canal in bold):

1. ***Mya arenaria***
2. *Nereis virens*

Of the 18 species established in German coastal waters of both the North and Baltic Seas, 12 species have also been found in the Kiel Canal (marked in bold above), indicating the likeliness of the Kiel Canal as an invasion pathway. However, it cannot be ruled out that some of these 12 may have migrated around Denmark. A few species have been found to be widespread along the coasts of both the North and Baltic Sea prior to the opening of the Kiel Canal.

The Kiel Canal turns out not to be a major migration pathway for non-native marine species. Some species, such as *Cercopagis pengoi*, are not able to reach the North Sea estuaries by spreading around Denmark as excess salinity limits their spread. The Kiel Canal is thus seen as a key migration pathway for only brackish water species between the Elbe River estuary and the Baltic Sea.

As Reise et al. (1999), we conclude that the German coastal waters are not impacted severely by introduced species and that most established non-native species can rather be seen as additions, without major consequences.

However, exceptions occur. A monetary impact assessment carried out for two species, the Chinese mitten crab and the ship-worm revealed that these two invaders, since their first records in German coastal waters, have caused a cost equivalent to 98.5 to 134.8 Million €. As both species were first recorded a long time ago, this monetary impact is comparatively small.

Acknowledgements

We express our sincerest thanks to Prof. Kartsen Reise, Alfred-Wegener-Institute, List, Sylt for the critical review of this chapter and the valuable comments he provided as well as to Dan Minchin, Ballina, Killaloe, Ireland and Henri Dumont, University of Gent, Belgium for improving the English.

11 References

Aleem, A.A. & Schulz, E. 1952. Über Zonierung von Algengemeinschaften (Ökologische Untersuchen im Nord-Ostsee-Kanal). Kieler Meeresforschungen 9, 70-76.

Aliani, S., Bianchi, C.N., De Asmundis, C., Meloni, R. 1995. Scanning electron microscope observations on the tube of the reef forming serpulid *Ficopomatus enigmaticus* (Fauvel) (Annelida, Polychaeta) Boll. Zool. 62, 4, 363-367.

Andresen, P. 1899. Versuchsfischerei im Kaiser Wilhelm Kanal. Naturwissenschaftliche Wochenschrift 15(19): 219-221.

Andresen, P. 1899. Versuchsfischerei im Kaiser-Wilhelm-Kanal. Naturw. Wochenschr. 14, 219-221.

Andresen, P. 1900. Die Resultate der im Jahre 1899 ausgeführten Versuchsfischerei im Kaiser-Wilhelm-Kanal. Ebg. 15, 210-211.

Anonymous 1958. Der Nordostseekanal war die Heimat. Fischwaid. 10: 325.

Arndt, H. & Schnese, W. 1986. Population dynamics and production of *Acartia tonsa* (Copepoda: Calanoida) in the Darss-Zingst estuary, southern Baltic. Ophelia, (4), 329-334 pp.

Arndt, W. 1931/32. Die Tierwelt des Nordostseekanals und ihr Lebensraum. Der Naturfoscher 8, 16 pp.

Augener, H. 1940. Beiträge zur Polychaetenfauna der Ostsee. Kieler Meeresforsch. 3, 133-147.

Ax, P. 1952. Eine Brackwasser-Lebensgemeinschaft an Holzpfählen des Nord-Ostsee-Kanals. Kieler Meeresforsch. 9, 229-243.

Backe-Hansen, P., Dahl, E. & Danielssen, D. 1998. On the bloom of Chattonella on the North Sea/ Skagerrak in April-May 1998. Poster presentation.

Barfod, H. 1903. Seltene Gäste unter den Fischen der westlichen Ostsee. Nerthus. 5, 601.

Barfod, H. 1904. Das Vordringen der Ostseeorganismen in den Kaiser-Wilhelm-Kanal mit besonderer Berücksichtigung der wichtigsten Nutzfische. Nerthus. 6, 101-109.

Barnes, R.S.K. 1994. The brackishwater fauna of northwestern Europe. Cambridge Univ. Press.

Bartsch, I. & Kuhlenkamp, R. 2000. The marine macroalgae of Helgoland (North Sea): an annotated list of records between 1845 and 1999. Helgoland Marine Research 54: 160-189.

Beseke, C. 1893. Der Nordostseekanal. Nachdruck der Originalausgabe 1982. Lühr & Dircks, Hamburg.

Betz, K.-H. 1979. *Sphaeroma rugicauda* Leach, 1814 (Isopoda: Flabellifera) im Nord-Ostsee-Kanal. Schriften Naturwiss. Ver. Schleswig-Holstein 49: 81-84.

Bick, A. & Burckhardt, R. 1989. Erstnachweis von *Marenzelleria viridis* (Polychaeta Spionidae für den Ostseeraum, mit einem Bestimmungsschlüssel der Spioniden der Ostsee. Mitt. Zool. Mus. Berlin. 65(2), 237-247.

Bick, A. & Zettler, M.L. 1997. On the identity and distribution of two species of *Marenzelleria* (Polychaeta, Spionidae) in Europe and North America. Aquat. Ecol. 31, 137-148.

Bley, O. 1989. Personal communication, Chair of the Sport-Fischerei-Verein (SFV) "Forelle" Todenbüttel, filed documentation at headoffice of the SFV. (Cited in Hartmann & Spratte 1995).

Boettger, C.R. 1933a. Über die Ausbreitung der Muschel *Congeria cochleata* Nyst in europäischen Gewässern und ihr Auftreten im Nordostseekanal. Zool. Anzeiger 77, 267-269.

Boettger, C.R. 1933b. Die Ausbreitung der Wollhandkrabbe in Europa. Sitzungberichte der Gesellschaft naturforschender Freunde, 399-415.

Boettger, C.R. 1932. Über die Ausbreitung der Muschel *Congeria cochleata* Nyst in europäischen Gewässern und ihr Auftraten im Nord-Ostsee-Kanal. Zoologischer Anzeiger 101: 43-48.

Boje, R. 1965. Die Bedeutung von Nahrungsfaktoren für das Wachstum von *Mytilus edulis* L. in der Kieler Förder und im Nord-Ostsee-Kanal. Kieler Meeresforschungen 21: 81-100.

Bothmann, O. 1998. Epifauna und Pfahlbewuchs im Salzwassergradienten des Nord-Ostsee-Kanals. Diplomarbeit, Institut für Meereskunde an der Universität Kiel, 102 pp.

Brammer, J. 1987. Markierte Karpfen im Nord-Ostsee-Kanal Kiel. Mitteilungen für Schleswig-holsteinische Jäger und Fischer 12: 4.

Brandhorst, W. 1955. Über Laichen und Aufwuchs des Frühjahrsherings im Nord-Ostsee-Kanal unter besonderer Berücksichtigung der Hydrographie. Dissertation, Universität Kiel, Germany, 116 pp.

Brandhorst, W. 1956. Über Laichen und Aufwuchs des Frühjahrsherings im Nord-Ostsee-Kanal. Kieler Meersforschungen 12: 186-199.

Brandt, K. 1896a. Die Tierwelt der Kieler Bucht und des Kaiser-Wilhelm-Kanals. Kiels Einrichtungen für Gesundheitspfl. u. Unterricht. Kiel. Festschr. XXI. Versamml. deutsch. Ver. f. öff. Gesundheitspfl. 27-35.

Brandt, K. 1896b. Das Vordringen mariner Tiere in den Kaiser-Wilhelm-Kanal. Zool. Jahrb. Abt. Syst. 9, 387-408.

Brandt, K. 1897. Das Vordringen mariner Thiere in den Kaiser-Wilhelm-Canal. Zoologische Jahrbücher Abt. f. Systematik, Geographie und Biologie der Thiere 9: 387-408.

Brandt, K. 1897. Ueber die Tierwelt und die Lebensbedingungen im Kaiser-Wilhelm-Kanal. Mitt. Deutsch. Seefisch. Ver. 13, 6.1-10.

Bulnheim, H.-P. 1976. *Gammarus tigrinus*, ein neues Faunenelement der Ostseeförde Schlei. Schlei Schr. Naturw. Ver. Schlesw.-Holst. 46, 79-84.

Bulnheim, H.-P. 1980. Zum Vorkommen von *Gammarus tigrinus* im Nord-Ostsee-Kanal. Arch. FischWiss. 30 (1), 67-73.

Carlton, J.T. & Geller, J.B. 1993. Ecological roulette: The global transport of nonindigenous marine organisms. Science 261: 78-82.

Carlton, J.T. 1992. Introduced marine and estuarine molluscs of North America: An end-of-the-20th-century perspective. Journal of Shellfish Research 11: 489-505.

Caspers, H. 1950. Die Lebensgemeinschaft der Helgoländer Austernbank. Helgoländer wiss. Meeresunters, 3, 119-169.

Cole, H.A. 1982. Status (1980) of Introductions of Non-indigenous Marine Species to North Atlantic Waters, ICES Cooperative Research Report 116, 87 pp.

Czerny, D. & Spratte, S. 1988. Kurzbericht über die Markierung von Regenbogenforellen *Salmo gairdneri* Richardson im Nord-Ostsee-Kanal in den Jahren 102(3-4): 68-76.

Czerny, D. 1987. Zander im Nord-Ostsee-Kanal. Mitteilungen für schleswig-holsteinische Jäger und Fischer 4: 3-5.

Czerny, D. 1990. Erträge der Heringsfischerei im Nord-Ostsee-Kanal. Die Heimat 10/11: 238-245., 12: 389.

Czerny, D. 1996. Vorkommen, Verbreitung und Häufigkeit der Fische und Neunaugen im Nord-Ostsee-Kanal. Fischökologie 9: 1-14.

Czerny, D. 1994. Wo bleiben die Fangmeldungen der Sportfischer für den Nord-Ostsee-Kanal? Jäger und Fischer 7: 32-36.

Dallmer, E. 1890. Veränderungen der Fischerei durch den Bau des Nord-Ostsee-Kanals. Unpublished position paper as of 01.12.1890 to the Royal Governmental President (Königlichen Regierungspräsidenten), Ministerium für Ernährung, Landwirtschaft, Forsten und Fischerei (MELFF), Kiel, 20 pp.

Dechow, F. 1920. Die Bodentiere des Kaiser-Wilhelm-Kanals. Dissertation Universität Kiel. 74 pp. (mimeographed).

Dehus, P. 1990. Die Verbeitung der Flußkrebse (Decapoda: Astacidae, Cambaridae) in Schleswig-Holstein. Faun. ökol. Mitt. 6, 95-105.

Dijkema, K.A. 1983. The salt-marsh vegetation of the mainland coast, estuaries and Halligen. In: Dijkema, K.A. & W.J. Wolff. (eds.). Flora and vegetation of the Wadden Sea islands and coastal areas. Stichtung, Leiden. 185-220.

Dobbrunz, W. 1976. Angeln zwischen Nord- und Ostsee von Brunsbüttel bis Kiel – mit Rute und Kescher am Kanal. Fischwaid (5): 252-255.

Dörjes, J. 1987. Die Biota des Sublitorals. In: Gerdes, G. & Krumbein, W.E. (eds.) Mellum, Portrait einer Insel. Kramer, Frankfurt am Main, 141-152.

Drake, L.A., Choi, K.-H., Ruiz, G.M. & Dobbs, F.C. 2001. Global redistribution of bacterioplankton and viroplankton communities. Biological Invasions 3: 193-199.

Dumoulin, E. 2004. Snelle areaaluitbreiding van het penseelkrabbetje *Hemigrapsus penicillatus* (de Haan, 1835) langs de kusten van de Zuidelijke Bocht van de Noordzee, status van haar opmars in de Westerschelde en beschouwingen over de ecologie en het gedrag van de soort. Journal De Strandvlo 24(1), 5-35.

Duncker, G. & Ladiges, W. 1960. Die Fische der Nordmark. Abhandlungen und Verhandlungen des Naturwissenschaftlichen Vereins in Hamburg, N.F. Volume 3, Supplement, Hamburg 1960. Kommissionsverlag Cram, De Gruyter, 432 pp.

Duncker, G. 1939. Hundsfische (*Umbra* spec.) in Schleswig Holstein? Heimat (Kiel) 49: 300-301.

Duncker, O. 1964. Angeln im Nord-Ostsee-Kanal. Die Fischwaid No 7: 235-237.

Elbrächter, M. 1999. Exotic flagellates of coastal North Sea waters. Helgoländer Meeresunters. 52, 235-242

Eliason, A. & Haahtela, I. 1969. *Polydora* (*Boccadria*) *redeki* Horst (Polychaeta, Spionidae) from Finland. Annales Zoologici Fennici 6, 215-218.

Elsner, H. 1994. Wikinger Museum Haithabu: Schaufenster einer frühen Stadt. Karl Wachholtz Verlag, Neumünster. 129 pp.

Essink, K. & Kleef, H.L. 1986. Establishment of a population of the Spionid worm *Marenzelleria* in the Ems Estuary (the Netherlands, Fed. Rep. of Germany). Proceedings of COST-647 Workshop "Intertidal Soft Sediments", Dec. 1986, St. Valery-sur-Somme, France. 5 pp.

Essink, K. 1985. On the appearance of the American jack-knife clam *Ensis directus* (Conrad, 1983) (Mollusca, Cultellidae) in the Dutch Wadden Sea. Basteria 49, 4-6, 73-80.

EU 2000. Directive 2000/60/EC of the European Parliament and of the Council 23 October 2000.

FAO 1995. Precautionary approach to fisheries. Part 1: Guidelines on the precautionary approach to capture fisheries and species introductions. FAO Fisheries Technical Paper, Rome, 350/1 52pp.

Fiedler, M. & Kils, U. 1990. Fischereibiologische Untersuchung im Nord-Ostsee-Kanal. Institut für Meereskunde, Universität Kiel (Project Report, unpublished), 123 pp.

Fitschen, B. 1986. Synökologische Untersuchungen an Brackwasserbenthos im olighalinen Bereich des Nord-Ostsee-Kanals (Schleswig-Holstein)I. Diploma Thesis, University Kiel, 76 pp.

Frier, J.O. 1976. Oxygen consumption and osmoregulation in the isopods *Spharoma hookeri* Leach and *S. rugicauda* Leach. Ophelia 15(2): 193-203.

German Federal Bureau of Maritime Casualty Investigation 2004. Jahresstatistik 2002 / 2003. 34 pp.

Gollasch, S. & Riedel-Lorje, J.-C. 2000. Warum war "Holland in Not"? Auswirkungen eingeschleppter Arten. SDN-Magazin 2000, 16-22.

Gollasch, S. 1996. Untersuchungen des Arteintrages durch den internationalen Schiffsverkehr unter besonderer Berücksichtigung nichtheimischer Arten. Diss., Univ. Hamburg, Verlag Dr. Kovac, Hamburg, 314 pp.

Gollasch, S. 1999. The Asian decapod *Hemigrapsus penicillatus* (de Haan, 1833) (Decapoda, Grapsidae) introduced in European waters, status quo and future perspective. Helgoländer Meeresunters. 52, 359-366

Gollasch, S. 2002. The importance of ship hull fouling as a vector of species introductions into the North Sea. Biofouling 18: 105-121.

Gollasch, S., Macdonald, E., Belson, S., Botnen, H., Christensen, J., Hamer, J., Houvenaghel, G., Jelmert, A., Lucas, I., Masson, D., McCollin, T., Olenin, S., Persson, A., Wallentinus, I., Wetsteyn, B. & Wittling, T. 2002. Life in Ballast Tanks. 217-231 pp. In: Leppäkoski, E., Gollasch, S. & Olenin, S. (eds.): Invasive Aquatic Species of Europe: Distribution, Impacts and Management. KLUWER Academic Publishers, Dordrecht, The Netherlands. 583 pp.

Gollasch, S., Rosenthal, H., Botnen, H., Crncevic, M., Gilbert, M., Hamer, J., Hülsmann, N., Mauro, C., McCann, L., Minchin, D., Öztürk, B., Robertson, M., Sutton, C. & Villac, M.C. 2003. Species richness and invasion vectors: sampling techniques and biases. Biological Invasions 5: 365-377.

Grabow, K., Eggers, T.O. & Martens, A. 1998. *Dikerogammarus villosus* Sovinsky (Crustacea: Amphipoda) in norddeutschen Kanälen und Flüssen. Lauterbornia, 33, 103-107.

Grahl, H. & Kelm, A. (eds.) 1992. Der blaue Tod – Die Cholera In Hamburg 1892. Museumspädagogischer Dienst Hamburg 1992, 94 pp.

Greve, W. 1994. The 1989 German Bight Invasion of *Muggiaea atlantica*. ICES J. mar. Sci. 51: 355-358.

Hagmeier, A. & Kändler, R. 1927. Neue Untersuchungen im Nordfriesischen Wattenmeer und auf den fiskalischen Austernbänken. Wiss. Meeresunters. Abt. Helgoland 16(6), 1-90.

Hahn, A. 1956. Die Bekämpfung der Bohrmuschel. Die Küste 5, 49-72.

Hartlaub, C. 1897. Die Hydromedusen Helgolands. Wiss. Meeresunters. Abt. Helgoland, II, 449-536.

Hartmann, U. & Spratte, S. 1995. Daten zur Fischfauna im Einzugsgebiet des Nord-Ostsee-Kanals. Ministerium für Ernährung, Landwirtschaft, Forsten und Fischerei des Landes Schleswig-Holstein (Project Report, mimeographed), 157 pp.

Hartmann-Schröder, G. & Stripp, K. 1968. Beiträge zur Polychaetenfauna der Deutschen Bucht. Veröffentl. Inst. Meeresforsch. Bremerhaven 11, 1-24.

Hasle, G.R. 1983. *Thalassiosira punctigera* (Castr.) comb. nov., a widely distributed marine planktonic diatom. Nordic Journal of Botany 3: 593-608.

Hasle, G.R. 1990. Diatoms of the Oslo Fjord and Skagerrak. Species new to the area: immigrants or overlooked in the past? Blyttia 48, 33-38.

Hauser, B. 1973. Bestandsänderungen der Makrofauna an einer Station im ostfriesischen Watt. Jber. Forschungsstelle Norderney, 24, 171-203.

Hayes, K.R., & Sliwa, C. 2003. Identifying potential marine pests – a deductive approach applied to Australia. Marine Pollution Bulletin 46: 90-98.

Henning, G. 1939. Was bietet der Kaiser-Wilhelm Kanal dem Sportfischer? Der Deutsche Sportangler 6: 170-172, vol 7: 197-199.

Hessland, I. 1946. On the quarternary *Mya* period on Eirope. Ark. Zool. 37A(8), 1-51.

Hinkelmann, A. 1897 Über die im Jahre 1896 ausgeführte Versuchsfischerei im Kaiser-Wilhelm-Kanal. Mitt. Deutsch. Seefischerei Ver. 13, 242-245.

Hinkelmann, A. 1898 Über die im Jahre 1896 ausgeführte Versuchsfischerei im Kaiser-Wilhelm-Kanal. Mitt. Deutsch. Seefischerei Ver. 14, 45-50.

Hinkelmann, A. 1899. Die Fische und sonstigen Nutztiere des Kaiser-Wilhelm-Kanals mit besonderer Berücksichtigung der Lebensverhältnisse des Herings. Die Heimat. 9, 173-178.

Hinkelmann, A. 1900. Über die im Jahre 1899 ausgeführte Versuchsfischerei im Kaiser-Wilhelm-Kanal. Mitteilungen des Deutschen Seefischereivereins 16: 40-43.

Hinkelmann, A. 1903. Über die im Jahre 1902 ausgeführte Versuchsfischerei im Kaiser-Wilhelm-Kanal. Mitteilungen des Deutschen Seefischereivereins 19: 88-91.

Hinkelmann, A. 1903. Über die neuesten Ergebnisse der Versuchsfischerei auf dem Kaiser-Wilhelm-Kanal. Die Heimat. 13, 153-164.

Hinkelmann, A. 1905. Über die im Jahre 1904 ausgeführte Versuchsfischerei im Kaiser-Wilhelm-Kanal. Mitteilungen des Deutschen Seefischereivereins 21: 16-18.

Hinkelmann, A. 1906. Über die im Jahre 1905 ausgeführte Versuchsfischerei im Kaiser-Wilhelm-Kanal. Mitteilungen des Deutschen Seefischereivereins 22: 14-18.

Hinkelmann, A. 1907. Über die im Jahre 1896 ausgeführte Versuchsfischerei im Kaiser-Wilhelm-Kanal. Mitt. Deutsch. Seefischerei Ver. 23, 5-7.

Hinkelmann, A. 1915. Versuchsfischerei im Kaiser-Wilhelm-Kanal nach seiner Erweiterung im Jahre 1915.Mitteilungen des Deutschen Seefischereivereins 37: 57-59.

Höpner Petersen, G. 1978. Life cycles and population dynamics of marine benthic bivalves from the Disko Bugt area of West Greenland. Ophelia 17: 95-120.

Höpner Petersen, G. 1999. Five Recent Mya species, including three new species and their fossil connections. Polar Biol.22: 322-328.

Holm, U., Neumann, M. 1990. Untersuchungen zur Besiedlung der Papenau mit Fischen und bach-typischen wirbellosen. Bericht für das Amt für Land-und Wasserwirtschaft Kiel, 30 pp. (unpublished report).

Hoppe, K. 2002. *Teredo navalis* - The Cryptogenic Shipworm. 116-119 pp. In: Leppäkoski, E., Gollasch, S. & Olenin, S. (eds.): Invasive Aquatic Species of Europe: Distribution, Impacts and Management. KLUWER Academic Publishers, Dordrecht, The Netherlands. 583 pp.

Hoppenrath, M., Saldarriaga, J., Schweickert, M. & Elbrächter, M. 2004. Description of *Thecadinium mucosum* sp. nov. (Dinophyceae), a new sand-dwelling marine dinoflagellate, and an emended description of Thecadinium inclinatum Balech. J, Phycol. 40, 945-951.

Hülsmann, N. & Galil, B.S. 2001. The effects of freshwater flushing on marine heterotrophic protists – implications for ballast water management. Marine Pollution Bulletin 42: 1082-1086.

ICES 2003. ICES Code of Practice on the Introductions and Transfers of Marine Organisms. International Council for the Exploration of the Sea (www.ices.dk).

Jaeckel, S. 1962. Die Tierwelt der Schlei – Übersicht einer Brackwasserfauna. Schr. Naturwiss. Vereins Schleswig-Holstein 33, 11-32.

Jazdzewski, K. & Konopacka, A. 2002. Invasive Ponto-Caspian Species in Waters of the Vistula and Oder Basins and the Southern Baltic Sea. In: Invasive aquatic species of Europe – distribution, impacts and management. Leppäkoski, E., Gollasch, S. & Olenin, S. (eds). Kluwer Academic Publ., Dordrecht, Boston, London: 384-398.

Jazdzewski, K. 1969. Biology of two hermaphroditic Crustacea, *Cyathura carinata* (Kröyer) (Isopoda) und *Heterotanais oerstedi* (Kröyer) (Tanaidacea) in waters of the Polish Baltic Sea. Zoologica Polonia 19: 5-25.

Jensen, K.R. & Knudsen, J. A summary of alien marine invertebrates in Danish waters. (submitted).

Jungbluth, J.H. 1996. Einanderer in der Molluskenfauna. In: Gebietsfremde Tierarten. Auswirkungen auf einheimische Arten, Lebensgemeinschaften und Biotope. H. Gebhardt. R. Kinzelbach & S. Schmidt-Fischer (eds.), 314 p.

Kafemann, R. 2000. Räumliche und zeitliche Veränderungen der Struktur einer Brackwasser-fischgemeinschaft und ihre Steuerung durch abiotischen und biotische Umwelteinflüsse. Eine fischbiologische Studie im Nord-Ostsee-Kanal. Dissertation Universität Hamburg, 278 pp. (Verlag dissertsation.de).

Kafemann, R., Thiel, R. & Finn, J.E. 1998. Die Bedeutung abiotischer Schlüsselfaktoren für die Struktur der Fischgemeinsschaft im Nord-Ostsee-Kanal. Fischökologie 11: 1-20.

Kafemann, R., Thiel, R. & Sepulveda, A. 1996. De fischökologische Bedeutung der Nebenstromgewässer der Unterelbe. Archive Hydrobiologie (Suppl.) 110: 199-214. (Untersuchungen zum Elbe-Ästuar).

Kardel, 1995. Fisherman, personal communication (mentioned in Hartmann and Spratte 1995).

Kerckhof, F. 2005. National Report for Belgium. ICES Working Group on Introductions and Transfers of Marine Organisms (WGITMO) form Meeting 2005, 18-19.

Kerstan, S. & Kerstan, M. 1989. Parasitenuntersuchungen, Längen-Altersbestimmung an drei Stintproben aus dem Ost- und einen aus dem Westteil des Nord-Ostsee-Kanals. Institut für Meereskunde, unpublished reports, 12 pp.

Kils, U. 1992. Untersuchungen zum Einfluß der Stein-Schüttungen und des Schiffsverkehrs im Nord-Ostsee-Kanal auf die Laich-Biologie des Herings, *Clupea harengus*. Institut für Meereskunde an der Christian-Albrechts-Universität Kiel, unpublished report, 20 pp.

Kinne, O. 1955. *Neomysis vulgaris* (Thompson), eine autökologisch-biologische Studie. Biol. Jb. Syst. 74: 160-202.

Kinne, O. 1956. Zur Ökologie der Hydroidpolypen des Nord-Ostsee-Kanals. *Laomedea loveni* (Allmann), *Cordylophora caspia* (Pallas), *Perigonimus megas* (Kinne). Zeitschr. zur Morphologie und Ökologie der Tiere 45, 217-249.

Kinzelbach, R. 1991. Die Körbchenmuscheln *Corbicula fluminalis*, *Corbicula fluminea* und *Corbicula fluviatilis* in Europa (Bivalvia: Corbiculidae). Mainer Naturwiss. Arch. 29, 215-228.

Kirchenpauer, J.U. 1862. Die Seetonnen der Elbmündung. Abhandl. Gebiete Naturwiss. 4(3), 1-59.

Klein, G. 1969. Amphipoden aus der Wesermündung und der Helgoländer Bucht, mit Beschreibung von *Talorchestia frisiae* n.sp. Veröff. Inst. Meeresforsch. Bremerhaven 11, 173-194.

Klein, K., Plötz, J. & Nöthlich, I. 1976. Bericht über den chemisch-biologischen Zustand des Nord-Ostsee-Kanals. Bundesanstalt für Gewässerkunde, Koblenz zu N/268/2536.

Klie, W. 1933. Neues zur Crustaceen-Fauna Nordwestdeutschlands. Abh. Nat. Ver. Bremen, 28, (4), 271-276.

Knudsen, J. 1989. Immigration of marine invertebrates to the Limfjord (Denmark) and the North Sea - Baltic transition area. [in:] Environmental Quality and Ecosystm Stability. Vol. IV-B, Environmental Quality. Spanier, E. Steinberger, Y. Luria, M. (eds.). ISEEQ Publication, Jerusalem, 135-145.

Köhn, J. & Gosselck, F. 1989. Bestimmungsschlüssel der Malakostraken der Ostsee. Mitt. Zool. Mus. Berlin, 65, 1-114.

Koske, H. & Szekielda, K.H. 1965. Untersuchungen zur Hydrographie, Schwebstoffzusammensetzung und Bakteriologie des Nord-Ostsee-Kanals. Teil 1. Kieler Meeresforschungen 21: 132-134.

Kothe, P. & Sindern, J. 1973. Die Verbreitung des Makro-Zopobenthos im Nord-Ostsee-Kanal und ihre Abhängigkeit vom Salzgehalt. II: Organismenverbreitung und biologische Indikation des Seewassereinflusses. Deutsche gewässerkundliche Mitteilungen. 17: 21-26.

Kothe, P. 1973. Die Verbreitung des Makrozoobenthos im Nord-Ostsee-Kanal und ihre Abhängigkeit vom Salzgehalt. II. Organismenverbreitung und biologische Indikation des Seewassereinflusses. Deutsche Gewässerkundl. Mitt. 17, 21-26.

Kraeplin, K. 1887. The German freshwater Bryozoa. A monography. Abhandl. Verhandl. Naturwiss. Vereins Hamburg 10, 1-168.

Krumm, H. & Rheinheimer, G. 1966. Untersuchungen zur Hydrographie, Schwebstoffzusammensetzung und Bakteriologie des Nord-Ostsee-Kanals. Teil 2. Kieler Meeresforschungen 22: 121-134.

Kuckuck, P. 1957. Der Strandwanderer. Lehmanns Verlag, München.

Kühl, H. 1954. Über das Auftreten von *Elminius modestus* Darwin in der Elbmündung. Helgoländer wiss. Meeresunters. 5, 53-56.

Kühl, H. 1962. Die Hydromedusen der Elbmündung. Abhandl. Verhandl. Naturwiss. Vereins Hamburg. 6, 209-232.

Kühl, H. 1965: Fang einer Blaukrabbe, *Callinectes sapidus* Rathbun (Crustacea, Portunidae) in der Elbmündung. Arch. Fischereiwiss., 15, (3), 225-227.

Kühl, H. 1977. Neue Tierarten an der deutschen Nordseeküste. Inf. Fischw. 24, 2, 61-62.

Kylin, H. 1930. Über die Entwicklungsgeschichte der Florideen. Lunds Univ. Årsskr. Swed., 2, 26 (6) : 1-104.

Lamp, F. 1990. Experimente zur Verpflanzung von Nordseekabeljau in die westliche Ostsee. Archiv für Fischereiwiss. 40(1/2): 69-86.

Landessportfischereiverband Schleswig-Holstein e.V. 1990. Hinweise zum Ausfüllen der Fangmeldungen für den Nord-Ostsee-Kanal. Mitteilungen für schleswig-holsteinische Jäger und Fischer 2: 6 pp.

Lenz, J., Andres, H.-G., Gollasch, S. & Dammer, M. 2000. Einschleppung fremder Organismen in Nord- und Ostsee: Untersuchungen zum ökologischen Gefahrenpotential durch den Schiffsverkehr. UBA Project Water: 102 04 250, Umweltbundesamt, Berlin, Texte 5, 273 pp.

Leppäkoski, E., Gollasch, S., Olenin, S. (eds.) 2002. Invasive aquatic species of Europe. Distribution, impacts and management. Kluwer Academic Publishers, Dordrecht. 583pp.

Lloyds Register- Fairplay 2002. Ports & Terminal Guide 2003. Lloyds Register- Fairplay. Lombard House, Surrey, UK (published as CD-ROM).

Lowe, S.J., Browne, M., & Boudjelas, S. 2000. 100 of the worlds worst invasive species. IUCN Invasive Species Specialist Group, Auckland, New Zealand 12pp.

Luther, G. 1987. Seepocken der deutschen Küstengewässer. Helgoländer Meeresunters. 41, 1-43.

Mallok, J. 1995. Fischfauna im Flemhuder See und im angrenzenden Bereich des Nord-Ostsee-Kanals: Jungfischwachstum und Altersrückberechnung. Diplomarbeit, Institut für Zoologie, Universität Frankfurt, Main, Germany, 84 pp. (mimeographed).

Marcus, G.J. 1980. The conquest of the North Atlantic. The Boydell Press, Woodbridge. 224 pp.

Marquard, O. 1926. Die Chinesische Wollhandkrabbe, *Eriocheir sinensis* Milne- Edwards, ein neuer Bewohner deutscher Flüsse. Fischerei, 24, 417-433.

McCarthy, S.A. & Khambaty, F.M. 1994. International dissemination of epidemic *Vibrio cholerae* by cargo ship ballast and other non-potable waters. Applied Environmental Microbiology 60: 2597-2601.

McKinley, R.S., Patrick, P.H., Matousek, J.A. & Wells, A.W. 1989. Field testing of behavioural barriers for fish exclusion at cooling-water intake systems: Ontario Hydro Pickering Nuclear Generating Station: Final report. Technical Report EPRI-GS-6246, 130 pp. http://www.osti.gov/energy citations/product.biblio.jsp?osti_id=6240492 (visited February 15th 2005).

Meixner, J. 1943. Über die Umbildung einer Turbellarienart nach Einwanderung aus dem Meere ins Süßwasser. Int. Rev. Hydrobiol, 43.

Meyer-Waarden, P.E. 1964 Mastversuche mit portugiesischen Austern im deutschen Wattengebiet. Inf. Fischw. 11, 77-78.

Minchin, D. & Gollasch, S. 2003. Fouling and ships' hulls: how changing circumstances and spawning events may result in the spread of exotic species. Biofouling (Supplement) 19: 111-122.

Minchin, D. & Rosenthal, H. 2002. Exotics for Stocking and Aquaculture, Making Correct Decisions. In: Leppäkoski, E., Gollasch, S. & Olenin, S. (eds.) 2002. Invasive Aquatic Species of Europe: Distribution, Impacts and Management. KLUWER Academic Publishers, Dordrecht, The Netherlands. 583 pp.

Minchin, D. 1996. Management of the introduction and transfer of marine molluscs. Aquatic Conservation: Marine and Freshwater Ecosystems 6: 229-244.

Möbius, K. 1872. Die Prosobranchia und Lamellibranchia der Kieler Bucht.

Möbius, K. 1873. Die wirbellosen Thiere der Ostsee. Jahresbericht der Commission zur wissenschaftlichen Untersuchung der deutschen Meere 1, 97-144.

Möbius, K. 1887. Schlußbericht über den Versuch des deutschen Fischereivereins, kanadische Austern in der Ostsee anzusiedeln. Deutscher Fischerei Verein. Mitt. Section Küsten- und Hochseefisch. 1.

Moy, P.B. 2000. Development of an aquatic nuisance species barrier in a commercial waterway. In: Perderson, J. (ed.) First National Conference on Bioinvasions, USA, Massachusetts Institute of Technology (MIT), MIT Sea Grant Program, Center for Coastal Resources, Cambridge, USA, 24.-27. Januar, 1999. Conference proceedings, 357-362.

Müllegger, S. 1921. *Heliactis bellis* bei Büsum gefunden. Schrift. Zool. Stat. Büsum, 2, 61.

Müllegger, S. 1937. *Portumnus latipes* (Pennant), ein neuer Krebs an der holsteinischen Westküste. Wschr. für Aquarien- und Terrarienkunde, 34, 35.

n.n. 2003. Ausbau des Nord-Ostsee-Kanals. Bericht der Landesregierung. Drucksache 15/1936. 29 pp.

Neb, E. 1952. Untersuchungen über Fortpflanzung und Wachstum an den Heringen der westlichen Ostsee mit besonderer Berücksichtigung der Kieler Förde. Dissertation, University Kiel, Germany, 153 pp.

Nehls, G. 2004. *Gracilaria vermiculophylla* (Gracilariales, Rhodophyta) – a Pacific algae spreading on tidal flats of the Wadden Sea. Presentation at MARBEF Workshop: Aquatic invasive species and the functioning of European coastal ecosystems. Jan. 2005, List, Sylt (Germany).

Nehring, S. & Leuchs, H. 2000. Neozoen im Makrobenthos der deutschen Brackgewässer an der deutschen Nordseeküste. Lauterbornia 39, 73-116.

Nehring, S. 1994. First living *Alexandrium minutum* resting cysts in Western Baltic. IOC Newsletter on toxic algae and algal blooms. 9, 2 pp.

Nehring, S. 1995. Ein Fossil kehrt zurück. Ruhestadien des Phytoplanktons als Anzeiger für Veränderungen im Ökosystem Nordsee. Wattenmeer International 3, 18-19.

Nehring, S. 1998. Non-indigenous phytoplankton species in the North Sea: supposed region of origin and possible transport vector. Archive of Fishery and Marine Research 46, 181-194.

Nehring, S. 2000. Neozoen im Makrobenthos der deutschen Ostseeküste. Lauterbornia 39, 117-126.

Nehring, S. 2001. After the TBT era: alternative anti-fouling paints and their ecological risks. In: Kröncke, I., Türkay, M., Sündermann, J. (eds) Burning issues of North Sea ecology. Senckenbergiana maritima 31.

Nehring, S. 2002. Different human-mediated vectors for non-indigenous macrozoobenthic species. In: Leppäkoski, E., Gollasch, S. & Olenin, S. (eds.) 2002. Invasive Aquatic Species of Europe: Distribution, Impacts and Management. KLUWER Academic Publishers, Dordrecht, The Netherlands. 583 pp.

Neubaur, R. 1926. Eine chinesische Wollhandkrabbe im Wittensee. Fischereizeitung. (Neudamm). 29, 1111.

Neubaur, R. 1931. Die Nutzung der Fischerei im Nord-Ostsee-Kanal. Untersuchung des Oberfischmeisters vom 6.11. 1931 aufgrund der Bereisung vom 24-25. 9. 1931. (unpublished Report).

Neubaur, R. 1936. Ein neuer Mitbewohner Schleswig-holsteinischer Fischgewässer. Fischereizeitung 39, 725-726.

Neudecker, T. (1990): Genutzte Muscheln und Schnecken. In: Lózan, J.L., Lenz, W., Rachor, E., Watermann, B. & Westernhagen, H. v. (eds.): Warnsignale aus der Nordsee. Paul Parey, Berlin, 165-176.

Neuhaus, E. 1933. Studien über das Stettiner Haff und seine Nebengewässer. Zeitschr. Fisch. u. deren Hilfswiss. 31, 427-489.

Neumann, M. 1993. Fischereibiologische Begleituntersuchung in der Papenau (Station 0+00 – 60+00) (2 Jahre nach dem naturnahen Ausbau). Im Auftrag des Amtes für Land- und Wasserwirtschaft Kiel. Labor für biologische Gewässeruntersuchungen, Kiel, 30 pp. (unpublished report).

Neumann, M. 1994. Fischereibiologische Begleituntersuchung in der Papenau – Untersuchungen zum Naturaufkommen der Forelle – (Nachweis von Jungforellen). Im Auftrag des Amtes für Land- und Wasserwirtschaft Kiel. Labor für biologische Gewässeruntersuchungen, Kiel, 11 pp. (unpublished report).

Noël, P.Y., Tardy, E. & D'Udekem D'Acoz, C. 1997. Will the crab *Hemigrapsus penicillatus* invade the coasts of Europe? C.R. Acad. Sci. Paris, Sciences de la vie/Life Sci. 320, 741-745.

Ockelmann, W.K. 1958. The Zoology of East Greenland. Marine Lamellibranchiata. Medd. Grønland 122. 4, 256 pp.

OFM 1927. Aus Schleswig-Holstein. Fischerei-Zeitung 30(3): 56-57 (OFM = Oberfischmeister, at the time Mr. R. Neubaur).

Ojaveer, H., Gollasch, S., Jaanus, A. & Kotta, J. Distribution of the Chinese mitten crab *Eriocheir sinensis* (Arthropoda, Malacostraca) in the northeastern Baltic Sea (submitted).

Olenin, S., Orlova, M. & Minchin, D. 1999. *Dreissena polymorpha* (Linneaus, 1758). 43-48 pp. In: Gollasch, S., Minchin, D., Rosenthal H. & Voigt, M. (eds.): Exotics Across the Ocean. Case histories on introduced species: their general biology, distribution, range expansion and impact. Logos Verlag, Berlin, Germany, 78 pp.

Ostenfeld, C.H. 1908. On the immigration of *Biddulphia sinensis* Grev. and its occurence in the North Sea during 1903-1907. Medd. Komm. Havunders., Ser. Plankton, 1, (6), 1-46.

Pankow, H. 1971. Algenflora der Ostsee I. Benthos (Blau-, Grün, Braun- und Rotalgen). Gustav Fischer, Jena, 419 pp.

Panning, A. 1938. The Chinese Mitten Crab. Annual Report Smithsonian Institute, 361-375.

Panning, A. 1952. Die chinesische Wollhandkrabbe. Die neue Brehm-Bücherei, 70:1-46.

Panning, A. & Peters, N. 1932. Wollhandkrabbe und Elbfischerei. Hamb. Nachr., (6), 1-16.

Panning, A. 1938. The Chinese Mitten Crab. Smithsonian Rep., 361-375.

Pax, F. 1920. Die Aktinienfauna von Büsum. Schr. Zool. Stat. Büsum Meeresk., 5, 1-24.

Pax, F. 1936. Anthozoa. In: Grimpe, G. & Wagler, E. (eds.): Die Tierwelt der Nord- und Ostsee. Bd. 30, Akad. Verlagsges. Becker & Erler, Leipzig, 317 pp.

Pelz, G.R., 1994. Fischereiliches Gutachten über den Fischbestand und –ertrag im Flemhuder See und dem angrenzenden Nord-Ostsee-Kanal. UVU Flemhuder See. Bundesanstalt für Gewässerkunde, Koblenz (unpublished Report) 64 pp.

Pelz, G.R., 1996. Fischbestand und –ertrag im Flemhuder See und dem angrenzenden Nord-Ostsee-Kanal. Fischökologie Aktuell 9: 76-85.

Peters, N. 1933. B. Lebenskundlicher Teil. In: Peters, N. & Panning, A. (eds). Die chinesische Wollhandkrabbe *(Eriocheir sinensis* H. MILNE-EDWARDS) in Deutschland. Leipzig: Akademische Verlagsgesellschaft GmbH. p 59-156.

Peters, J. 1934. Heringsfänge im Nordostseekanal. Die Heimat No. 7: 245-246.

Petersen, K.S., Rasmussen, K.L., Heinemeier, J. & Rud, N. 1992. Clams before Columbus?. Nature, 359, 679.

Por, F.D. 1978. Lessepsian migration: the Influx of Red Sea Biota in the Mediterranean by way of the Suez Canal. Springer Verlag – Ecological Studies, Berlin, Heidelberg, Germany. Retrospect and viewpoints. 228 pp.

Post, D. & Jandmann, M. 1994. Verbreitungsatlas der Fließgewässerfauna in Ostfriesland. Staatl. Amt Wasser Abfall. 141 pp.

Rady, G.A. 1913. Anleitung zu den Grundzügen für die Entwicklung einer Austernkultur an unserer deutschen Nordseeküste. Lehren aus der zeit zur Hebung der Küstenfischerei. Hans Hofmann, Norden.

Redeke, H.C. 1937. Über die Vebreitung einiger Malakostraken in niederländischen Gewässern. Int. Rev. ges. Hydrobiol. Hydrogr. 35, 217-228.

Reibisch, J. 1926. Über Änderungen in der Fauna der Kieler Bucht. Schr. naturwiss. Ver. Schleswig-Holstein 17, 227-232.

Reinhardt, F., Herle, M., Bastiansen, F. & Streit, B. 2003. Economic impacts of the spread of alien species in Germany. UBA Texte, 80/03, 229 pp.

Reinke, J. & Darbishire, O.V. 1898. Notiz über die marine Vegetation des Kaiser-Wilhelm-Kanals im August 1896. Wiss. Meeres-Unters. Abt. Kiel, 3, 33-36.

Reise, K. 1998a. Exoten der Nordseeküste. Wattenmeer Internat. 1, 21-22.

Reise, K. 1998b. Pacific Oysters invade mussel beds in the European Wadden Sea. Senckenbergiana maritima, 28(4/6), 167-175.

Reise, K., Gollasch, S. & Wolff, W.J. 1999. Introduced marine species of the North Sea coasts. Helgoländer Meeresunters. 52, 219-234.

Rheinheimer, G. 1997. The influence of natural salinity gradients on bacteria communities in flowing waters. Limnologica 27(1): 29-35.

Rigby, G. & Taylor, A. 1995. Ballast water: its impacts can be managed. ICES Annual Science Conference, Aalborg, 8 pp.

Rudolph, K. 1997. Zum Vorkommen des Flohkrebses *Pontogammarus robustoides* im Peenemündungsgebiet. Natur und Museum 127, 306-312.

Schlienz, K. 1922. Eine Süßwasser-*Orchestia* in der Außenalster in Hamburg. Arch. Hydrobiol. 14, 144-150

Schlienz, W. 1923. Verbreitung und Verbreitunsgbedingungen der höheren Krebse im Mündungsgebiet der Elbe. Arch. Hydrobiol. 14, 429-452.

Schmitz, W. 1960. Die Einbürgerung von *Gammarus tigrinus* Sexton auf dem europäischen Kontinent. Arch. Hydrobiol. 57, 223-225.

Schnakenbeck, W. 1933. *Leander longirostris* (H. M.-Edw.) in der Unterelbe. Zool. Anz., 102, (5/6), 129-135

Schulz, W. 1995. Der Nord-Ostsee-Kanal. Eine Fotochronik der Baugeschichte. Westholsteinische Verlagsanstalt Boyens & Co, Heide in Holstein. 173 pp.

Schulze, F.E. 1981. Über den Bau und die Entwicklung von *Cordylophora lacustris* Allmann nebst Bemerkungen über Vorkommen und Lebensweise dieses Tieres. Engelmann, Leipzig 52 pp.

Schütt, E.C. 1991. Die Chronik Hamburgs. Chronik Verlag in der Harenberg Kommunikation Verlags- und Mediengesellschaft mbH & Co. KG, Dortmund. 639 pp.

Schütz, L. & Kinne, O. 1955. Über die Mikro- und Makrofauna der Holzpfähle des Nord-Ostsee-Kanals und der Kieler Förde. Kieler Meeresforsch. 11(1), 110-135.

Schütz, L. 1961. Verbreitung und Verbreitungsmöglichkeiten der Bohrmuschel *Teredo navalis* L. und ihr Vordringen in den Nord-Ostsee-Kanal bei Kiel. Kieler Meeresforsch. 17, 228-236.

Schütz, L. 1963a. Die Beziehung zwischen Ei-, Embryonenzahl und Körpergröße der Weibchen einiger Peracaridae aus dem Nord-Ostsee-Kanal (Mesohalinikum). Zool. Anzeiger 171(5/8): 291-303.

Schütz, L. 1963b. Ökologische Untersuchungen über die Benthosfauna im Nordostseekanal – I: Autökologie der sessilen Arten. Int. Rev. ges. Hydrobiol. 48, 361-418.

Schütz, L. 1966. Autökologie der vagilen und hemisessilen Arten im Bewuchs der Pfähle: Mikro- und Mesofauna. Internationale Revue der gesamten Hydrobiologie 51: 633-685.

Schütz, L. 1966. Die Fauna der Fahrrine des NO-Kanals. Kieler Meeresforschungen 22: 104-115.

Schütz, L. 1969. Ökologische Untersuchungen über die Benthosfauna im Nordostseekanal - III. Autökologie der vagilen und hemisessilen Arten im Bewuchs der Pfähle: Makrofauna. Int. Rev. ges. Hydrobiol. 54, 553-592.

Schütz, L. 1983a. Die Fauna der Fahrrinne des NO-Kanals. Kieler Meeresforschungen 19: 104-115.

Schütz, L. 1983b. Ökologische Untersuchungen über die Benthosfauna im Nord-Ostsee-Kanal. I: Autökologie der sessilien Arten. Internationale Revue der gesamten Hydrobiologie 38: 361-418.

Seemann, 1957. Anregung zum Bau einer Fischtreppe an der Wehrausmündung zum Nord-Ostsee-Kanal. Wasserbauamt Kiel-Holtenau. Schriftverkehr, Aktennummer 244/4. (cited in Hartmann & Spratte, 1995).

Sellius, G. 1733. Historia naturalis teredinis seu xylophagi marini tubulo conchoidis speciatim Belgici. Trajecti ad Rehenum, [xxxiv], 356, 1 pp.

Seuront, L. 2005. First record of the calanoid copepod *Acartia omorii* (Copepoda: Calanoida: Acartiidae) in the southern bight of the North Sea. Journal of Plankton Research 27(12): 1301-1306.

Sexton, E.W. 1939. On a new species of *Gammarus* (*G. tigrinus*) from Droitwich District. J. mar. biol. Ass. U.K. 23, 543-551.

Sikorski, A.V, & Bick, A. 2004. Revision of *Marenzelleria* Mesnil, 1896 (Spionidae, Polychaeta). Sarsia, 89: 253-275.

Sindermann, C.J. 1992. The introduction of marine species: What the future holds. In: DeVoe, R. (ed) Proceedings of the Conference & Workshop. Introductions & Transfers of Marine Species. Achieving a Balance between Economic Development and Resource Protection. Session VII, pp 143-148. South Carolina Sea Grant Consortium, USA.

Sordyl, H., Bönsch, R., Gercken, J., Gosselck, F., Kreuzberg, M. & Schulze, H. 1998. Verbreitung und Reproduktion des Schiffsbohrwurms *Teredo navalis* L. an der Küste Mecklenburg-Vorpommerns. DGM 42(4), 142-149.

Spratte, S. & Hartmann, U. 1992. Daten zur limnischen Fischfauna im Eidergebiet. Ministerium für Ernährung, Landwirtschaft, Forsten und Fischerei und Landessportfischereiverband Schleswig-Holstein e.v. (Hrsg). 137 pp.

Spratte, S. & Hartmann, U. 1997. Fischartenkataster: Süßwasserfische und Neunaugen in Schleswig-Holstein. Ministerium Landwirtschaft, Ernährung und Tourismus, Kiel Germany. 183 p.

Steusloff, U. 1909. *Paludestrina jenkinsi* Smith an der deutschen Ostseeküste. Arch. Ver. Fr. Nat. Ges. Meckl. 63, 82-93.

Steusloff, U. 1927. Die Bedeutung der *Paludestrina jenkinsi* E.A. Smith für unsere Vorstellungen über Art-Entstehung und Art-Verbreitung. Verhandl. Internat. Ver. f. theor. u. angew. Limnol. 3, 454-459.

Szaniawska, A & Dobrzycka-Krahel, A. 2004. National Report for Poland. ICES Working Group on Introductions and Transfers of Marine Organisms (WGITMO) form Meeting 2004, 64-74.

Thienemann, A. 1950. Verbreitungsgeschichte der Süßwassertierwelt Europas. In Thienemann, A. (ed.): Die Binnengewässer. Schweizbartsche Verlagsbuchhandlung, Stuttgart, 809 pp.

Tittizer, T., Schöll, F., Banning, M., Haybach, A. & Schleuter, M. 2000. Aquatissche Neozoen im Makrozoobenthos der Binnenwasserstraßen Deutschlands. Lauterbornia, 39. 1-72.

Törlitz, H. 1921. Über die fischereilichen Verhältnisse im Kaiser-Wilhelm-Kanal. Wasser- und Schiffahrtsamt Brunsbüttel (unpublished Report).

Wägele, J.W. 1979. Der Fortpflanzungszyklus von *Cyathura carinata* (Isopoda, Anthuridae) im Nord-Ostsee-Kanal. Helgoländer wiss. Meeresuntersuchungen 32: 295-304.

Walford, L. & Wicklund, R. 1973. Contribution to a world-wide inventory of exotic marine and anadromous organisms. FAO Fish. Tech. Pap., 121, 1-49.

Wallentinus, I. 1992. Dispersal of *Sargassum muticum* in the ICES countries. ICES WGITMO Report, 6 pp.

Wallentinus, I. 1999. Introduction and transfer of plants. In: Munro ALS, Uting SD & Wallentinus I (eds) Status of introductions of non-indigenous marine species to North Atlantic waters 1981-1990. ICES Cooperative Research Report 231: 1-43.

Wasser- und Schifffahrtsdirektion Nord. 2001. Der Nord-Ostsee-Kanal. Wasser- und Schifffahrtsdirektion Nord, Kiel. 6 pp.

Weber, H. 1971. Die Laichplätze des Herings (*Clupea harengus* L.) der westlichen Ostsee. Kieler Meeresforschungen 27: 194-208.

Werner, B. 1950. Die Meduse *Gonionemus murbachi* MAYER im Sylter Wattenmeer. Zool. Jb., 78, 471-505

Wiese, R. (no year). Der Nord-Ostsee-Kanal. Blinker, Sonderheft (special issue) no. 40: 56-59.

Wilke, H. 1979. Kann die Eider in Schleswig-Holstein durch die Ableitung des geklärten Abwassers in den Nord-Ostsee-Kanal verbessert werden? Fischwaid 12: 7199.

Williamson, M. 1996. Biological Invasions. Chapman & Hall, London, 244 pp.

Wiltshire, K.H. & Dürselen, C.-D. 2004. Revision and quality analysis of the Helgoland Reede long-term phytoplankton data archive. Helgol. Mar. Res. 58, 252-268.

Wolff, T. 1977. The Horseshoe Crab (*Limulus polyphemus*) in North European waters. Vidensk. Meddr Dansk Naturh. Foren., 140, 39-52.

Wolff, W.J. 2005. Non-indigenous marine and estuarine species in The Netherlands. Zoologische Mededelingen Leiden 79, 1-116.

The Panama Canal

Chapter I

Cutting a Canal Through Central America

ANDREW N. COHEN

San Francisco Estuary Institute, 7770 Pardee Lane, Oakland, CA 94621-1424, USA

1 Introduction and prehistory

Men had dreamed of a ship canal through Central America for three centuries before the first shovelful of dirt was dug, and three decades more would pass before the first ship sailed through the completed canal. This marked the reconnection of two bodies of water and two biotas that had been separating for perhaps 10 million years, and had been fully separated for at least 2 million years. The reconnection would affect not only these two regions, but would have at least indirect impacts on coastal marine biotas throughout the world.

The locations and times of existence of ancient seaways connecting the Atlantic and Pacific oceans have been the subject of lively scientific debates. These rely on interpretations of sedimentary deposits; inferences based on fossil, taxonomic and molecular genetic evidence of exchanges of terrestrial organisms between North and South America and of marine organisms between the Atlantic and Pacific oceans; and paleogeochemical evidence of the timing of water exchanges between the oceans. Such seaways may have existed as early as the late Cretaceous and throughout most of the Tertiary (Darlington 1957; Whitmore & Stewart 1965; Briggs 1974). As the Central American isthmus rose during the late Tertiary, water gaps of various sizes remained or opened at different periods. For land animals, these gaps acted sometimes as a complete barrier and at other times as a selective filter, allowing the passage of some "island hopping" species but not others. Similarly, when these gaps were deep, wide and common, marine plants and animals may have passed freely between the oceans, but narrower or shallower gaps would have acted as filters, restricting the flow of certain species through them.

During the middle Miocene, one or more straits connected Atlantic and Pacific waters through the developing Central American isthmus, as shown by the close similarity of the fossil marine faunas on either side of it (Dickerson 1917; Woodring 1966). Dickerson (1917) believed these openings were at Tehuantepec and in Panama. Woodring (1966) concluded that the main opening was through the Bolivar Trough, where the Atrato River now runs in western Colombia (which Briggs (1974) refers to as the "Panama Portal"), based on the extent of Miocene marine deposits found there and the discovery of early or early middle Miocene fossils of North American land mammals in central Panama (Whitmore & Stewart 1965). Smaller or intermittent straits may also have opened during the Miocene through the Nicaraguan Depression in southern Nicaragua (now occupied by Lake Nicaragua), or in central Costa Rica or central Panama (Whitmore & Stewart's (1965) "Strait of Panama" in the vicinity of the Canal Zone) where there are significant Miocene marine and brackish-water deposits. Woodring (1966) suggested that the North Equatorial Current would have carried planktonic larvae from the Atlantic through these straits and into the Pacific.

It was Wallace (1876) who first realized that there were extensive migrations of land animals between North and South America during the Tertiary. These occurred in four distinct periods, based on fossils recovered primarily in western North America, Florida and Argentina (Webb 1976; Marshall et al. 1979). The first exchange involved a few genera of ground sloths moving north, and procyonids (racoons and their relatives) and probably the earliest sigmodontinid rodents moving south. These animals, dubbed the "New Island Hoppers," were capable of crossing water gaps in the island arcs that were the precursors of the central American land bridge in Hemphillian (North American stage) and Huayquerian (South American stage) time, between 6 and 9.5 million years ago (Marshall et al. 1979). The second and third interchanges involved a larger and more diverse group of genera, suggesting that the land bridge was then more-or-less complete. These occurred in late Blancan to early Irvingtonian stages in North America, and Chapadmalalan to early Uquian stages in South America, roughly 3 to 2 million years ago (Webb 1976; Marshall et al. 1979). Finally, a few further genera migrated to South America in the Ensenandan stage, around one million years ago (Webb 1976). Webb (1976) notes that an absence of exchanges over long periods does not necessarily mean that the land bridge was broken. Instead ecological factors, including tropical habitats and discontinuities in savanna habitat across the Central American isthmus, could have blocked migration entirely at times, and acted as a selective filter, blocking some families or groups of organisms, at other times (Marshall et al. 1982).

An arcuate, east-west belt of thin sediment in the central Caribbean is thought to mark the course of a persistent ocean current that was strong enough to inhibit sediment deposition, with two breaks in the sediment record indicating periods when the current was strong enough to stop deposition altogether (Holcombe & Moore 1977). The later of these occurred from Eocene to early Miocene time when the Atlantic equatorial current apparently flowed through to the Pacific. The resumption of sedimentation starting in early-to-middle Miocene indicates a weakening of the current in the central Caribbean, which was probably due to the constriction and closure of straits that had connected to the Pacific. Marine sediments east of Florida show a gradual coarsening from late Miocene up to mid-Pliocene time (about 3.8 million years ago), which has been interpreted as evidence of stronger winnowing of sediments due to an increase in the average velocity of the Gulf Stream (Kaneps 1979). This is thought to have also been caused by the gradual emergence of the Central American isthmus, forcing flows that had formerly entered the Pacific to turn north and strengthen the Gulf Stream. This had the further effect of warming northern waters so that a subtropical fauna became established around Chesapeake Bay (Stanley & Campbell 1981).

Keigwin (1982) compared the carbon and oxygen isotope ratios in benthic and planktonic foraminifera recovered from seabed cores in the western Caribbean and eastern Pacific, and concluded that the initial shoaling of the isthmian region occurred in the Miocene, that deep to intermediate water exchange between the Atlantic and Pacific ended around 6 million years ago, and that surface water exchange became increasingly restricted between 4 million and 3 million years ago. It has been suggested that the restriction of water exchange at that time led to a sustained change in regional climate, resulting in a period of rapid speciation in shallow water ostracods (Cronin 1985). Keller et al. (1989) compared foraminifer faunas and isotope values in the western Caribbean and eastern Pacific and concluded that the closure of the strait occurred in four phases: first, an increase in coldwater upwelling foraminifer species at intermediate depths in the Caribbean around 6.2 million years ago, indicating a restriction of westward flow and its deflection northward, corroborated by divergence of oxygen isotope values in benthic foraminifers; a second faunal change accompanied by divergence of oxygen isotope values in planktonic foraminifers at about 4.2 million years ago, probably caused by increasing surface salinity in the Caribbean related to further restriction of westward flow; third, a major differentiation of Caribbean and Pacific faunas beginning around 2.4 million years ago as high salinity surface species of foraminifers became relatively more abundant in the Caribbean, with high surface salinities corroborated by other faunal and isotope data; and finally, an acceleration of faunal divergence starting at 1.8 million years ago, indicating final closure of the strait. Further comparisons of fossil foraminifera assemblages suggest that a barrier prevented exchange between the oceans between 12.9 million and 7.0 million years ago (Knowlton et al. 1993), facilitated by a low sea-level stand at about 10.5 million years ago (Banford et al. 2004); followed by restricted shallow water exchange that may have extended to 150 meters depth 6-7 million years ago (Lessios 1998), and to less than 50 meters depth by 6.3 million years ago (Knowlton et al. 1993).

Knowlton et al. (1993) reported a pattern of divergence in behavior, allozymes and mitochondrial DNA in sibling pairs of alpheid shrimp species collected on the Atlantic and Pacific coasts of Panama that indicated separation of the species pairs at markedly different times. Calibrating by mitochondrial DNA divergence and assuming the least differentiated pairs were separated 3.5 million years ago (see estimates of the time of isthmian closure in the next paragraph), yields separation times for the more differentiated pairs of 6.1, 6.3 and 9.1 million years ago, which is consistent with the temporal pattern of circulatory barriers and shoaling derived from foraminiferal assemblages and isotope changes.

The limited distribution of Pliocene marine deposits in Costa Rica and the limited distribution of Pliocene marine deposits in Costa Rica and Panama suggests that the southern part of the isthmus was complete during the Pliocene (Woodring 1966). Based on the fossil record of a major interchange of North and South American land animals, the closure of the straits and completion of the land bridge is estimated to have occurred by 2-3 million years ago (Woodring 1966), 3 million years ago (Marshall et al. 1979, 1982; Keigwin 1982), 2.5-2.8 million years ago (Bermingham & Lessios 1993), or 2.8-3.1 million years ago (Banford et al. 2004). Other estimates of the time of closure as reviewed or reported by various authors are about 1 million years ago (Olsson 1972), 1-2 million years ago (Voss 1972), 2-3 million years ago (Springer & Gomon 1975), around 3.5 million years ago (Holcombe & Moore 1977; Stanley & Campbell 1981; D'Croz & Robertson 1997), 1-3 million years ago (Voss 1978), around 5.7 million years ago (Gunter 1979), between 3.5-5.7 million years ago and 2 million years ago (Lessios 1979, 1981), 1-5 million years ago (Leschine 1981), 3.1-3.5 million years ago (Lessios 1984), 3-4 million years ago (Cronin 1985), around 3.1 million years ago (Vermeij 1991), 3.0-3.5 million years ago (Knowlton et al. 1993; Lessios 1998), and 2.9-3.5 million years ago (Bermingham & Lessios 1993). Some evidence also suggests that a final opening in the isthmus may have occurred in association with a high sea level stand 2.8-3.1 million years ago (Knowlton & Weigt 1998), or a more limited breach allowing limited exchange around 1.8-2 million years ago (Lessios 1998; Knowlton & Weigt 1998; Banford et al. 2004).[1]

With the closure or even the partial closure of the isthmus, the biota of the Caribbean and Panamic[2] regions began to diverge. Woodring (1966) noted a much greater loss of Tertiary molluscan genera and subgenera in the Caribbean than in the Panamic region, and Vermeij (1991) similarly calculated that the Caribbean lost 32% and the Panamic 15% of their respective Miocene and Pliocene subgenera. In contrast, nearly all of the Panamic coral genera but less than half of the Caribbean coral genera were lost since the closure (Vermeij 1991).

Later analysis of fossil assemblages indicated that the highest rate of extinction of molluscan subgenera in the Caribbean occurred somewhat later, around 2.4 million years ago (Vermeij 1993) or around the end of the Pliocene (Jackson et al. 1993), and that speciation and invasions from other regions more than made up for the losses, so that the number of Caribbean subgenera continuously increased from the late Miocene to the end of the Pliocene (Jackson et al. 1993). Similarly, an analysis of fossil assemblages in Florida found that the number of mollusk species in the western Atlantic has remained more or less constant from the late Pliocene to the present (Allmon et al. 1993). A species-level analysis of strombinid mollusks from both sides of the isthmus found a steady increase on the Pacific side from 3 species in the early/middle Miocene to 33 species currently, whereas Caribbean strombinids peaked at 23 species in the early Pliocene and then declined to 3 species. As with the Caribbean subgenera, both Caribbean and Pacific strombinid species suffered a high rate of extinction around the end of the Pliocene, but the Pacific fauna gained more species through speciation and invasion than it lost (Jackson et al. 1993). Vermeij (1993) summarized the overall pattern portrayed by these and other studies as being characterized by a greater degree of speciation and diversification among large suspension-feeding animals and shallow-water mollusks occurring on mud or sand in the eastern Pacific than in the western Atlantic, and greater speciation

---

1 As noted, "closure," "completion" or "emergence" of the isthmus did not happen all at once or in a single event, and these terms may mean somewhat different things to researchers variously concerned with changes in hydrology, migrations of land animals, or migration of marine organisms. This accounts for some of the range and variation in these time estimates, along with the incomplete and still accumulating stock of evidence, and differences in interpretations of the evidence.

2 "Caribbean" and "Panamic" are used in this chapter to refer to the modern coastal marine regions in the tropical western Atlantic and tropical eastern Pacific, respectively.

and diversification in reef- and rock-dwelling organisms in the western Atlantic than in the eastern Pacific.

2 Construction of the canal

Dreams of cutting a canal through the Central American isthmus have been recorded since the 16th century, but no attempt was made until a French company began digging in the 1880s.[3] After 8 years and the deaths of tens of thousands of workers, the French effort ended in failure, bankruptcy and scandal. After another 10 years of labor, the U.S. government completed a canal in the early 20th century. Apart from the monumental physical effort and medical challenges, and the accompanying political and financial drama, the canal project is primarily a tale of repeatedly determining, reconsidering and re-determining two key decisions: where to build the canal, and whether to construct it as a sea-level or a lock canal. The latter decision in particular would have a huge effect on the canal's biological consequences.

2.1 Plans and explorations

In 1513, Vasco Nuñez de Balboa crossed the isthmus of Panama and reached the shore of the Pacific, demonstrating that a mere strip of land separated the Atlantic and Pacific oceans.[4] In 1534, King Charles I of Spain ordered his governor in Panama to survey a route for a ship canal through the isthmus along the valley of the Chagres River. The governor concluded that a canal could not be dug there, but the priest Francisco López de Gómara thought otherwise. Writing in 1552, he described four possible routes, concluding that "there are mountains, but there are also hands, and for a king of Castile, few things are impossible" (McCullough 1977: 27). By then, however, the king had other priorities.

In 1811 the German naturalist and explorer Alexander von Humboldt published a report that considered several routes and recommended one through Nicaragua. Several events over the following decades would encourage serious consideration of a transisthmian canal. The Erie Canal and the Caledonian

---

[3] The following account of the building of the canal is largely based on McCullough (1977), updated with information from the Panama Canal Authority (ACP 2005).

[4] The shortest straight-line distance is at San Blas, where the two oceans are less than 50 kilometers apart. At the site of the canal, from Limon Bay to Balboa, the straight-line distance across is about 60 kilometers.

Canal both opened in the 1820s. The 98 kilometer-long Caledonian Canal, designed by Thomas Telford and built in 1803-1822, cuts across Scotland to link the Atlantic and the North Sea. Among its 29 locks is a set of eight known as "Neptune's Staircase," which the poet Robert Southey described as Britain's greatest work of art. DeWitt Clinton's Erie Canal, built in 1808-1825, was the longest canal in the world, with a lock system that lifts barges over an elevation of nearly 700 feet. The successful completion of these massive engineering projects made a transisthmian canal seem more feasible, and in later years both Telford and Clinton would consider plans for a Central American canal.

Gold was discovered in California in January of 1848. A year later, 200 men landed at the mouth of the Chagres River on the Caribbean side of the Panama region and beat their way through the jungle to meet up with a California-bound ship on the Pacific side, thus becoming the first gold-seekers to reach California via the "Panama Route." More would follow, so many that work on a railroad would begin in the following year. The 48-mile-long, single-track Panama Railroad was completed in 1855, and in the next 10 years some 400,000 travelers would buy a $25 one-way ticket for a 3-hour ride from ocean to ocean (Fig. 1).

Fig. 1. The Panama Railroad. From: Illustrated History of the Panama Railroad, by F.N. Otis, Harper & Brothers, Publishers, New York, 1862.

In 1850 Dr. Edward Cullen announced that he'd found a path through the Darien region that rose no more than 150 feet above sea level. Such a route did not, in fact, exist, but Cullen's claim spurred others to examine the isthmus more carefully. One of these was Navy Lieutenant Isaac Stain, who in 1854 entered the Darien wilderness with 27 men in search of the canal route and, finding no route, reached the Pacific seven weeks later with 20 survivors described as "living skeletons, covered with foul ulcers."

Between 1870 and 1875, U.S. President Ulysses S. Grant, who as an Army captain had led the Fourth Infantry across the isthmus in 1852, dispatched seven expeditions to Central America to survey possible canal sites. These included routes in Colombia (along the Atrato and Napipi rivers, the site of an ancient seaway, the Bolivar Trough), in the Darien region (from Caledonia Bay along the Sucubti River, and from the Gulf of San Blas along the Mandinga River), along the Chagres River and the line of the Panama Railroad (the site of another ancient seaway), in southern Nicaragua through Lake Nicaragua, and across the isthmus of Tehuantepec in southern Mexico. To evaluate the results, Grant appointed an Interoceanic Canal Commission, which in 1876 concluded that the best route was through southern Nicaragua. The value of linking the oceans was more evident than ever, and the geographical and engineering knowledge needed to build the canal seemed increasingly within reach.

2.2 The French effort

The Suez Canal was completed in 1869. Six years later its triumphant creator, Ferdinand de Lesseps, announced his interest in constructing a canal through Central America at a special meeting of the Société de Géographie de Paris (Fig. 2). Even at that early stage, de Lesseps declared that the canal must be a waterway dug through at sea level, like the canal at Suez, rather than a lock canal.

De Lesseps and others formed a company, La Société Civile Internationale du Canal Interocéanique de Darien, and sent an expedition to Panama to search for a route, under the leadership of Naval Lieutenant Lucien Napoléon Bonaparte Wyse. Panama was then a province of Colombia, which gave the expedition permission to search only in the Darien region, east of the Panama Railroad. Lieutenant Wyse returned in April 1877 with a plan and route for a canal that included both ship locks and tunnels, but de Lesseps rejected the plan.

While a second expedition in 1877-78 examined a route along the Panama Railroad, Wyse traveled to Bogotá and negotiated a treaty that granted an exclusive right to the Société Civile to build and operate a canal through Panama. This time the expedition returned to Paris with a plan for a sea-level canal along the railroad route, with an 8 kilometer-long tunnel under the Continental Divide. De Lesseps found the new plan acceptable. Working with the Société de Géographie de Paris, he organized an international scientific congress to consider the construction of an inter-oceanic canal. De Lesseps expected the congress to give legitimacy to the Société Civile's plan and thereby assist it in obtaining financing.

Fig. 2. Ferdinand de Lesseps, fresh from his triumph at Suez. From: Vanity Fair, Nov. 27, 1869.

The congress met in Paris in May 1879, and considered 14 different routes across Central America. Baron Godin de Lépinay, the chief engineer for the French Department of Bridges and Highways, proposed a lock canal with a dam at Gatun to block the Chagres and Gatun Rivers and create an artificial lake that would serve as the main waterway, with a cut through the Culebra Gap. Lépinay's plan was very similar to the canal that was built 35 years later, but the congress gave it little attention. There was more interest in an American plan for a lock canal in southern Nicaragua. But after an impassioned presentation by de Lesseps, the congress endorsed the Société Civile's plan for a sea-level canal along the line of the Panama Railroad.

Following the congress, the Société Civile was reorganized as the Compagnie Universelle du Canal Interocéanique de Panama with Ferdinand de Lesseps as its president. De Lesseps' son Charles was hired to supervise the work, which officially began with an explosion at Culebra on January 10, 1880. The work, however, did not go smoothly. The company had vastly underestimated the amount of excavation needed to create a sea-level waterway; there were frequent landslides that stopped work, destroyed equipment and had to be laboriously dug out; and workers died in huge numbers from tropical fevers and other diseases. As progress slowed and the death toll mounted, it became harder and harder to raise the financing needed to continue the work.

In 1887 the plan to dig a sea-level waterway - which had been the central, unwavering component of Ferdinand de Lesseps' Central American vision for the past twelve years - was discarded in favor of a plan to first construct and open a lock canal, which would later and gradually be excavated down to sea level by floating dredges. The change, though a sensible one, came too late to save the effort. The Compagnie Universelle went bankrupt in December 1888, a liquidator was appointed, and work was halted in May 1889. In the aftermath of the company's failure, claims emerged that government officials had been bribed to support the issuing of public bonds to finance the work. Some of the Compagnie Universelle's officers and contractors were tried and found guilty of fraud and bribery, and Charles de Lesseps went to jail.

2.3 America takes over

Two events around the turn of the 19th century set the stage for the United States to take on the role of canal builder. In 1898, the Spanish-American War left the United States in possession or control of new territories, including the islands of Puerto Rico, Guam and the Philippines, and a naval base in Cuba. And in 1901, an anarchist shot and killed President William McKinley, putting Theodore Roosevelt - a participant in and ardent supporter of the Spanish-American War - in the White House (Fig. 3).

Roosevelt was also a disciple of Alfred Mahan (Fig. 5A), a former naval officer and the author of "The Influence of Sea Power Upon History." In this widely-read and highly-regarded book, Mahan argued that a nation's success in commerce and war depended on its dominance at sea, and that a Central American canal was a commercial and military necessity for the United States. Mahan's lessons gained force with the sudden expansion of U.S. possessions in both the Atlantic and Pacific at the close of the Spanish-American War, along with the annexation of Hawaii in 1898. Meanwhile, an incident during the war vividly demonstrated the perils of having the U.S. fleet divided between two oceans, when it took the battleship *USS Oregon* more than two months to sail from San Francisco around the South American continent to provide a military presence in the waters off Cuba.

Fig. 3. Theodore Roosevelt, Colonel of the 1st Volunteer Cavalry (the "Rough Riders") in the Spanish-American War. From: The Rough Riders, by Theodore Roosevelt, Charles Scribner's Sons, New York, 1899.

Fig. 4. Foredeck of the battleship USS Oregon. Photo courtesy of the Library of Congress.

More than any other modern U.S. president, Roosevelt supported the expansion of the nation's territorial and military authority, with the expectation that the United States would become the dominant power in the Pacific. Armed with Mahan's sense of the importance of the Navy, Roosevelt considered a U.S.-controlled isthmian canal to be essential to the country's future, allowing it to quickly merge its fleets while denying its enemies the same swift passage between oceans in time of war. By lopping nearly 8,000 miles off the distance between the Atlantic and the Pacific, a canal would grant the United States an enormous military advantage.

Fig. 5. (A) Alfred Thayer Mahan. (B) Secretary of State John Hay. Photos by J.E. Purdy, courtesy of the Library of Congress.

Roosevelt began preparing the United States for the pursuit of a Central American canal even before he assumed the presidency. In 1850 the United States and Great Britain had signed the Clayton-Bulwer Treaty, agreeing to joint control of any canal built across Central America. This forestalled the worrisome possibility of Britain building and controlling its own canal in the New World. Forty years later, however, the United States was prepared to consider building a canal on its own, and so in 1890 the U.S. Secretary of State, John Hay (Fig. 5B), and the British ambassador negotiated a new treaty which gave the United States the right to build and operate an *unfortified* canal that would be free and open to the ships of all nations, both in peace and in war.

To Roosevelt, these terms were unacceptable. "If that canal is open to the war ships of an enemy it is a menace to us in time of war," he wrote to Hay. "If fortified by us, it becomes one of the most potent sources of our possible sea strength." Roosevelt had allies in the U.S. Senate, and when they blocked ratification of the treaty, Hay was forced to renegotiate it. The new treaty, which deleted the prohibition against fortifying the canal and gave the United States the right to do whatever was necessary to protect it, was signed in November 1901 after Roosevelt became President.

During these negotiations, it had been assumed by nearly all the parties that any U.S. canal would be built in Nicaragua. The Nicaraguan route had been selected by Grant's Interoceanic Canal Commission in 1876, championed by the U.S. delegation to the Paris congress in 1879, and confirmed by the Isthmian Canal Commission, also known as the Walker Commission, in November 1901. A careful reading of the Walker Commission's report, however, revealed that the selection of a route through Nicaragua rather than Panama was based on cost, and a large portion of that was the $109 million estimated price for acquiring the canal properties and equipment that had belonged to the Compagnie Universelle.[5] In early January of 1902, after a flurry of activity in Paris, the asking price was slashed to $40 million; and by the end of the month, at Roosevelt's request, the Walker Commission issued a supplemental report that favored the Panama route.

The debate then moved to the U.S. Congress and carried on for weeks, when the earth itself appeared to step in and have its say. Among the many differences between the proposed routes, the threat posed by Nicaragua's several volcanoes had been only a minor point of argument. Then on May 2, 1902, Mt. Pelée, a little known volcano on the island of Martinique in the eastern Caribbean, began erupting. Six days later it exploded, killing 30,000 people and leveling the town of St. Pierre (Fig. 6). On May 14, Momotombo in Nicaragua erupted, and then on May 20 Mt. Pelée erupted again, along with an eruption of Souffriere that devastated the northern portion of the island of St. Vincent.

Though the proponents of the Panama route made much of the volcanic threat, it's not clear how much it affected the outcome of the debate. In any event, in late June the Senate and House passed bills that provided for an initial attempt to construct a canal in Panama, but to revert to the Nicaragua route if the

---

5 The purchase would actually be made from the Compagnie Nouvelle du Canal de Panama, a company that had been formed to take over the holdings of the Compagnie Universelle when it went bankrupt.

country could not obtain clear title to the Panama canal properties within a reasonable time.

Fig. 6. (A) Mt. Pelée erupting. (B, C) The destruction of St. Pierre. Photos courtesy of the Library of Congress.

In January 1903, John Hay negotiated a treaty with Colombia granting the United States a 100-year franchise to build and operate a canal in Panama in exchange for $10 million. The terms of the treaty were quite favorable to the United States, and included U.S. control of a six-mile-wide canal zone across Panama and renewal of the franchise at the United States' sole option. The U.S. Senate quickly ratified the treaty, but Colombia balked. The Roosevelt administration, through both diplomatic communications and public statements, expressed increasing impatience with Colombia, even suggesting publicly that it might support a Panamanian revolt.

That support was provided just eight months later. On November 3, 1903, a small group of revolutionaries took over Panama City in a bloodless coup, and declared Panama's independence. On November 5, U.S. gunboats landed at Colon on the Caribbean coast and took control of the Panama Railroad. When a Colombian ship reached Colon that evening with a large detachment of soldiers, the U.S. forces stopped them from traveling to Panama City, and eventually negotiated their return to Colombia. Over the following months eight U.S. gunboats would patrol both coasts of Panama, preventing Colombian troops from landing.

On November 6 the United States formally recognized Panama, and twelve days later John Hay signed a treaty with Panama granting the canal concession to the United States. This agreement was similar to the one that Colombia had rejected, but was even more favorable to the United States. For example, the canal zone would be ten rather than six miles wide, it would be held by the United States in perpetuity, and the United States could expropriate any additional land or water needed for the construction, operation or defense of the canal. In return, the United States would pay Panama $10 million and guarantee its independence. After the treaty was ratified by the two countries, the United States purchased the Compagnie Universelle's canal properties for $40 million.

2.4 Construction

The U.S. canal project began in earnest with the arrival of Colonel William Gorgas, the Chief Sanitary Officer, at Colon in June of 1904 (Fig. 7B). Gorgas recognized the critical importance to the construction effort of controlling tropical fevers and other diseases. During the building of the Panama Railroad, thousands had died from cholera, dysentery, malaria, yellow fever and smallpox. So frequent and regular were these deaths that the railway company did a steady side-business in supplying pickled cadavers to medical schools for dissection and study. The high toll continued during the Compagnie Universelle's canal work. Later estimates indicated that 20,000 to 22,000 workers died in the 8-year effort.

Until nearly the start of the U.S. canal project, the prevailing medical and public opinion was that malaria and yellow fever were caused by "miasmal mists" that arose from swampy ground or disturbed soils. For a long time, however, there had been hints that mosquitoes were involved. By 1881 a Havana physician, Carlos Finlay, had become convinced through 20 years of observations that yellow fever was transmitted by the mosquito *Aedes aegypti* (then known as *Stegomyia fasciata*) (Fig. 8A). He tried to prove it by allowing mosquitoes to suck blood from yellow fever patients, and then having them attack the skin of volunteers, but the volunteers did not come down with yellow fever. The solution to this puzzle began to emerge when a Mississippi physician, Henry Rose Carter, initiated a statistical study of yellow fever among rural patients, and discovered that a 12-20 day "extrinsic incubation period" was needed after an individual developed yellow fever before another person could catch the disease by visiting the infected individual's home. Carter published his results in 1900, just as he and Walter Reed (Fig. 7A) were sent to Havana to deal with a yellow fever outbreak, working alongside Finlay. Reed focused on *Aedes aegypti* and demonstrated through a series of experiments and observations that it was the agent of transmission of yellow fever, and that Carter's extrinsic incubation period was the time needed for the disease organisms to develop in their mosquito hosts. As final proof, William Gorgas, though not yet fully persuaded by Reed's work, undertook a massive effort to rid Havana of *Aedes aegypti*, thereby eliminating yellow fever from the city in eight months.

Meanwhile, in 1897 the English physician Ronald Ross, working in India, discovered *Plasmodium falciparum*, the protozoan that causes malaria, multiplying in the stomach of an *Anopheles* mosquito that had recently fed on a malarial patient. For discovering the mosquito vector of malaria, Ross won the Nobel Prize in 1902.

Fig. 7. (A) Walter Reed. (B) William Gorgas. (C) John Stevens.

Gorgas now arrived in Panama, with the job of controlling both malaria and yellow fever in the Canal Zone, and to reduce the incidence of pneumonia, tuberculosis, chronic diarrhea and dysentery, which took nearly as many lives. Although Gorgas understood what needed to be done to control these two tropical fevers, his superiors still believed that they were caused by bad air rising from wet or disturbed ground and exacerbated by immoral life styles, and they failed to support his efforts to control mosquitoes.

In July of 1905 a new director arrived in the Canal Zone: John Stevens, formerly an engineer with the Great Northern Railroad (Fig. 7C). He immediately saw the importance of managing these diseases, accepted Gorgas' plans, and provided unstinting support for a medical campaign that included controlling mosquitoes. Gorgas virtually eliminated yellow fever from the Canal Zone in a year and half, and greatly reduced the incidence of malaria (Fig. 8B). In all, during the ten years it took to construct the canal there were 5,609 deaths from disease and accidents out of a total work force of 56,300 people - a very large number, but corresponding to an annual death rate that was only one-fifth of the rate during the years of the French effort.

When Stevens took over, one key decision still to be made was whether to build a sea-level or a lock canal. A review board appointed by Roosevelt recommended a sea-level canal in November 1905. But later that winter, Stevens observed the Chagres River in flood. He decided that a sea-level canal, which required containing or diverting the Chagres' raging flood waters before they reached the waterway of the canal, was impractical; and that the best solution was to allow the Chagres' flows to enter and fill a lake that would form the major part of the waterway in a lock canal.

Fig. 8. (A) *Aedes aegypti*, the vector for yellow fever. Photo courtesy of the U.S. Department of Agriculture. (B) One of the drainage and fumigation brigades, similar to those he organized in Havana, that William Gorgas used to rid the Canal Zone of yellow fever. Photo courtesy of CanalMuseum.com.

The next advisory body to take up the question was the Isthmian Canal Commission. In February 1906 Stevens convinced first the commissioners, and then Roosevelt himself, to endorse a lock canal. Then the issue advanced to a U.S. Senate committee, which voted in favor of a sea-level canal, before the full Senate voted for a lock canal by a narrow margin and finally settled the matter.

With Gorgas' medical campaign underway and the form of the canal decided, Stevens, a former railroad engineer, now reorganized the canal excavation as a railroad project. In his view, the fundamental challenge of the canal's construction was the coordinated movement of men, food, supplies and - most importantly - dirt, throughout the Canal Zone. Enormous quantities of dirt had to be moved quickly and efficiently out of the diggings at Culebra Cut and the other excavation sites and transported to Gatun where it was needed to dam the Chagres and Gatun rivers, and to other fill and disposal sites. To Stevens, the obvious solution to the problem was the Panama Railroad, and so he rebuilt it with heavier rails and a second track, ordered heavy locomotives and cars to replace the lightweight French stock, and hired railroad engineers to run the project. "The digging is the least of it," he said, and the efficient operation of the project after his reorganization showed that he was right (Fig. 9).

Fig. 9. Moving the dirt. Photo courtesy of the Panama Canal Authority.

In November 1906, at the height of the rainy season, Roosevelt made a famous tour of the canal project - the first time a U.S. president had left the country. He slogged through mud, posed at the controls of a massive steam shovel, asked innumerable questions, and by his energy and enthusiasm greatly increased the visibility and the popularity of the project in the United States. However, a few months after Roosevelt's return to the United States, Stevens, for reasons that were never clear, suddenly resigned. Roosevelt turned the canal project over to Major George Washington Goethals and other officers from the Army Corps of Engineers, and from then on the project was run under a command structure of military engineers working in a civilian capacity and reporting to the Secretary of the War Department until it was completed seven years later.

The main excavation through Culebra Gap was plagued by a series of massive landslides. These forced repeated downward adjustments in the estimate of the final slope that would stabilize the sides of the excavation, which increased the amount of rock and sediment that had to be removed from Culebra to 80 million cubic meters, compared to a 1906 estimate of 45 million cubic meters of excavation for the entire canal. Despite these setbacks, the dry excavation at Culebra was completed in June 1913, when two steam shovels met nose-to-nose at the bottom of the cut. On October 10, 1913, President Woodrow Wilson pressed a button in Washington, sending a telegraph signal to Panama which detonated an explosion that broke open the Gamboa Retention Dam and allowed the rising waters of Gatun Lake to flood the cut. Dredges were then floated in to complete the digging (Fig. 10).[6]

Fig. 10. Floating dredges at work in the Culebra Cut. Photo courtesy of CanalMuseum.com.

On January 7, 1914, the *Alexandre La Valley*, an old Compagnie Universelle crane boat that had been brought up from Limon Bay to work on the canal some time earlier, passed down through the locks to the Pacific side, becoming the first ship to pass through the canal from ocean to ocean. Hardly anyone noticed. The canal was officially opened on August 15, 1914, to little ceremony, as World War I had just broken out in Europe.

The following year, however, the completion of the canal was celebrated at an international exposition in San Francisco (Fig. 11). The United States had spent about \$375 million building the canal (including a \$40 million payment for the Compagnie Universelle's property and equipment and a \$10 million payment to

[6] Though in truth, the digging is never really over. In the first 13 months after the canal was opened, slides at Culebra blocked the canal three times, the last of which closed it to traffic for seven months while dredges dug it out again. Slides, repairs, and improvements have continued the digging to the present day.

the Republic of Panama), which is about $175 million more than the 1906 cost estimate. In all, some 200 million cubic meters of rock and sediment were excavated, or over four times the amount that was estimated as necessary in 1906 and nearly three times the amount excavated at Suez.

Fig. 11. "The Thirteenth Labor of Hercules" - Perlham W. Nahl's poster for the 1915 Panama Pacific International Exposition in San Francisco, showing Hercules parting the isthmus.

3 References

ACP. 2005. Panama Canal Website, Authoridad del Canal de Panama, Balboa-Ancon, Republic of Panama, at http://www.pancanal.com/index.html (accessed in December 2005).

Allmon, W.D., G. Rosenberg, R.W. Portell and K.S. Schindler. 1993. Diversity of Atlantic coastal plain mollusks since the Pliocene. Science 260: 1626-1628.

Banford, H.M., E. Bermingham and B.B. Collette. 2004. Molecular phylogenetics and biogeography of transisthmian and amphi-Atlantic needlefishes (Belonidae: *Strongylura* and *Tylosaurus*): perspectives on New World marine speciation. Mol. Phylogenet. Evol. 31: 833-851.

Bermingham, E., and H.A. Lessios. 1993. Rate variation of protein and mitochondrial DNA evolution as revealed by sea urchins separated by the Isthmus of Panama. Proc. Natl. Acad. Sci. 90: 2734-2738.

Briggs, J.C. 1974. Operation of zoogeographic barriers. Systematic Zool. 23(2): 248-256.

Cronin, T.M. 1985. Speciation and stasis in marine ostracoda: climatic modulation of evolution. Science 227: 60-63.

Darlington, P.J., Jr. 1957. Zoogeography. Wiley, New York.

D'Croz, L. and D.R. Robertson. 1997. Coastal oceanographic conditions affecting coral reefs on both sides of the Isthmus of Panama. Proc. 8th Int. Coral Reef Sym. 2: 2053-2058.

Dickerson, R.E. 1917. Ancient Panama Canals. Proc. Calif. Acad. Sci. 7(8): 197-205.

Gunter, G. 1979. Marine fishes of Panama as related to the Canal. Gulf Research Reports 6(3): 267-273.

Holcombe, T.L and W.S. Moore. 1977. Paleocurrents in the eastern Caribbean: geologic evidence and implications. Marine Geology 23: 35-56.

Jackson, J.B.C., P. Jung, A.G. Coates and L.S. Collins. 1993. Diversity and extinction of tropical American mollusks and emergence of the isthmus of Panama. Science 260:1624-6.

Kaneps, A.G. 1979. Gulf Stream: velocity fluctuations during the late Cenozoic. Science 204: 297-301.

Keigwin, L. 1982. Isotopic paleoceanography of the Caribbean and east Pacific: role of Panama uplift in Late Neogene Time. Science 217:350-353.

Keller, G., C.E. Zenker and S.M. Stone. 1989. Late Neogene history of the Pacific-Caribbean gateway. Journal of South American Earth Sciences 2(1): 73-108.

Knowlton, N., and L.A. Weigt. 1998. New dates and new rates for divergence across the Isthmus of Panama. *Proc. R. Soc. Lond.* 265: 2257-2263.

Knowlton, N., L.A. Weigt, L.A. Solórzano, D.K. Mills, and E. Bermingham. 1993. Divergence in proteins, mitochondrial DNA, and reproductive compatibility across the Isthmus of Panama. Science 260:1629-1632.

Leschine, T.M. 1981. The Panamanian sea-level canal. Oceanus 24(2): 20-30.

Lessios, H.A. 1979. Use of Panamanian sea urchins to test the molecular clock. Nature 280:599-601.

Lessios, H.A. 1981. Divergence in allopatry: molecular and morphological differentiation between sea urchins separated by the Isthmus of Panama. Evolution 35(4): 618-634.

Lessios, H.A. 1984. Possible prezygotic reproductive isolation in sea urchins separated by the Isthmus of Panama. Evolution 38(5): 1144-1148.

Lessios, H.A. 1998. The first stage of speciation as seen in organisms separated by the Isthmus of Panama. Pages 186-201 in: D. Howard and S. Berlocher (eds.), Endless Forms: Species and Speciation. Oxford University Press, Oxford.

Marshall, L.G., R.F. Butler, R.E. Drake, G.H. Curtiss and R.H. Tedford. 1979. Calibration of the Great American Exchange. Science 204: 272-279.

Marshall, L.G., S.D. Webb, J.J. Sepkoski, Jr. and D.M. Raup. 1982. Mammalian evolution and the Great American Interchange. Science 215: 1351-1357.

McCullough, D. 1977. The Path Between the Seas: The Creation of the Panama Canal 1870-1914. Touchstone Books, Simon & Schuster, New York.

Olsson, A.A. 1972. Origin of the existing Panamic molluscan biotas in terms of their geologic history and their separation by the isthmus land barrier. Bull. Biol. Soc. Wash. 2: 117-123.

Springer, V.G., and M.F. Gomon. 1975. Revision of the Blenniid fish genus *Omobranchus* with descriptions of three new species and notes on other species of the tribe Omobranchini. Smithsonian Contributions to Zoology 177: 1-135.

Stanley, S.M. and L.D. Campbell. 1981. Neogene mass extinction of Western Atlantic molluscs. Nature 293: 457-459.

Vermeij, G.J. 1991. When biotas meet: understanding biotic interchange. Science 253:1099-1104.

Vermeij, G.J. 1993. The biological history of a seaway. Science 260:1603-1604.

Voss, G.L. 1972. Zoogeographical considerations of the impact of a sea-level canal across Panama. Proc. International Congress of Zoology. 17(3): 1-13.

Voss, G.L. 1978. Panama Sea-Level Canal-II. Biological catastrophe or grand experiment? *Sea Frontiers* 24: 206-13.

Wallace, A.R. 1876. The Geographical Distribution of Animals. Macmillan, London.
Webb, S.D. 1976. Mammalian faunal dynamics of the great American interchange. Paleobiology 2: 220-234.
Whitmore, F.C., Jr. and R.H. Stewart. 1965. Miocene mammals and Central American seaways. Science 148: 180-185.
Woodring, W.P. 1966. The Panama land bridge as a sea barrier. Proc. Amer. Phil. Soc. 110(6): 425-433.

Chapter II

Shipping Patterns Associated with the Panama Canal: Effects on Biotic Exchange?

GREGORY M. RUIZ[1], JULIO LORDA[2], ASHLEY ARNWINE[1] & KELLY LION[1]

[1] *Smithsonian Environmental Research Center, P.O.Box 28, Edgewater, Maryland 21037, USA*
[2] *Department of Ecology, Evolution, and Marine Biology, University of California, Santa Barbara, California 93106, USA*

1 Introduction

Shipping has been a primary mode of trade for millennia and is undergoing constant change (Couper 1972; see Suez Canal chapter). Vessel size and speed have certainly increased through time. Worldwide, the number of recipient and source ports engaged in international commerce have increased in recent history, as have the cumulative number of vessel arrivals across these ports. Together, these changes in scale and tempo of shipping are driving the increased globalization of economies.

Trade routes have also shifted through time. Changes in vessel characteristics such as motorization (speed), size, and refrigeration have overcome earlier physical or temporal constraints associated with some routes. New commodities and markets have emerged, and older ones have sometimes declined. Opening of new passages has resulted from discovery and the creation of canals. World events such as wars and trade embargos or agreements have limited use of pre-existing routes. In addition, trade routes have also responded at various timescales to environmental changes (e.g. ice cover or water level surrounding passages) and storm events.

Although it is evident that the scale, tempo, and routes of shipping are highly dynamic, the temporal and spatial pattern of changes and not been well documented to date. Many of the changes in shipping are punctuated rather than a gradual shift over time. Such shifts are exemplified by the advent of steamships or the opening of canals as new passage ways, which rapidly changed shipping on a global scale (Couper 1972, see Suez Canal chapter).

Changes in shipping patterns affect not only transport of cargo but also transfer of organisms to new geographic regions. It is well known that many species are transferred unintentionally in the cargo of ships and by the hulls and ballasted materials of ships (Visscher 1928, Carlton 1985, Carlton and Geller 1993, Coutts 1999, Gollasch 2002). Upon release to a new geographic region, many species have established self-sustaining populations. Due to the magnitude of shipping and the extensive species pool associated with ships' ballast and hulls, shipping is a leading source of biological invasions in coastal ecosystems throughout the world (Cohen and Carlton 1995, Reise 1998, Ruiz et al. 2000, Fofonoff et al. 2003, Hewitt et al. 2004).

In this chapter, we begin to explore some patterns of shipping associated with the Panama Canal (see Cohen, Panama Canal chapter I for history and description of the canal). The opening of this passage in 1914 was indeed a punctuated event, causing a change in commercial shipping on a global scale. We compiled historical records from the Panama Canal Authority to (a) describe changes in the magnitude of shipping through the Panama Canal from 1914-2004, (b) examine the directional flux of different vessel types, including the frequency of ballasted versus cargo laden transits, through the canal and (c) compare the magnitude of shipping through the canal to that of the largest port systems in the United States. Based upon this background, we consider the implications of creating this new passageway, and its expanding use, for biological invasions.

2 Magnitude and tempo of commercial shipping in the Panama Canal

Since its opening in 1914, an estimated 781,363 ocean-going commercial vessels have passed through the Panama Canal. This estimate excludes all ships (a) operated by Panama and Colombia, (b) operated by the United States through 2000, or (c) under 500 tons displacement.

The number of transits by these ocean-going vessels shows a strong increase through time, exhibiting two periods of rapid increase followed by relatively little inter-annual change (Figure 1A). The first period of increase occurred from 1915 (1,075 transits) to 1928 (6,456 transits), where the number of annual transits are reported by fiscal year ending in June. The number of annual transits did not exceed this range until 1952. There was a marked decline in traffic during World War II (1942-1945), when the number of transits ranged from 1,562 to 2,688. The second period of increase occurred from 1952 (6,562 transits) to 1971 (14,020 transits). Since this time, annual transits have remained relatively stable, having a mean of 12,625 transits (sd = 1,030) and a range of 9,936 to 15,194.

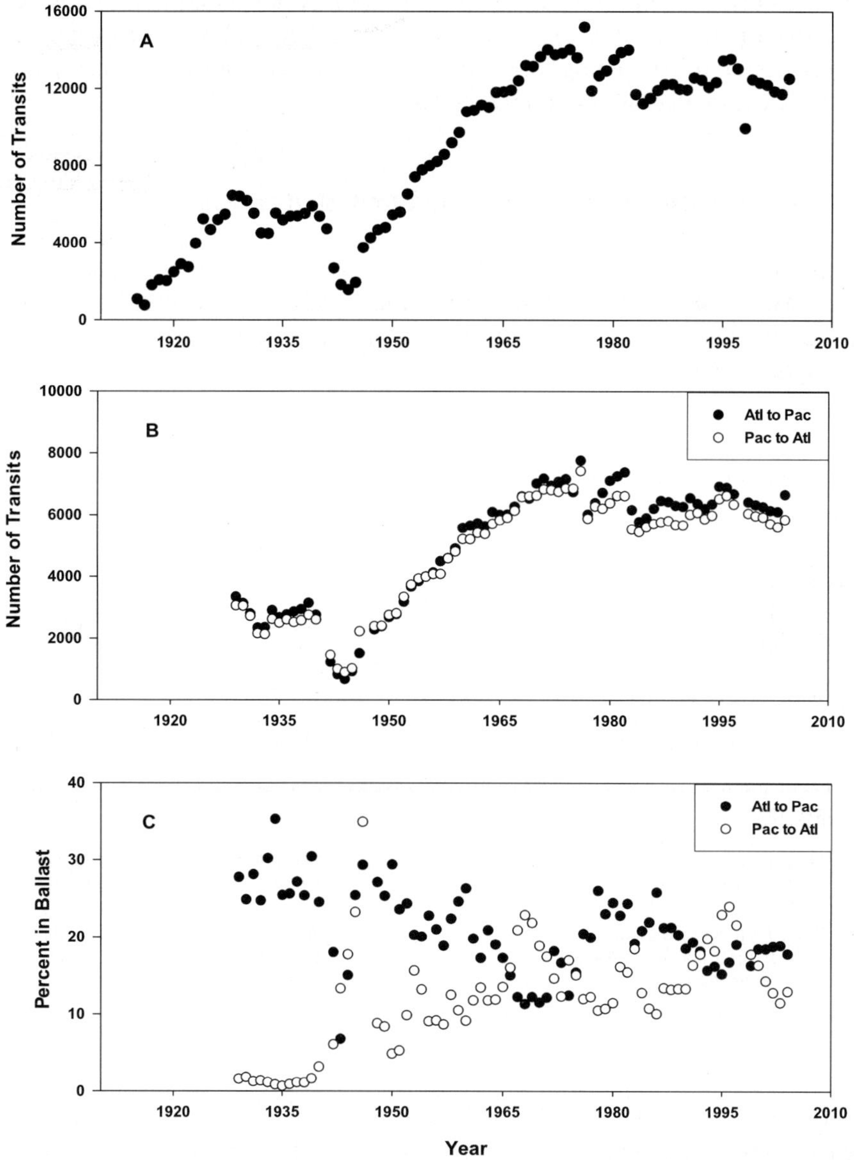

Fig. 1. Number of transits by year. Shown for each year is the number of transits through the Panama Canal by ocean-going commercial vessels. (A) Total number of transits, (B) Number of westbound transits (dark circles) and eastbound transits (open circles), and (C) Percent of westbound and eastbound transits reported to be ballasted. Data as reported by the Panama Canal Authority for each fiscal year (ending in June).

In recent years, small commercial vessels (<500 tons) added approximately 10% more transits. For example, the number of reported annual transits for small vessels ranged from 1,323 to 1,517 for the years 2002 - 2004. Today, we therefore estimate total commercial traffic from ocean-going and small vessels to be approximately 13,000 to 14,000 transits per year. This is the current scope of international traffic using the canal, as these estimates exclude domestic (local) traffic and also recreational vessels.

3 Overall direction of traffic and ballasted transits

Beginning in 1929, data were available on the annual number of transits in each direction and whether these ships were laden with cargo versus in ballast. As might be expected, the number of transits in each direction is similar among years (Figure 1B).

Table 1. Cumulative statistics for total and ballasted transits by direction. Summary statistics are shown for two different time periods: (A) 1929 - 2004; (B) 1929 - 1991.

	Atlantic to Pacific	Pacific to Atlantic
A. 1929 - 2004		
Total Transits	**366,693**	**348,914**
% of Total Transits	51.20%	48.80%
Total Transits in Ballast	**72,979**	**47,271**
% of Total Transits in Ballast	60.70%	39.30%
% of Directional Transits in Ballast	19.90%	13.50%
B. 1929-1991		
Total Transits	**289,286**	**276,325**
% of Total Transits	51.10%	48.90%
Total Transits in Ballast	**59,450**	**34,432**
% of Total Transits in Ballast	63.30%	36.70%
% of Directional Transits in Ballast	20.60%	12.50%

However, there appears to be a strong directional bias in the percentage of transits in ballast (Figure 1C). Such directional data were available for the 73 of 76 years from 1929-2004, as shown in Figure 1C. For 56 (77%) of these years, the percentage of transits in ballast from Atlantic to Pacific exceeded those in the opposite direction, often by a large margin. Interestingly, the difference in annual percentage of ballasted eastbound versus westbound voyages was greatest before 1965, suggesting temporal change and convergence in directional ballast operations through time.

These patterns are equally evident when comparing cumulative data across all years (Table 1A). From 1929-2004, 51.2% of all transits were westbound (Atlantic to Pacific) and 48.8% were eastbound (Pacific to Atlantic). During this same period, the westward traffic accounted for 60.7% of all ballasted voyages for both directions. Since the cumulative number of transits was similar in each direction, this indicates that a higher percentage of westbound traffic was in ballast (19.9%) compared to eastbound traffic (13.5%), as shown in Table 1A.

4 Direction of traffic and ballasted transits by vessel type

To gain a better understanding of directional patterns of ballasted transits, we examined the frequency of transits classified as "in ballast" by vessel type and direction from 1929-1991. During this time period (1929-1991), new types of vessels were added to the classification scheme used by the Panama Canal Authority, probably reflecting changes in specialization, design, and size of vessels (Couper 1972). General Cargo and Tanker vessels were present in the classification scheme for the entire period, whereas other vessel types were represented for only part of the period (Ore Carriers 1938-1972; Dry Bulk 1968-1991; Refrigerated Cargo 1968-1991; Containers 1968-1991; Passenger vessels were not consistently reported until 1938-1991). It is likely that some of these latter vessel types are included earlier as General Cargo vessels and were not classified separately until they began to increase in frequency, but any lag-time in reporting new vessel types in the transit records has not yet been evaluated. After 1991, the number of different commercial vessel types included in the classification scheme for transits in the Panama Canal doubled, expanding from seven to over fourteen. To simplify our analysis, and to avoid the confounding effects of adding additional vessel types through time, we examined transits for the 62-year period prior to 1992.

For the period 1929-1991, most (63.3%) of all ballasted transits occurred from the Atlantic to Pacific (Table 1B) and a higher percentage of westbound transits (20.6%) were ballasted than eastbound transits (12.5%). A similar directional bias exists in ballasted transits for the period 1929-2004 (Table 1A), but this pattern is weighted strongly by the early years and changed dramatically through time (as noted above and seen in Figure 1C); the cause of such temporal change is the focus of current study.

Figure 2A shows the composition by vessel type for all transits from 1929-1991. General cargo vessels were the most frequent vessel type (50% of total), followed by dry bulk carriers (15%), tankers (14%), refrigerated cargo vessels (8%), containerships (5%), ore carriers (1%), and passenger vessels (1%). Other vessel types in combination contributed the remaining 6%, and included military vessels, but little information was available to characterize these further.

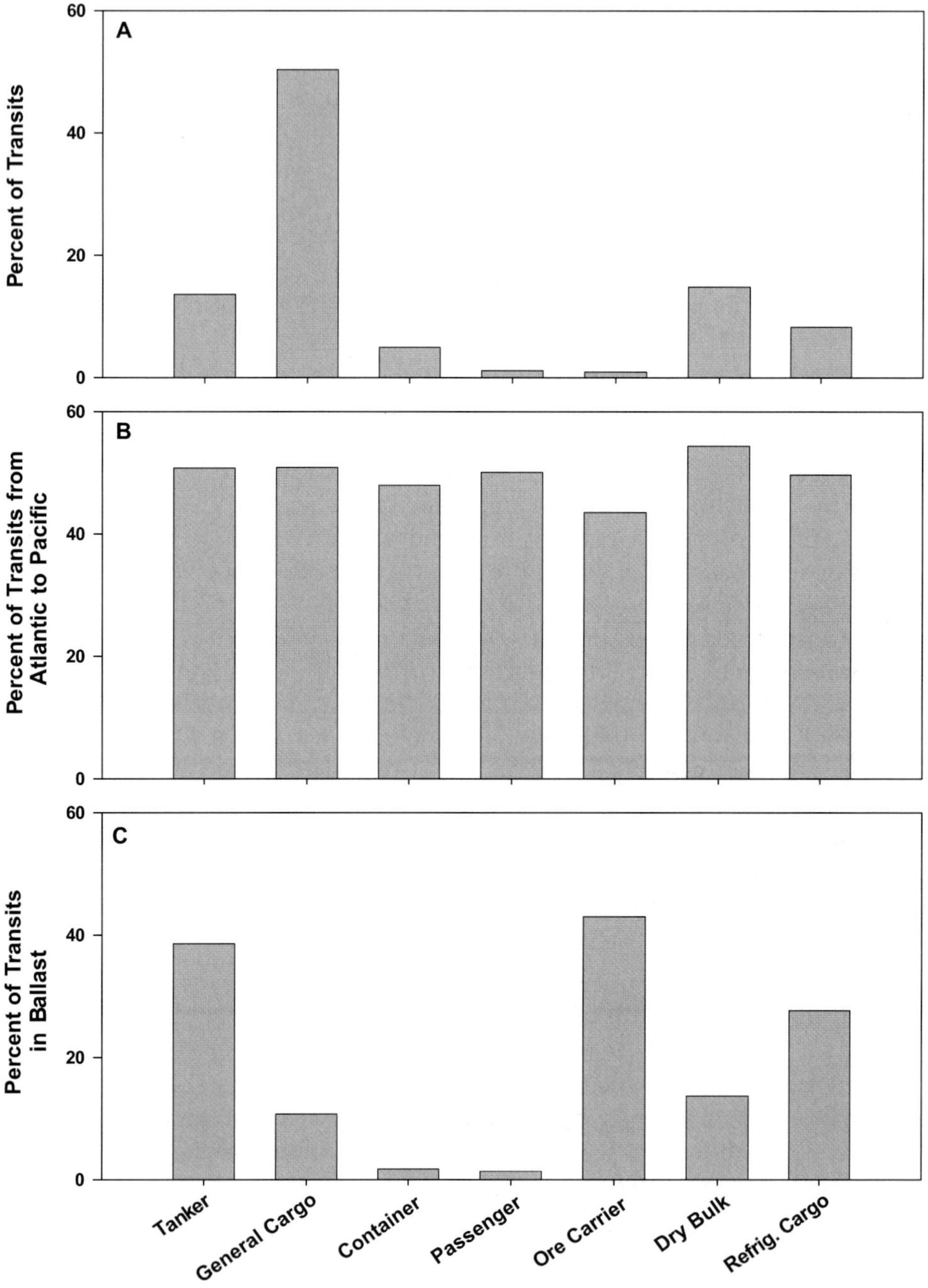

Fig. 2. Cumulative transit statistics by vessel type, 1929 - 1991. Shown are (A) percent of total transits by vessel type, (B) percent of transits within vessel type that was westbound, Atlantic to Pacific, and (C) percent of transits within each vessel type that was ballasted.

In general, the flux of vessels eastbound was equivalent to that westbound for each vessel type. As shown in Figure 2B, the percentage of total traffic that was westbound (Atlantic to Pacific) ranged from a low of 43.5% (for ore carriers) to a high of 50.9% (for general cargo vessels).

Five of the seven vessel types were classified as having arrived to the Panama Canal "in ballast" for at least 10% of their total transits (Figure 2C). Ore Carriers, tankers, and refrigerated cargo vessels arrived most frequently in ballast (43%, 38%, and 27% of transits, respectively). Dry bulk and general cargo vessels arrived in ballast much less frequently (14% and 10% of transits). Less than 2% of container and passenger vessels arrived in ballast.

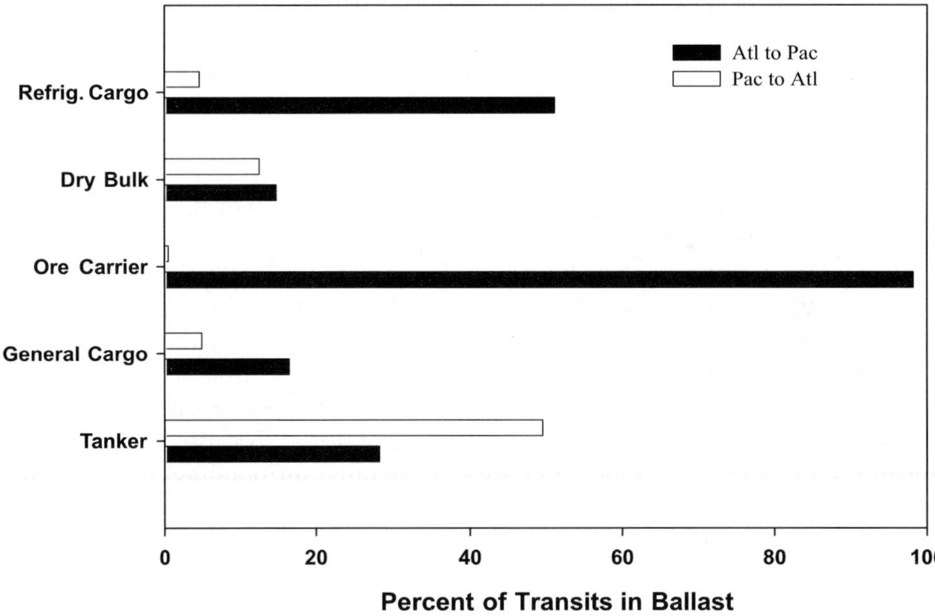

Fig. 3. Percent of ballasted transits by vessel type and direction, 1929 - 1991. Shown separately for each vessel type is the percent of eastbound and westbound transits in ballast.

For these five vessel types most often arriving in ballast, Figure 3 contrasts the percent of ballasted transits for eastbound versus westbound traffic. For four of the five vessel types, a higher percentage of in ballast transits occurred in the westbound direction. This was most pronounced for ore carriers and refrigerated cargo vessels. Nearly all ore carriers were reported in ballast from the Atlantic to Pacific (98%) and almost none (2%) were classified in ballast when eastbound, suggesting transport of cargo in only the latter direction.

Approximately 51% of westbound refrigerated cargo vessels were in ballast compared to 5% of eastbound transits by this vessel type.

Although much less pronounced, the percentage of cargo vessels in ballast was still 3-fold greater in the westbound versus eastbound direction. The percentage of dry bulk carriers in ballast was very similar for eastbound (12.5%) and westbound transits (14.8%). In contrast, tankers exhibited a bias in the opposite direction for this time period. The percentage of vessels in ballast for eastbound transits was approximately twice that reported for westbound transits.

Such directional bias may have important consequences for species transfers and biological invasions. In general, ships laden with cargo can carry much less ballast water than those considered in ballast. For this reason, the frequency of ballasted voyages may provide a useful, albeit coarse, proxy for the net direction of transfer for biota associated with ballast tanks within particular vessel types.

At the present time, we are not able to go beyond this coarse-level analysis and estimate actual volumes of ballast water transferred through the Panama Canal or compare these volumes across vessel types, source regions, and recipient regions. The frequency distribution and average for ballast water volumes carried by any one vessel type (e.g. tankers) when 'in ballast' will certainly differ from the others (Carlton et al. 1995, Verling et al. 2005). Even within vessel type, there are also likely to be differences through time that result from changes in vessel size, design, and cargo. While it is evident that many factors influence ballast water use (volume) by ships, and that this affects transfer and dynamics of associated biota, the specific details surrounding ships in the Panama Canal remain unresolved (see also Effects of the Panama Canal on Biotic Exchange).

5 Relative scale of shipping associated with the Panama Canal

It is important to consider the current scale and nature of shipping in the Panama Canal in the context of other major port systems, underscoring the broad-scale shift in commercial shipping patterns that resulted from opening of this passageway. For this purpose, we compare the number of transits in the Panama Canal to commercial ship arrivals to ports of the United States at the present time.

From 2000-2004, the mean number of annual transits in the Panama Canal by commercial vessels was 12,121 (sd = 325). All of these are commercial vessels arriving from outside the country, as transits from Panama and Colombia are excluded from these estimates. Most of these vessels are not undergoing any cargo operations and pass through the canal as quickly as possible, with a transit time of 6-12 hours. The last port and next port of call of these vessels, defining the trade routes through the canal, is the topic of ongoing investigation.

For this same time period, all ports in the United States received approximately 100,000 ship visits per year (U.S. Maritime Administration, unpublished data). While this surpasses the number of transits in the Panama Canal, these arrivals are distributed across scores of ports and thousands of miles, along both the Atlantic and Pacific coasts. A more direct comparison to the Panama Canal, where ships pass the same geographic location, would be to consider individual port systems in the U.S. Each of the largest port systems in the U.S. received annually less that 50% of the total transits in the Panama Canal for the same years, 2000-2004 (Figure 4). The mean number of annual arrivals to each of the six largest U.S. port systems ranged from 2,118 to 5,358.

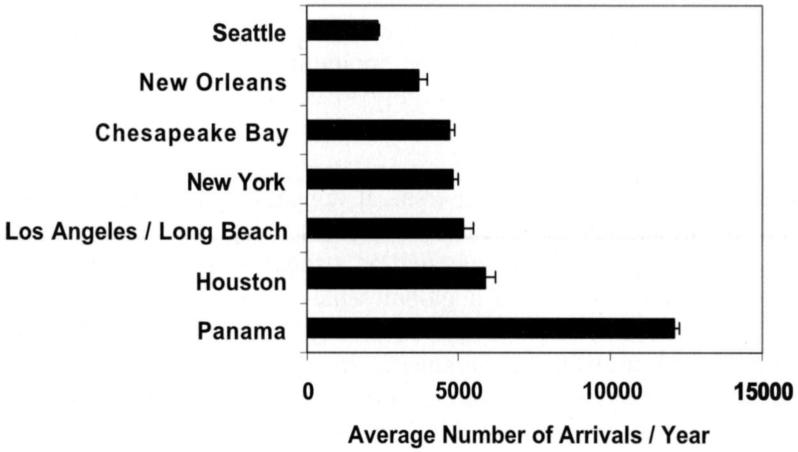

Fig. 4. Mean number of annual arrivals by location, 2000 - 2004. Shown is the mean number (+ s.e.) of arrivals to each of six major port systems in the United States compared to mean number (+ s.e.) of total transits for the Panama Canal.

It is also noteworthy that the number of arrivals to U.S. ports included both foreign and domestic (coastwise) traffic, whereas those reported for the Panama Canal are only foreign arrivals. Approximately 50% of arrivals to the U.S. ports are domestic in origin, and domestic arrivals to the six U.S. ports in Figure 4 ranged from 23-79% of the total.

6 Effects of the Panama Canal on biotic exchange

The creation of the Panama Canal as a major passageway for commercial shipping, where none existed previously, has affected the movement of aquatic organisms in two general ways: (1) unaided movement of organisms into or through the canal from adjacent waters and (2) transfers of organisms by ships. Either mode of biotic exchange can breach historical barriers to dispersal and result in biological invasions, allowing organisms to establish self-sustaining populations in new geographic locations.

For unaided dispersal of organisms via the canal, especially coast-to-coast movement across Panama, a great deal depends upon the organisms' environmental tolerance. Cohen (Panama Canal chapter III) reviews examples of several fish and invertebrate taxa that appear capable of making the transit through the canal. However, the transition from marine to freshwater may be very restrictive, creating a dispersal barrier for many species (Rubinoff 1970). Certainly some species may raft (Sheffey 1968, Thiel and Gutow 2005), possibly avoiding full and prolonged exposure to freshwater by attaching to floating materials. Others may be have environmentally tolerant resting stages (e.g. seeds, eggs, and cysts) or be associated with fast-moving organisms, perhaps as commensals or parasites, facilitating survivorship and transport. The capacity for dispersal unaided by ships, including especially salinity tolerance and the extent of rafting, remains largely to be studied in the Panama Canal system (Rubinoff 1970).

With respect to ship-mediated dispersal, it appears that a massive movement movement of organisms has occurred via ships using the Panama Canal. For almost a century, the canal has operated as a focal point for international ship traffic, shaping trade routes on a global scale. Approximately 800,000 commercial ship transits have now occurred, having many potential implications for the ship-mediated transfer of organisms in a regional and global context. However, to a large extent, we can only draw inferences (outlined below) about the transfer of organisms by these ships, and these await further data for testing.

6.1 Ship-mediated transfer: Regional perspective

In general, the coasts of Panama and surrounding region have been exposed to an increased propagule supply of non-native organisms, resulting from the large and continuous flux of transiting ships. These ships arrive from many parts of the world, and they would not otherwise have come to this region except for the canal. There exists a vast body of research from around the world describing abundant and taxonomically diverse assemblages of organisms that are transported in and on ships (see below). As a result, it *is* certain that ships arriving to the canal transport organisms and that some of these organisms are released to surrounding waters (Rubinoff 1970, Dawson 1973; see also Cohen, Panama Canal chapter III for discussion).

Ship-mediated transfer of aquatic organisms occurs primarily through association with ships' ballasted materials and ships' hulls. Organisms are routinely entrained in ships' ballast tanks, which are filled with surrounding waters at one port or location (to maintain stability, under rough conditions or often in lieu of cargo) and discharged at subsequent ports of call (Carlton 1985, Carlton and Geller 1993). Nearly all ballast tanks contain living organisms. It is not unusual to find concentrations of organisms in the water itself in the range of 10^0 - 10^2 zooplankton per liter, 10^3 - 10^6 phytoplankton per liter, 10^8 - 10^9 bacteria per liter, and 10^9 - 10^{10} viruses per liter (Smith et al. 1999, Zhang and Dickman 1999, Drake et al. 2001, Minton et al. 2005). Organisms also reside at the bottom of ballast tanks, and microorganisms form biofilms on the inner surfaces of these tanks (Bailey et al. 2005, Drake et al. 2005). Moreover, organisms associated with tank bottoms and surfaces can form resting stages or cysts that can remain viable for relatively long periods of time, even with little overlying water (Bailey et al. 2003). Thus, when ballast is discharged, organisms are released to the surrounding waters.

Organisms are also frequently found on the exposed, underwater surfaces of ships. Contemporary with the operation of the canal, a wide variety of species have been reported from around the world on the hulls, rudders, and other underwater surfaces (Visscher 1928, Coutts 1999, Gollasch 2002, Minchin & Gollasch 2003). Organisms also appear to be common in the sea chests of ships; although part of ballast intake systems, these are protected recesses along the outer surface and therefore easily colonized by a diverse array of organisms (Coutts et al. 2003).

Based upon existing information across many ship types and global regions, we surmise that (a) most ships arriving to the Panama Canal have living aquatic organisms in their ballast tanks and outer surfaces, (b) the cumulative number of these organisms passing through the region through time must be great, and (c) viable organisms (propagules) have frequently been released from the ballast tanks and hulls of vessels to local waters. Given the importance of ship-mediated transfers as a source for biological invasions in many parts of the world (Cohen and Carlton 1995, Ruiz et al. 2000, Hewitt et al. 2004), we might also expect many non-native species to be established along the coasts of

Panama due to shipping. However, there is a paucity of data available to characterize the quantity and species diversity of biota associated with arriving vessels to Panama or the extent to which propagules are released to the surrounding waters (see discussions by Rubinoff 1970 and Cohen, Panama Canal, chapter III). The capacity of such organisms to tolerate and colonize local waters, or the extent to which invasions have already occurred, also has not been adequately tested, despite some earlier surveys (see Panama Canal chapter III for review). Thus, a robust assessment of the relationship of shipping to propagule delivery and invasion dynamics in the region is not yet available.

6.2 Ship-mediated transfer: Global perspective

In addition to any regional effects on biotic exchange, the Panama Canal has also affected the global flux of biota associated with transiting ships. As a minimum, opening of this passageway resulted in different trade routes, altering transit times as well as surrounding environmental conditions and voyage conditions. In addition, access to the canal likely affected the source and recipient ports for some commodities. Each of these changes, operating alone and in combination, can affect the biota transferred by ships.

Any changes in geographic route or ports will obviously affect the species assemblage that can be moved by ships, either in ballast tanks or on outer surfaces. Opening a new trade route or adding a new port is likely to result in changes not only to species composition but also relative abundances encountered by ships, affecting the initial colonization of ships and possibly the fate of organisms during transit. While the Panama Canal has surely caused a shift in both species composition and abundance of ship borne biota, the scope of such change has not been evaluated to date.

Independent of colonization of ships, the condition and survivorship of organisms will be strongly affected by voyage duration and environmental conditions encountered. The Panama Canal was built to reduce transit time between ports. Survivorship in ballast tanks is time-dependent, and past studies have shown that species from many taxonomic groups decline in abundance during voyages (Gollasch et al. 2000, Wonham et al. 2001, Verling et al. 2005). Thus, survivorship in ballast tanks should increase with reduced voyage duration, resulting in a higher density of organisms at the end of voyages, controlling for other factors (see Cohen, Panama Canal chapter III for further discussion). Although we are not aware of similar studies for time-dependent mortality for biota on the hulls of vessels, we expect a similar pattern to exist.

The rate of decline during voyages will depend upon surrounding environmental conditions, which are also affected by route. For ballasted communities, large differences in survivorship rates may exist between transits in tropical water versus high latitudes, especially if ambient conditions approach thermal tolerance limits. Although physically isolated from the environment, ballast tanks often acclimate to outside conditions experienced during voyages (Wonham et al. 2001, Gollasch et al. 2000, Ruiz unpubl. data).

The same general principles apply to organisms on the outer surfaces of vessels, and it may be that exposure to freshwater within the Panama Canal is particularly stressful for many organisms. Unlike biota in ballast tanks, these organisms are exposed to a change from full seawater to freshwater upon entering the canal. They must be able to withstand freshwater exposure for the period of transit (6-12 hours) and then acclimate again to full seawater. These rapid changes in salinity may serve as a strong biocide, actually removing many species from hulls (see however Rubinoff and Rubinoff 1969). On the other hand, there is some evidence that rapid changes in environmental conditions (especially temperature) can induce spawning (Minchin & Gollasch 2003), which may increase propagule supply to surrounding areas.

To our knowledge, the dynamics of biota associated with ships transiting the Panama Canal have not been measured. It would be especially informative to take such measures along multiple voyage routes, especially to examine the effect of pre- and post-canal routes on (a) the initial biotic content and (b) survivorship functions for both ballast and hull communities.

On balance, it is difficult at the present time to quantify exactly how the Panama Canal has affected the global flux of biota. The nature of organism transfers certainly changed with associated shifts in route, transit time, and environmental conditions. Furthermore, invasions inevitably resulted from post-canal shipping traffic, given the overall importance of ship-mediated transfers (see discussion in Panama Canal chapter III). Whether the magnitude of species movement and invasions is greater than would have occurred without the canal remains a challenging question.

7 References

Bailey S.A., Duggan I.C., van Overdijk C.D.A., Jenkins P.J. & MacIsaac H.J. 2003. Viability of invertebrate diapausing eggs collected from residual ballast sediments. Limnol. Oceanogr. 48:1701-1710.

Bailey S.A., Duggan I.C., Jenkins P.J. & MasIsaac H.J. 2005. Invertebrate resting stages in residual ballast sediments of transoceanic ships. Can J Fish Aquat. Sci. 62:1090-1103.

Carlton J.T. 1985. Transoceanic and interoceanic dispersal of coastal marine organisms: the biology of ballast water. Oceanogr. Mar. Biol. Ann. Rev. 23:313-71.

Carlton J.T. & Geller J.B. 1993. Ecological roulette: the global transport of nonindigenous marine organisms. Science 261:78-82.

Carlton J.T., Reid D.M. & van Leeuwen H. 1995. The role of shipping in the introduction of nonindigenous aquatic organisms to the coastal waters of the United States (other than the Great Lakes) and an analysis of control options, Report to U. S. Coast Guard, Washington D.C.

Cohen A.N. & Carlton J.T. 1995. Nonindigenous species in a United States estuary: A case study of the biological invasions of the San Francisco Bay and Delta, U.S. Fish and Wildlife Service and National Sea Grant College Program (Connecticut Sea Grant).

Couper A.D. 1972. The geography of sea transport. Hutchinson and Company LTD, London.

Coutts A.D.M. 1999. Hull fouling as a modern vector for marine biological invasions: investigation of merchant vessels visiting northern Tasmania. Masters thesis, Australian Maritime College.

Coutts A.D.M., Moore K.M. & Hewitt C.L. 2003. Ships' sea chests: an overlooked transfer mechanism for non-indigenous marine species. Mar. Poll. Bull. 46:1510-1512.

Dawson C.E. 1973. Occurrence of an exotic eleotrid fish in Panama with discussion of probable origin and mode of introduction. Copeia 1:141-4.

Drake L.A., Choi K.-H., Ruiz G.M., & Dobbs F.B. 2001. Microbial inventories in ballast water of ships arriving in Chesapeake Bay. Biol. Invasions 3:193-199.

Drake L.A., Meyer A.E., Forsberg R.L., Baier R.E., Doblin M.A., Heinemann S., Johnson W.P., Kock M., Rublee P.A. & Dobbs F.C. 2005. Potential invasion of microorganisms and pathogens via 'interior hull fouling': biofilms inside ballast tanks. Biol. Invasions 7:969-982.

Fofonoff P.W., Ruiz G.M., Steves B. & Carlton J.T. 2003. In ships or on ships? Mechanisms of transfer and invasion for non-native species to the coasts of North America. In: Ruiz GM & Carlton JT (eds), pp 152-182, Invasive species: vectors and management strategies. Island Press, Washington.

Gollasch S. 2002. The importance of ship hull fouling as a vector of species introductions into the North Sea. Biofouling 18:105-121.

Gollasch S., Rosenthal H., Botnen H., Hamer J., Laing I., Leppakoski E., Macdonald E., Minchin D., Nauke M., Olenin S., Utting S., Voigt M. & Wallentinus I. 2000. Fluctuations of zooplankton taxa in ballast water during short-term and long-term ocean-going voyages. Internat. Rev. Hydrobiol. 85:597-608.

Hewitt C.L., Campbell M.L., Thresher R.E., Martin R.B., Boyd S., Cohen B.R., Currie D.R., Gomon M.F., Keough M.J., Lewis J.A., Lockett M.M., Mays N., MacArthur M.A., O'Hara T.D., Poore G.C.B., Ross D.J., Storey M.J., Watson J.E. & Wilson R.S. 2004. Introduced and cryptogenic species in Port Philip Bay, Victoria, Australia. Mar. Biol. 144:183-202.

Minchin D. & Gollasch S. 2003. Fouling and ships hulls: how changing circumstances and spawning events may result in the spread of exotic species. Biofouling 19:111-122.

Minton M.S., Verling E., Miller A.W. & Ruiz G.M. 2005. Reducing propagule supply and coastal invasions via ships: effects of emerging strategies. Front. Ecol. Envir. 3:304-308.

Reise K., Gollasch S. & Wolff W.J. 1999. Introduced marine species of the North Sea coasts. Helgoländer Meeresunters. 52:219-234.

Rubinoff I. 1970. The sea-level canal controversy. Biological Conservation 3(1): 33-36.

Rubinoff R.W. & Rubinoff I. 1969. Observations on the migration of a marine goby through the Panama Canal. Copeia 2: 395-397.

Ruiz G.M., Fofonoff P., Carlton J.T., Wonham M.J. & Hines A.H. 2000. Invasions of coastal marine communities in North America: Apparent patterns, processes, and biases. Ann. Rev. Ecol. Syst. 31: 481-31.

Sheffey J.P. When Caribbean and Pacific waters mix. Science 162:1329.

Smith L.D., Wonham M.J., McCann L.D., Ruiz G.M., Hines A.H. & Carlton J.T. 1999. Invasion pressure to a ballast-flooded estuary and an assessment of inoculant survival. Biol. Invasions 1:67-87.

Thiel M. & Gutow L. 2005. The ecology of rafting in the marine environment. II. The rafting organisms and community. Oceanogr. Mar. Biol. Ann. Rev. 43:279-418.

Verling E., Ruiz G.M., Smith L.D., Galil B., Miller A.W. & Murphy K. 2005. Supply-side invasion ecology: characterizing propagule pressure in coastal ecosystems. Proc. Royal Soc. Lond. B, 272:1249-1256.

Visscher J.P. 1928. Nature and extent of fouling of ships' bottoms. Bull. Bureau Fisheries 43:193-252.

Wonham M.J., Walton W.C., Ruiz G.M., Frese A.M. & Galil B. 2001. Going to the source: role of the invasion pathway in determining potential invaders. Mar. Ecol. Prog. Ser. 215:1-12.

Zhang F. & Dickman M. 1999. Mid-ocean exchange of container vessel ballast water. 1. Seasonal factors affecting the transport of harmful diatoms and dinoflagellates. Mar. Biol. 176:243-51.

Chapter III

Species Introductions and the Panama Canal

ANDREW N. COHEN

San Francisco Estuary Institute, 7770 Pardee Lane, Oakland, CA 94621-1424, USA

1 The canal and its environs

1.1 Structure, operations and environmental characteristics

The Panama Canal has been described as being more of a "bridge of water" over the Central American isthmus than an excavation through it (McCullough 1977).[1] The bridge was built by damming the Chagres and Gatun rivers to create Gatun Lake, 26 meters above sea level. The lake covers 425 square kilometers and crosses the Continental Divide through the 14-kilometer-long Culebra Cut. When it was impounded, it was the largest artificial water body in the world.[2]

The canal, including its approach channels, runs about 80 kilometers from Limon Bay on the Caribbean side to near Balboa on the Pacific side, with a distance of about 57 kilometers from tidewater to tidewater (Fig. 1). Because of the curve in the isthmus, an Atlantic-to-Pacific transit through the canal follows a compass path from the northwest to the southeast. A vessel first enters the Atlantic approach channel through Limon Bay. At the end of the channel it is lifted up through three lock chambers at Gatun Locks to Gatun Lake (Fig. 2). The vessel then follows a 51 kilometer-long shipping lane through the lake and Culebra Cut to Pedro Miguel Locks. Here it is lowered 9.5 meters in a single lock chamber to Miraflores Lake. At the far end of the 2.1 kilometer-long lake it drops through two lock chambers at Miraflores Locks into the Pacific shipping

[1] The description of the canal in this section is based primarily on information from Jones & Dawson 1973, McCullough 1977 and ACP 2005.

[2] It remained the largest artificial lake until 1936 when the completion of Hoover Dam created Lake Mead. Until Hoover Dam the Panama Canal's locks were also the largest concrete structures in the world.

channel, which runs past the Port of Balboa, under the Bridge of the Americas, along the Naos Breakwater and out to Panama Bay.

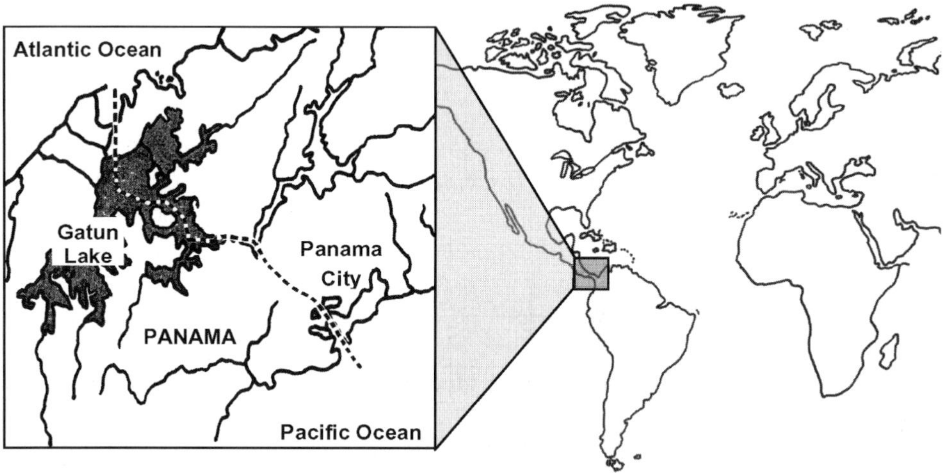

Fig. 1. The canal route. Insert = Central Panama with Gatun Lake (grey) and location of the Panama Canal (dotted line). Drawing by Stephan Gollasch, Hamburg, Germany.

Fig. 2. Three views looking up Gatun Locks from the Caribbean approach channel toward Gatun Lake.

At each end of the canal there are thus three lock steps, each of which consists of a pair of lock chambers built side-by-side (so there are six pairs or 12 chambers in all). This provides two lanes of traffic, so that vessels can be simultaneously moved up or down the locks in opposite directions or in the same direction, depending on traffic needs. Each chamber is essentially an enormous concrete box, 305 meters long by 33.5 meters wide, and averaging 26 meters deep. The chambers are closed at each end by two gates, except at the lower entrance to the Gatun Locks where there is only one gate. Each gate is made of two swinging steel doors, 20 meters wide and 2 meters thick, and from 14 to 25 meters tall. The largest weighs nearly 750 tons, but they are hollow and buoyant so that each one can be opened and closed by a 40-horsepower motor (Fig. 3).

Fig. 3. Miraflores Locks drained for maintenance. The view is toward the lower end of the upper lock chamber. In the center of the image are the two slightly gaping doors of the chamber's lowermost gate: the doors of the gate just above them are folded back into the lock walls. In the foreground are five of the circular wells set into the floor of the chamber through which water enters and drains. Photo courtesy of Mark Torchin.

About 100 million liters of water are needed for each filling of a chamber. Water enters and leaves the chambers though 5.5-meter diameter culverts built into the side walls, and through smaller culverts running laterally beneath the floors of the chambers. Each chamber is served by 20 lateral culverts, and each of these drains and releases water though five 1.4-meter diameter wells in the floor of the chamber, so there are 100 such wells per chamber. With these, a chamber can be filled or drained in only eight minutes (Fig. 4).

To run the locks, the impounded water of the Chagres River is released progressively downward from one water body or lock chamber to the next and on into the saline coastal waters of the Atlantic and Pacific Oceans. Each plug of downward flow out of an upper chamber lowers its surface, and each plug flowing into a lower chamber raises its surface, to bring the water on either side of a lock gate to the same level before the gate is opened. However, when a lock gate is opened between an upper chamber or water body containing fresher water, and a lower chamber or water body containing saltier water, the lighter, fresher water flows from the upper to the lower chamber on the surface while the denser, saltier water flows from the lower to the upper chamber along the bottom, mixing the water between the two chambers in a process called gravitational circulation. This pulls some salt water (along with drifting or swimming organisms) up into the locks, so that incrementally greater salinities are encountered as one progresses down the locks from Gatun Lake.

Fig. 4. The first flooding of Miraflores Locks, with water entering through wells in the floor of the chamber. Photo courtesy of CanalMuseum.com.

The strength of the gravitational circulation depends on the density differences between the waters above and below each set of lock gates, and this is mainly a function of salinity differences, which are greater in the lower chambers. When the gates between lower chambers or between the lowest chambers and the ocean at Gatun or Miraflores are opened, a surface current of about 0.25 meters per second has been observed flowing out from the upper chamber accompanied by a "readily perceptible wave of ripples," indicating strong gravitational circulation (Jones & Dawson 1973). Saltier water may also be pushed or drafted into higher chambers by vessels moving upward through the locks.

Table 1 shows some surface salinity measurements in the canal system. The measurements indicate that there is generally fully fresh water in Gatun Lake, Gatun Upper Locks and Pedro Miguel Locks, slightly brackish water (0-3 ppt) in Gatun Middle Locks, Miraflores Lake and Miraflores Upper Locks, brackish water (5-26 ppt) in Gatun Lower Locks and Miraflores Lower Locks, and somewhat saltier water (10-30) in the approach channels. However, Dawson's (1973) record of 20 ppt in the sump areas after dewatering of the upper east lock at Miraflores suggest that even the locks at this level (the middle of the three steps) can sometimes attain relatively high salinities at least near the bottom of the chambers. This is consistent with records of the estuarine barnacle *Fistulobalanus* (=*Balanus*) *pallidus* in Miraflores Upper Locks (Jones & Dawson 1973; Spivey 1976), and a collection of the stenohaline goby *Gobiosoma nudum* in a sump of the Miraflores Upper Locks (Fishbase 2005). Jones and Dawson (1973) reported some vertical salinity stratification in the approach channels and Miraflores Lower Locks in November, but fully mixed conditions otherwise. They noted that the rapid, turbulent flow of water into the bottoms of the chambers during filling, turbulence generated by rotating propellers, and the "piston effect" of a large ship moving into a tight-fitting lock chamber all help to mix the water in the locks. No substantial differences in salinity are apparent between the end of the wet season (November 1972) and the end of the dry season (April 1972).

Table 1. Measurements of surface salinity (parts per thousand) in the Panama Canal. References 1 = Hildebrand 1939, 2 = Menzies 1968, 3 = Abele 1972a, 4 = Dawson 1973, 5 = Jones & Dawson 1973, 6 = Jones & Rützler 1975, * = Values estimated from graph, ** = In sump areas after dewatering.

Date	Pacific Approach	Miraflores Lower Locks	Miraflores Upper Locks	Miraflores Lake	Pedro Miguel Locks	Gatun Lake	Gatun Upper Locks	Gatun Middle Locks	Gatun Lower Locks	Atlantic Approach	Ref.
Jun 1935							0	≈0-1	10-16	18-20	1
<1939	16-20			0.1-3.0		≤0.02				18-20	1
≤1968	30*	26*		1.0*	0.0*	0.0*			24*	28.5*	2
Feb 1969					0.0-0.4						3
Jan 1972			20**								4
Apr 1972	27*		1-3*	0*	0*	0*	0.1*	0-1*	5-11*	17*	5
Nov 1972	23*	14-15*	3*	0-1*	0*	0*	0.2-0.3*	1-2*	6-8*	10-17*	5
Mar 1974						0.2	0.02-0.8				6

Jones and Dawson (1973) also noted that in some cases there was significant horizontal variation in salinity within a lock. Their investigations were sparked by an earlier observation in January 1972 of a slanting upper distributional boundary for the estuarine barnacle *Fistulobalanus pallidus* on the side of an upper lock chamber at Miraflores, slanting from 7 meters above the chamber floor at the seaward end down to floor of the chamber at 25 meters in,[3] suggesting a salt wedge within the chamber, though their salinity measurements provided little evidence of vertical stratification. Jones and Rützler (1975) similarly reported a slanting lower distributional boundary for the freshwater sponge *Trochospongilla leidii* on the side of an upper Gatun Lock chamber in

[3] Spivey (1976) noted the same slanting distribution on the lock wall in August 1974.

1974 that suggested the presence of a salt wedge, and several estuarine species (the mussel *Mytilopsis sallei,* the isopod *Uromunna* (=*Munna*) *reynoldsi* and the amphipod *Gitanopsis tortugae*?) restricted to the seaward half of the chamber, suggesting differences in salinity from one end of the chamber to the other. Their measurements did show slightly higher salinities toward the bottom and toward the lower end of the chamber. Abele and Kim (1989) also noted the walls of one Gatun Lock chamber covered with the bivalve *Isognomon* sp. "along a line apparently following a salinity gradient."

Jones and Dawson (1973) found no vertical temperature stratification anywhere in the canal system. In April, water temperatures were about 29-30°C in Gatun Lake, dropping to about 28°C in the Atlantic approach and about 24-25°C in the Pacific channel. In November, temperatures were about 27-30°C throughout the system. This is consistent with other temperature records with means between 27 and 29°C, and extreme ranges in the approaches of 21-30°C (Hildebrand 1939; Abele 1972; Jones & Rützler 1975).

An additional component of the canal system that is relevant to biological studies is the "Miraflores Third Lock," a brackish water lagoon that partially fills an excavation alongside the approach to the Miraflores Locks, where a number of Atlantic species have been found. The excavation was part of an effort, abandoned in the 1940s, to build a third and larger set of locks.[4] The lagoon is about 100 meters wide and 1,400 meters long, and averages 20 meters deep. The lagoon receives a mix of freshwater runoff and sea water that enters at high tides several times a month. The sea water comes in through five culverts that connect to the canal's approach channel a short distance below Miraflores Locks, and surface water from the lagoon drains back to the approach channel through a small (2-3 meter wide) surface creek (Rubinoff & Rubinoff 1968; McCosker & Dawson 1975).

The nearly vertical sides of the lagoon are covered to a depth of 5.5 meters by shells of the oyster *Ostrea palmula*. Red mangroves, *Rhizophora mangle*, which are native to both coasts of Panama, grow in the occasional shallow areas (Rubinoff & Rubinoff 1968; McCosker & Dawson 1975). The tide range in the lagoon is only several centimeters. Salinities have been measured at 6-12 ppt at the surface and 14-18 ppt at 15-18 meters depth (Rubinoff & Rubinoff 1968; Dawson 1970; McCosker & Dawson 1975), and pH declined from 7.8 at the

---

[4] The new locks were to be nine meters wider and 60 meters longer than the old ones (Challinor 1972). The project was authorized in 1939 and abandoned in 1942 according to Challinor (1972) and Leschine (1981); begun in 1941 and abandoned in 1943 according to Rubinoff and Rubinoff (1968); and excavated in 1940 and filled with water after work was abandoned in 1946 according to McCosker and Dawson (1975).

surface to 7.3 at 18 meters depth in February-March 1971 (McCosker & Dawson 1975). The temperature profile is an inverted thermocline, with a layer of warmer but saltier and denser water beneath cooler and lighter brackish water; in March 1971 temperatures were 29.6°C at the surface, 29.1-29.2°C at 1-4 meters depth, and 30.7-30.8°C at 7 meters and below (McCosker & Dawson 1975). Dissolved oxygen at that time was 6-7 ppm between the surface and 4 meters depth, declining to an anoxic zone of hydrogen sulfide and suspended detritus below 12 meters. The decline in oxygen is presumably responsible for a drop in the concentration of macroscopic organisms at around 5 meters depth (McCosker & Dawson 1975).

The canal's construction and operation has no doubt altered some of the physical conditions at least in the near vicinity of the ends of the canal. The dredging of approach channels has deepened some sections and probably resulted in the intrusion of some higher salinity water along the bottom. The long breakwater at Naos protecting the approach on the Pacific side has presumably altered longshore currents and sedimentation dynamics. The damming of the Chagres and Gatun rivers that formerly emptied into Limon Bay on the Atlantic shore and the use of the impounded water for lockage on both the Atlantic and the Pacific side has reduced the amount and the temporal variation in water and sediment supplied to the Atlantic side and increased the discharge of fresh water to the Pacific side.

1.2 Environmental and biological characteristics at either end of the canal

The canal connects two tropical marine regions that differ significantly in their environmental characteristics and biological composition, and these differences affect which organisms can invade the other region through the canal. The Gulf of Panama on the Pacific side of the canal has a much greater range of tides, temperature and salinity than the western Caribbean, and is generally more productive and supports more individuals and a larger biomass than the western Caribbean (Rubinoff 1968; Martin et al. 1970; Graham 1971; Glynn 1972; Rubinoff 1972; D'Croz & Robertson 1997). The annual occurrence of upwelling in the Gulf of Panama during the dry season (January to April) which brings cooler, saltier and more nutrient-rich water to the surface, and the lack of such upwelling on the Caribbean side, is responsible for some of these differences (D'Croz & Robertson 1997). For example, nitrogen concentrations are roughly the same on both sides of the isthmus during the wet season but are about five times greater in the Gulf of Panama than on the Caribbean side during the dry season; and phosphate concentrations in the Gulf of Panama are four times greater than in the Caribbean in the wet season and ten times greater in the dry season, though Caribbean phosphate levels were similar to those measured in a non-upwelling section of Panama's Pacific coast (D'Croz & Robertson 1997). The Pacific side is also subject to episodic warming from El Niño/Southern Oscillation events at 4-9 year intervals (D'Croz & Robertson 1997).

The average daily tidal range is 0.2-0.3 meters at the Atlantic entrance to the canal compared to 3.9 meters at the Pacific entrance, the annual range between the highest and lowest tide is 0.7-0.8 meters at the Atlantic entrance and 6.5-7.5 meters at the Pacific entrance, and sea level is about 0.3 meters lower on the Atlantic side (Martin et al. 1970; Glynn 1972, 1982; Porter 1972). During the upwelling season, surface temperatures drop and surface salinities rise in the Gulf of Panama, so that the annual ranges on the Pacific side (18-28°C and 25-36 ppt) are greater than on the Atlantic side (26-28°C, 33-36 ppt) (Glynn 1972; Rubinoff 1972), and the average annual temperature is about 1°C lower (Graham 1971; Sheffey 1972; Porter 1972). This has led some researchers to conclude that environmental conditions on the Pacific side are generally more variable and more challenging to organisms. However, Glynn (1972) points out that the weather is more variable and extreme on the Caribbean side - windier and stormier, with twice as much rainfall and greater seasonal variation in cloud cover - so that shallow reef and shore species are more heavily buffeted by high seas and subjected to greater changes in turbidity and sediment load. The Caribbean tides, though much smaller, are also more variable due to the greater influence of winds and weather; and abrupt seasonal shifts in the timing of the tides produces sudden, lengthy mid-day low tides that stress and kill intertidal and shallow reef organisms in many parts of the Caribbean (Glynn 1972, 1982).

Some types of habitats also vary greatly across the isthmus. Coral reefs are common on the Caribbean side on both protected and exposed coasts, range from shallow water to 60 meters depth, often cover tens to hundreds of hectares, grow up to 33 meters thick, are composed of a large number of different frame-building coral species, and comprise a wide variety of habitat zones. In contrast, coral reefs are rare in the Gulf of Panama and uncommon elsewhere on the Pacific side, are restricted to protected areas in water shallower than 15 meters, usually cover no more than a few hectares and grow no larger than 12 meters thick, include only a few frame-building coral species, and exhibit only a few habitat zones[5] (Glynn 1972, 1982; Porter 1972; D'Croz & Robertson 1997). In general, reefs on the Pacific side are made of volcanic rock from extensive lava

---

[5] Comparing the coral diversity, Porter (1972) reported a maximum Shannon diversity of 1.81 in the eastern Pacific compared to 3.42 in the Caribbean.

flows, a substrate that is uncommon on the Caribbean side (Porter 1972). Caribbean beaches often consist of coarse coral sands and coral fragments, while Pacific-side beaches are generally composed of finer quartz sands and mud (Martin et al. 1970; Glynn 1972). Seagrass beds are common on the Caribbean coast, rare on the Pacific (Glynn 1972; Earle 1972).

Biological differences accompany these physical and habitat differences. During the dry season, primary productivity is 3-5 times greater[6] and net plankton density is more than five times greater on the Pacific than the Atlantic side (Martin et al. 1970; Rubinoff 1972), apparently because of Pacific-side upwelling. In the Gulf of Panama, D'Croz and Robertson (1997) measured chlorophyll concentrations that were double and zooplankton concentrations that were ten times the concentrations on the Caribbean side, differences that were more or less consistent over the year. Seaweed communities have been described as well-developed and stable on the Caribbean side and sparse and seasonally varying in the Gulf of Panama, especially in intertidal and shallow subtidal waters, possibly due to differences in tide range, upwelling, or grazing by fish and invertebrates (Earle 1972). Hay and Gaines (1984) reported that seaweed community cover tends to be dominated by larger, upright species on the Caribbean side and by crustose and small filamentous forms ("algal turf") on the Pacific side. They argue that this difference is due to differences in grazing pressure, in combination with the larger tide range and lack of significant reef flat habitat on the Pacific side. Beach macrofaunal species diversity and density are six times greater and biomass is nine times greater on the Pacific than the Atlantic side (Glynn 1972), perhaps due to differences in upwelling, tidal range or sediment size. Mud-bottom benthic biomass is 3-5 times greater on the Pacific side (Bayer et al. 1970; Martin et al. 1970). On coral reefs, invertebrate diversity is lower but fish abundance is higher on the Pacific side (Glynn 1982). There appears to be greater grazing pressure by fish on corals and under-rock encrusting fauna, and greater bioeroding of corals by invertebrates, on the Pacific side (Glynn 1972, 1982). Shallow water fish are more tolerant of high temperatures on the Atlantic side, and of lower temperatures on the Pacific side (Graham 1971). Bayer et al. (1970) found the fish and invertebrate assemblages to be more varied on the Caribbean side, which they felt reflected a greater variety of habitats.

Extinction, immigration and speciation have increasingly differentiated the biological composition of the Caribbean and Panamic regions since the closure of the Central American seaway, but there are a still a notably large number of

---

[6] Primary productivity of 210-650 mg C/m^2 on the Pacific side and 78-120 mg C/m^2 on the Atlantic side (Martin et al. 1970 at p. 62).

shared morphospecies[7] and pairs of morphologically similar, closely related species across the isthmus (Table 2). The latter were dubbed "geminate pairs" by Jordan (1908) based on a review of the shallow-water fish species. Strictly speaking, the term geminate pair - also called sibling species,[8] cognate pair, homologues or analogues[9] - refers to the pair of species that resulted from the most recent speciation in their lineage.[10] Trans-isthmian geminate pairs have been frequently studied to gain an understanding of the pace and process of speciation (e.g. Mayr 1954; Lessios 1979, 1981, 1984, 1998; Bermingham & Lessios 1993; Knowlton et al. 1993; Lessios & Weinberg 1994; Knowlton & Weigt 1998).

Table 2. Assessments of biotic diversity across the isthmus. a Collection = Comparison based on a particular collection of organisms from sites within the indicated areas or regions. Review = Comparison based on a more-or-less comprehensive review of records of organisms collected within the indicated areas or regions, b Oceans = Atlantic and Pacific oceans. Regions = Usually Caribbean and Panamic Regions, sometimes different but comparable regions. Coasts = Atlantic and Pacific coasts of the Republic of Panama. Ends = Areas near the Atlantic and Pacific ends of the Panama Canal, c Compares species collected on buoys in the eastern U.S. and Bahamas versus the western U.S. including Hawaii, d Compares the Eastern Atlantic to the Western Pacific, e Includes species assumed to be in Panama because of records north and south, f Based on genera occurring in the West Indies, and omitting one genus (Clypeaster) requiring taxonomic revision, g Jordan later (1908) reported that according to the latest authority it was doubtful that any fish species occurred on both sides of the isthmus, h Compares the western Caribbean to Panama Bay and adjacent waters, i Compares southern Caribbean coast to Pacific Panama coast, j Compares one beach on each coast, k Compares the mainland coasts of Central America.

Organism Group	Type of Study[a]	Regions Compared[b]	Atlantic Species	Pacific Species	Total Species	Shared Species	% Shared	References
Diatoms	Collection	Regions	118	97	135	80	59%	Voss 1967, 1972
Macroalgae	Review	Regions	–	265	–	≈93	–	Martin et al. 1970, citing Hume 1969
Green Algae	Review	Regions	–	45	–	16	–	Martin et al. 1970, citing Dawson 1962
Green Algae	Review	Coasts	38	19	52	5	10%	Earle 1972

7 The determination of a morphospecies is based entirely on its form, without considering its reproductive continuity or isolation.

8 The term "sibling species" is used more broadly by some authors to refer to "species that are difficult or impossible to distinguish based on morphological characters," without regard to the species' genetic relationship (Knowlton 1993).

9 Coan (1984) argues that "analogue" implies a lack of close genetic relationship, and is therefore an improper term for a geminate species.

10 Thus some morphologically similar species thought to be geminate pairs may turn out not to be, when examined genetically.

Table 2. continued.

Organism Group	Type of Study[a]	Regions Compared[b]	Atlantic Species	Pacific Species	Total Species	Shared Species	% Shared	References
Green Algae	Collection & Review	Coasts	78	43	107	14	13%	Wysor 2004
Green Algae	Collection	Ends	13	10	21	2	10%	B. Wysor & A.N. Cohen unpublished data 2002
Brown Algae	Review	Regions	–	22	–	8	–	Martin et al. 1970, citing Dawson 1962
Brown Algae	Review	Coasts	20	9	27	2	7%	Earle 1972
Brown Algae	Collection & Review	Coasts	38	10	44	4	9%	B. Wysor pers. comm. 2002
Brown Algae	Collection	Ends	4	0	4	0	0%	B. Wysor & A.N. Cohen unpublished data 2002
Red Algae	Review	Regions	–	167	–	44	–	Martin et al. 1970, citing Dawson 1962
Red Algae	Review	Coasts	60	51	103	8	8%	Earle 1972
Red Algae	Collection & Review	Coasts	102	48	134	16	12%	B. Wysor pers. comm. 2002
Red Algae	Collection	Ends	21	7	26	2	8%	B. Wysor & A.N. Cohen unpublished data 2002
Blue-green Algae	Review	Coasts	3	6	7	2	29%	Earle 1972
Sea Grasses	Review	Coasts	4	3-5	6	1-3	17-50%	Earle 1972
Sponges	Collection	Ends	21	16	31	6	19%	De Laubenfels 1936
Siphonophores	Collection & Review	Regions	41	39	52	28	54%	Alvariño 1974
Hydrozoans	Collection	Ends	24	17	34	7	21%	D. Calder & A.N. Cohen unpublished data 2002
Corals	Review	Regions	–	–	–	0	0%	Verrill 1866, cited in Dickerson 1917
Corals	Review	Coasts	67-73	20-23	86-95	1	1%	Porter 1972
Scleratinian Corals	Review	Regions	74	49	–	–	–	Voss 1972
Hermatypic Corals	Review	Oceans	–	–	≈800	1	0.1%	Porter 1972

Table 2. continued.

Organism Group	Type of Study[a]	Regions Compared[b]	Atlantic Species	Pacific Species	Total Species	Shared Species	% Shared	References
Hermatypic Corals	Review	Regions	49	20	67-69	0-2	0-3%	Glynn 1972
Octocorals	Review	Regions	184	89	–	–	–	Voss 1972
Intertidal Polychaetes	Review	Coasts	73	136	179	30	17%	Fauchald 1977
Sabellid and Serpulid Polychaetes	Collection	Ends	20	11	25	6	24%	S.I. Salazar-Vallejo, J.R. Bastida-Zavala & A.N. Cohen unpublished data 2002
Pelecypods & Gastropods	Review	Regions	≈1,000	≈4,500	–	–	–	Olsson 1972
Gastropods	Review	Regions	799	1,818	–	–	–	Voss 1972
Bivalves	Review	Regions	378	564	–	–	–	Voss 1972
Barnacles	Collection	Ends	11	14	22	3	14%	F.B. Pitombo & A.N. Cohen unpublished data 2002
Buoy-fouling Isopods & Tanaids	Collection	Oceans[c]	14	16	29	1	3%	Miller 1968; Glynn 1972
Sphaeromatid Isopods in fouling	Collection	Ends	–	–	12	1	8%	Glynn 1972:25
Isopods	Collection	Ends	6	4	9	1	11%	J.W. Chapman & A.N. Cohen unpublished data 2002
Amphipods	Collection	Ends	20	13	30	3	10%	J.W. Chapman & A.N. Cohen unpublished data 2002
Sandy Beach Decapods	Collection	Ends	8	17	25	0	0%	Abele 1972b
Mangrove Decapods	Collection	Ends	17	20	35	2	6%	Abele 1972b
Rocky Intertidal Decapods	Collection	Ends	67	78	145	0	0%	Abele 1972b
Porcelain Crabs	Review	Oceans[d]	34	88	–	–	–	Gore & Abele 1976
Porcelain Crabs	Review	Regions	31	65	–	–	–	Gore & Abele 1976
Porcelain Crabs	Review	Coasts[e]	21	48	65	4	6%	Gore & Abele 1976

Table 2. continued.

Organism Group	Type of Study[a]	Regions Compared[b]	Atlantic Species	Pacific Species	Total Species	Shared Species	% Shared	References
Porcelain Crabs	Collection	Coasts	12	29	38	3	8%	Gore & Abele 1976
Warm-water Brachyuran Crabs	Review	Oceans	–	–	–	–	2%	Menzies 1968
Brachyuran Crabs	Review	Regions	362	233	574	21	4%	Bayer et al. 1970
Crabs	Collection	Ends	18	14	29	3	10%	E. Campos-González & A.N. Cohen unpublished data 2002
Stomatopods	Review	Regions?	50	28	78	0	0%	Bayer et al. 1970
Bryozoans	Collection	Ends	7	25	30	2	7%	Powell 1971
Echinoderms	Review	Regions	–	–	–	0	0%	Ortmann, cited in Dickerson 1917
Echinoderms	Review	Oceans	–	–	–	–	0.3%	Menzies 1968
Echinoids	Review	Regions	–	–	–	0	0%	Agassiz 1869, cited in Dickerson 1917
Shallow-water Echinoids[f]	Review	Regions	19	20-23	39-42	0	0%	Mayr 1954
Shallow-water Echinoids	Review	Regions	24	27	48-50	1-3	2-6%	Chesher 1972
Shallow-water Asteroids	Review	Regions	18	37	55	0?	0%	Martin et al. 1970; Chesher 1972
Crinoids	Review	Regions	≈50	2	–	–	–	Chesher 1972
Ophiuroids	Collection	Regions	51	10	–	–	–	Chesher 1972
Crinoids	Collection	Regions	23	20	–	–	–	Chesher 1972
Tunicates	Collection	Ends	16	6	17	5	29%	G. Lambert & A.N. Cohen unpublished data 2002
Fish	Review	Regions	–	–	–	–	6%	Jordan 1895[g], cited in Dickerson 1917

Table 2. continued.

Organism Group	Type of Study[a]	Regions Compared[b]	Atlantic Species	Pacific Species	Total Species	Shared Species	% Shared	References
Fish	Review	Regions	–	1,307	–	72	–	Evermann & Jenkins 1891, cited in Dickerson 1917
Fish	Review	Regions	–	–	–	–	≈1%	Topp 1969
Fish	Review	Regions	≈2,000-2,500	≈1,000	–	–	–	Martin et al. 1970
Shelf & Shore Fish	Review	Regions	≈1,400	≈800	–	–	–	Martin et al. 1970
Shallow Water Fish	Review	Regions[h]	600	400	–	–	–	Briggs 1972b
Fish (minus circumtropical species)	Review	Regions	–	–	≈1,000	12	1%	Rubinoff & Rubinoff 1969
Fish	Review	Coasts[i]	≈750	≈600	–	–	–	Martin et al. 1970
Shelf & Shore Fish	Review	Coasts[i]	≈500	403-≈500	–	–	–	Martin et al. 1970
Gobies	Review	Regions	–	–	–	0	0%	Rubinoff & Rubinoff 1969
Invertebrates	Review	Regions[h]	7,800	5,200	–	–	–	Briggs 1972b
Invertebrates from Locks	Collection	Ends		126	164	19	12%	Jones 1976
Mainly Fouling Organisms	Collection	Ends	204	174	348	30	9%	A.N. Cohen unpublished data 2002
Sandy Beach Macroscopic Infauna	Review	Coasts[j]	≈14	41	53-55?	0-2?	0-4%	Glynn 1972
Shallow Water Invertebrates & Fish	Review	Regions[k]	>8,000	>6,000	–	–	–	Briggs 1968
Shallow Water Invertebrates & Fish	Review	Regions	≈8,400	≈5,600	–	–	–	Briggs 1972a
Biotas	Review	Regions	7,000	8,000	–	–	–	Newman 1972

The morphospecies that occur on both sides of the isthmus probably comprise several elements (Table 3). Some may be true, native, genetically-interconnected species, with ongoing, natural genetic exchange occurring between the populations on either side of the isthmus. Such exchange could occur, for example, by individuals swimming or drifting around the South American continent; by relay through populations distributed around South America or around the world through the Pacific, Indian and South Atlantic Oceans; or by seeds, spores, cysts or other reproductive or resting stages carried over the isthmus by wind or birds (e.g. Martin et al. 1970 at p. 76). Other morphospecies may be closely-related populations that are reproductively and genetically isolated from each other by the isthmus. Although morphologically indistinguishable, appropriate genetic analysis would detect differences, and depending on the degree of difference and the definition employed, the populations would be classified either as separate subspecies or separate species.

Table 3. Interpretations of single morphospecies found on both sides of the isthmus.

Native populations
- a single genetic species that is present on both sides, with ongoing natural genetic exchange between the two populations.
- two morphologically indistinguishable genetic species, one on each side.

One-way introduction
- simple introduction: with no morphologically similar species present on the side receiving the introduction.
- complex introduction: with a morphologically similar species present on the side receiving the introduction, with or without resulting hybridization.

Two-way introductions
- morphologically similar species present on both sides, with introductions in both directions, with or without resulting hybridizations.

Double introduction
- an introduction to both sides from some other part of the world.

The presence of a morphospecies on both sides of the isthmus could also result from or involve a migration or transfer of organisms across the isthmus associated with human activities. In the simplest case there is no pre-existing, morphologically or genetically similar population on the second (invaded) side of the isthmus, resulting in a native population on the first side, and an introduced population the second side. A more complicated situation arises if there is initially a pair of similar populations (subspecies or closely related species) on either side of the isthmus, and there is an introduction from one of these populations across the isthmus. The result then is a native population on the first side, and a separate native population on the second side with a similar, perhaps morphologically indistinguishable introduced population living alongside it. It starts to get complicated if the two populations on the second side can hybridize, with additional genetic complexity if over time there are multiple introductions, each bringing a different sample of the genetic diversity present in the population on the first side. Things can get even messier if, as before, there are initially a pair of similar populations on either side of the isthmus, but now there are introductions in both directions (two-way introductions). Finally, a single morphospecies occurring on both sides of the isthmus can arise from a double introduction of a single exotic species, that is, a species native to some other part of the world that is introduced to both coasts. This may well have happened with some circumtropical or cosmopolitan species.

The geographic proximity and geologically recent separation of the Caribbean and Panamic regions has prompted a line of scientific inquiry involving taxonomic, ecological and evolutionary comparisons of the biotic communities on either side of the Central American isthmus. These studies have, among other goals, sought to determine the rates and patterns of evolutionary change in populations separated by the closing of the Central American seaway (reviewed in Lessios 1998), and to understand how differences in habitat and environmental conditions in the coastal waters on either side of the isthmus have influenced the composition and structure of these biological communities.

The existence of the canal can affect these cross-isthmian comparative studies. Organisms that migrate or are transported through the canal may provide a test of conclusions regarding reproductive isolation, differences in community structure, and the like. However, they can also confuse and complicate taxonomic and genetic comparisons (e.g. Lessios & Weinberg 1994). Chesher (1968), for example, suggested that because of the volume of ballast water transported through the canal since its opening in 1914, comparative systematic and ecological surveys "are 50 years too late to describe the uncontaminated condition." Voss (1972) concluded that "an unknown but considerable introduction of foreign elements has already mixed the faunas to such an extent that the list of original twins or analogues may be hopelessly confused." Lessios (1998), however, argued that genetic studies can avoid the problem by a "judicious choice of organisms."

2 Effects of the canal on species introductions

2.1 Modes of biotic transport

Organisms can theoretically move or be transported through the Panama Canal from one ocean to another by a variety of mechanisms that are summarized in Table 4. These mechanisms fall into two major categories. First, organisms can potentially swim, crawl, drift or float through the canal, a process here called "migration." Second, organisms may be carried through the canal in or on vessels.

The locks do not pose a significant physical barrier to migration by swimming organisms, since these can pass freely from one chamber to the next whenever the lock gates between them are open, regardless of whether vessels are in the process of being raised or lowered. That many fish, at least, do so is indicated by observations and collections of large numbers of fish at all levels in the lock system (e.g. Hildebrand 1939). Crawling organisms face the challenge of ascending the roughly 9-meter high wall at the inland end of each lock chamber, but in the nearly weightless underwater environment, many organisms can climb up vertical faces almost as easily as they crawl over horizontal surfaces. However, the low salinity water in the upper locks and the fresh water of Gatun Lake do block organisms that cannot tolerate extended exposure to hyposaline conditions.

Table 4. Modes of biotic transit through the canal.

Migration
Individual Travel (individuals traversing the entire length of the canal)
Gradual Dispersal (multigenerational range extension through the canal)
Transport by Vessels
As Hull Fouling (including in borings and crevices)
In Ballast Tanks (and other seawater system components)
Subaerial Transport (on deck, in chain lockers, in nets or other equipment)

While strong swimming or crawling abilities no doubt increase an organism's potential to migrate through the canal, it is at least theoretically possible for drifting or floating organisms to do so as well. Organisms drifting in the lower parts of the water column could be carried from one lock chamber to the next higher one by the gravitational currents described in the preceding section. Drifting organisms could also be drafted along in currents created by the movement of large vessels into the next chamber upward, or by the prop wash or return currents from vessels moving downward. Floating organisms could be carried by such currents, or propelled from chamber to chamber by the wind. Once in Gatun Lake, water currents or wind could disperse drifting and floating organisms throughout the lake. On reaching the other lock system, organisms could be carried downward through the locks with the water passed from one chamber to the next through the culvert system, be drawn along in vessel-created currents, drift or float in surface gravitational currents, or be pushed along by the wind.

Migration through the canal can occur either by individual organisms traversing the entire route from one end of the canal to the other; or by gradual dispersal, wherein a population becomes established within the canal system and then gradually extends its range, over one or more generations, to the other end of the canal. An individual organism's capacity to traverse the canal, if not constrained by environmental sensitivities, depends on the distance to be covered (about 57 kilometers from tidewater to tidewater), its traveling speed and the length of its life. A fish's sustained swimming speed depends upon its body form and its size (larger fish are generally faster). For fish that are not built for continuous fast swimming and that are less than about a third of a meter long, the speeds reported in the literature typically range from about 0.3 to 0.9 meters per second, or about 1-3 kilometers per hour (Fishbase 2005),[11] so theoretically the minimum time needed for a fish of this type to swim through the canal is less than 1-3 days. Deviations from the shortest route, and time spent resting or feeding would of course make any actual passage considerably longer, but still virtually any fish species not deterred by fresh water should be capable of swimming through the canal in less than one fish's lifetime.[12]

Organisms, including very slow-moving ones, could also migrate by gradually extending a population through the canal over more than one generation. This would require the capacity to pass an entire life cycle, including the often sensitive reproductive and early developmental stages, in fresh water, not just tolerate it during one life stage for a limited period of time.

Ships, boats, barges and other vessels frequently carry a variety of organisms attached to their hulls, collectively known as fouling, which can include seaweeds, sponges, hydroids, serpulid and sabellid worms, mussels, oysters, barnacles, bryozoans and tunicates (Woods Hole Oceanographic Institution 1952). Wooden hulls may also harbor various species of wood-boring organisms including shipworms, pholad clams, limnoriid and sphaeromatid isopods, and the amphipod *Chelura terebrans*. Snails, sea slugs, worms, tanaids, isopods, caprellid and gammarid amphipods, crabs and other crawling, clinging and crevice-nestling organisms can be carried along in fouling or in bored cavities, including some types of fouling-associated fish and shrimp. Modern cargo ship practices, including the brief time spent in port and the use of toxic hull coatings to deter fouling growth (which by increasing hull friction reduces a ship's speed and increases its fuel consumption), have presumably reduced the density of hull fouling organisms relative to earlier times. However, some fouling organisms are carried even by modern, well-maintained cargo ships in full operation, and the hulls of vessels that have been poorly maintained or moored in one spot for a long period of time may carry a remarkably dense fouling cover (DeFelice 1999; Coles et al. 1999; Pauley et al. 2002; Coutts 2002).

---

[11] For example, a 7-centimeter-long sand goby (*Pomatoschistus minutus*) swims at a speed of 0.3 meters per second, and 6- and 14-centimeter-long goldfish (*Carassius auratus auratus*) swim at sustained speeds of 0.4 and 0.8 meters per second, respectively (Bainbridge 1960; Sambilay 1990). In contrast, 2- to 2.5-meter-long bluefin tuna (*Thunnus thynnus*) swim at sustained speeds of 2.2-3.5 meters per second (Sambilay 1990).

[12] Internal parasites of such fish, as well as external parasites that can tolerate the freshwater exposure, could travel along with them.

Cargo ships may also transport and release marine organisms in ballast water. Over the course of a voyage, and especially at the beginning and end, a cargo vessel may take on and discharge large volumes of water to adjust its draft and trim. The water enters through ports in the hull located below the waterline. Metal covers with holes or slots that are typically about a centimeter wide block the entrance of large objects that could damage the pumps, though corrosion or a missing cover can provide a larger opening. The ports empty into a large compartment or "sea chest" immediately inside the hull, which acts as a sump for the ballast pumps to draw on. Pipes from the sea chest lead to the pumps, and thence either to the ballast tanks or to empty cargo compartments that are used to carry ballast on some bulkers and tankers. To drain the tanks the water is either pumped back along the same route, or in some cases (such as wing tanks located above the waterline) emptied though ports in the side of the ship. Studies in the last few decades have demonstrated that ballast water routinely transports and releases large numbers of marine organisms to new regions around the world (e.g. Carlton & Geller 1993; Gollasch et al. 2000a,b; and studies summarized in Cohen 1998). While most of the research and management effort has focused on organisms carried in ballast tanks and cargo holds, organisms can also travel in sea chests, ballast system pipes, firemain systems, etc. (Shelton et al. 2002). In this chapter I refer to ballast tanks and these other pipes and components as parts of the ship's seawater system.

Little is known about the volume of ballast water transferred between or into the Caribbean and Panamic regions, or where ballast water is taken up or discharged by vessels transiting the canal. We do know that a regulation adopted by the Panama Canal Company in 1956 required large vessels to maintain a minimum draft in order to reduce the "windage" or the portion of the vessel sticking out above the water, in order to prevent vessels from becoming unmanageable in narrow sections of the canal on windy days[13] (Chesher 1968; Constant 1978; Sheffey 1978). When the regulation was first enacted, nearly every ship that arrived in ballast (that is, without cargo) was required to take on additional ballast water in the harbor before it was allowed to enter the canal; in later years, most ships loaded this additional ballast while en route to the canal in order to avoid delays, so that only a few ships each month had to take on more

---

[13] The Panama Canal Authority's Marine Directive No. D-1-2005 (Jan. 1, 2005) states that "All vessels transiting the Canal should have sufficient ballast to permit safe handling during transit" (Section 10, Paragraph d at page 21), and sets minimum drafts based on vessel length (Table V on page 21). Vessels unable to attain these drafts may have constraints placed on their transit (Section 10, Paragraph e at page 21). Chesher (1968) reports that this requirement was adopted in 1956, but Sheffey (1978) states that ships have been loading ballast before entering the canal and discharging it after exiting the canal ever since the canal was opened.

ballast at the canal entrance. Most ships discharge the extra ballast water after transiting the canal, to reduce the hull cross-section below the water and thereby minimize fuel consumption (Sheffey 1968, 1978). Thus, a substantial amount of additional ballast water has been transported through the canal since 1956, with a good part of it loaded from and discharged to Panamanian waters near the canal.

In contrast to this requirement of loading ballast before entering the canal, Dawson (1973) stated that a ballasted ship "will usually discharge at least some ballast before traversing the Isthmus," and that Canal Company employees reported that ships often discharge both bilge and ballast water within the locks, despite regulations prohibiting such discharge in waters of the Canal Zone. He noted the "presence of considerable oil on surface waters and lock walls" at both ends of the canal as possible evidence of this practice. Springer and Gomon (1975), similarly note that "ships intending to clear their bilges and ballast tanks must do so before entering the Canal."

While none of these authors state any reason for ships to discharge ballast water before entering or on entering the canal, four possibilities can be imagined: (1) to reduce draft to meet canal regulations[14] and to safely enter shallow locks or channels;[15] (2) to prevent an increase in draft when entering freshwater, which provides less buoyancy than salt water;[16] (3) to reduce draft and hull resistance in order to reduce fuel consumption during passage through the canal, where less ballast is needed for stability than in the ocean; or (4) to save time in port by starting the discharge of ballast water before arriving, for vessels that will

---

[14] "A vessel's initial transit draft may not exceed 35.5 feet (10.82 m) six inches TFW (deepest point of immersion) unless a deeper transit draft for the vessel is authorized in conformity with *ACP Navigation Regulations, article 52*" (Panama Canal Authority Marine Directive No. D-1-2005 (Jan. 1, 2005), Section 10, Paragraph c at page 21). "Vessels may be denied transit when the vessel's maximum point of immersion exceeds its maximum authorized transit draft, or when the vessel's maximum point of immersion exceeds the published TFW maximum draft then in effect" (Panama Canal Authority Marine Directive No. D-1-2005 (Jan. 1, 2005), Section 10, Paragraph f at page 21).

[15] The permissible draft of transiting ships is controlled by the level of Miraflores Lake and the depth of water over the south sill elevation at Pedro Miguel (Panama Canal Authority Marine Directive No. D-1-2005 (Jan. 1, 2005), Section 18, Paragraph b(1) at page 43). If sufficient water is available, Miraflores Lake can be raised to provide a depth of 12.56 m over this sill to accomodate deep draft vessels. Leschine (1981a), however, reported that "low rainfall in 1976 and 1977 led to...severe draft restrictions on vessels transiting the canal."

[16] "Transition to fresh water frequently alters the trim of large vessels 3 to 4 inches (8 to 10 cm) by the head" (Panama Canal Authority Marine Directive No. D-1-2005 (Jan. 1, 2005), Section 1, Paragraph k at page 1). Cheshire (1968) noted that a tanker 131 m long has to load 18 metric tons of ballast water to increase its draft by 1 cm, while a 202-meter-long tanker needs 46 metric tons of ballast water.

after transiting the canal, load cargo after transiting the canal at Balboa or Panama City near the Pacific entrance, or at Cristobal, Colon or Manzanillo near the Atlantic entrance. Vessels discharging ballast water before entering the canal or in the canal for any of the first three reasons would presumably be likely to take on ballast water after leaving the canal in order to return to proper ocean ballast conditions.

Bayer et al. (1970) additionally noted that ships that had passed through the canal from the Pacific to the Atlantic often stopped at Colon to take on fuel, and would discharge ballast water into the harbor. Glynn (1982; also Carlton 1985) reported that an increasing number of oil tankers that were too large to transit the canal carried Alaskan oil to Parita Bay in the western Gulf of Panama, where they transferred the oil to small tankers for transport through the canal. To accommodate the oil, each small tanker discharged into Parita Bay an estimated 20,000-30,000 tons of Caribbean ballast water.

Finally, besides transport in hull fouling or in ballast tanks or other seawater system components, marine organisms that can survive out of water for the short trip through the canal (typically 8-10 hours) could be carried on the deck or the hull of a ship above the waterline, in among nets, lines or other damp equipment, or in damp compartments such as anchor chain lockers (Schormann et al. 1990). While organisms such as ligiid isopods and grapsid crabs that commonly scurry about above the waterline on rocks, docks, pilings or mangrove roots may be the likeliest candidates, subaerial transport is not necessarily restricted to such supralittoral organisms. For example, it has been suggested that the Japanese seaweed *Codium fragile tomentosoides* may have been transported along the New England coast from Nantucket Sound to fishing grounds north of Cape Cod in damp nets piled on the decks of fishing boats (Carlton & Scanlon 1985), and the invasive colonial tunicate *Didemnum* sp. may have recently been transported by similar means from the coastal harbors of New England to offshore fishing grounds on Georges Bank. McMahon (1996) has also suggested that the freshwater zebra mussel, *Dreissena polymorpha*, could have traveled from Europe to North America as adults attached to an anchor chain in the damp environment of a chain locker, rather than in ballast water as is commonly reported.

An organism's method of transit through the canal will determine the length and intensity of its exposure to fresh water (Table 5). Migration, by individual organisms or by gradual dispersal of a population, involves exposures of probably weeks up to several generations. Hull fouling organisms are exposed first to steps of decreasing salinity as they are raised through the locks to the level of Gatun Lake, then to full fresh water for the 5-8 hours it typically takes for a vessel to travel the 51 kilometer route across the lake to the other locks,[17] then to steps of increasing salinity as the vessel is lowered to sea level on the other side. It is not known how quickly water exchange occurs between the ambient water outside the hull and the water retained inside bore holes and crevices, or the water in sea chests, but organisms in these situations are presumably buffered to some degree from the full effect of freshwater exposure. The freshwater exposure of organisms carried subaerially is limited to the effects of splash and rain, while organisms loaded in seawater into ballast tanks are fully protected against freshwater exposure.

Table 5. Transit modes and freshwater exposure.

Transportation Mode	Duration of Freshwater Exposure
Individual Travel	weeks? to <1 lifespan
Gradual dispersal	generations
Hull fouling	≈ 6-8 hrs
In bored wood/crevices	buffered for ≈ 6-8 hrs
In sea chests	buffered for ≈ 6-8 hrs
On deck, in chain lockers, entangled in nets, etc.	none (except for splash or rain)
In ballast tanks	none

2.2 Pathways

Organisms that are introduced to new regions via voyages that include a passage through the Panama Canal may travel on a variety of generalized pathways (Table 6). The introduction pathway that people primarily associate with the canal is the transport of an organism from the tropical waters on one side of the isthmus to the tropical waters on the other side, here called (somewhat awkwardly) the Near-Canal-Near Pathway. Organisms can also be picked up in a region distant from the canal, carried through it and then released into the tropical waters on the other side, the Distant-Canal-Near Pathway, or may be carried on the reverse route, from American tropical waters through the canal and transported to some distant region, the Near-Canal-Distant Pathway. In some of these cases the role and significance of the canal will be relatively obvious, since traveling through the canal will be the shortest route from the source area to the release point by a substantial margin, and the reduction in the duration of that voyage (and the corresponding increase in the probability that organisms will survive the voyage) will be great.

---

[17] Hay and Gaines (1984) calculated that 12% of the 13,087 ships that used the canal in 1977 spent less than 6 hours in fresh water during transit, and that 98% spent less than 12 hours in fresh water. Some vessels may, of course, take longer, and for various reasons may even stop in Gatun Lake for a while.

Table 6. Generalized introduction pathways via vessels using the canal.

Pathway	Description
Near-Canal-Near	Between nearby waters through the canal, *i.e.* from the Caribbean to the Panamic region, or the reverse.
Distant-Canal-Near	From a distant region through the canal to nearby waters.
Near-Canal-Distant	From nearby waters through the canal to a distant region.
Distant-Canal-Distant	Between distant regions through the canal.
Distant-Near-Canal	From a distant region to nearby waters before a ship passes through the canal.
Canal-Near-Distant	From nearby waters to a distant region after a ship passes through the canal.
Freshwater Introduction	Introduction of exotic freshwater species into the canal and its tributaries.

Organisms can also be transported between two distant regions, passing through the canal en route, the Distant-Canal-Distant Pathway. In these cases alternate routes connecting the source and release points may not be much longer than, and may even be shorter than, the route through the canal, and the argument that transport was through the canal will depend largely on other evidence, such the relative strength of different trade routes.

A vessel that is en route to the canal can also pick up an organism in one region and release it in another before the vessel passes through the canal (the Distant-Near-Canal Pathway); similarly, a vessel that has already passed through the canal and is heading for its destination can also pick up an organism in one region and release it in another (the Canal-Near-Distant Pathway). These cases too will require evidence other than the directness of the route to show that they involved a vessel that traveled through the canal.

Finally, vessels can introduce freshwater organisms into the canal's waters, from which they can spread into the rivers and reservoirs that are tributary to the canal.

2.3 Direct and indirect effects

Since its opening in 1914, the Panama Canal has affected the transport and introduction of organisms to new regions of the world in a variety of ways. I have classified these as direct when they involve either the migration of organisms through the canal or the transport of organisms on vessel voyages where the vessel transits the canal. I classify the effects as indirect if they involve the transport of organisms on voyages that do not include a canal transit. These effects may result from the canal's alteration of local and global shipping patterns, or from the canal's influence on the design and operation of ships. The most important of these direct and indirect effects are listed in Table 7.

Table 7. Effects of the canal on species introductions.

Direct Effects
Migrations of some marine organisms between the Caribbean and Panamic regions, and of freshwater species across the continental divide.
Increased ship-borne transfers of organisms by voyages that include a transit through the canal, involving both near and distant regions, related to greater trade volume, new trade routes and shorter voyages.
Ship-borne transfers of freshwater species from other regions into the canal system.
Indirect Effects
Increased transfers of organisms in ballast water and hull fouling, due to an increase in the ballast water transported and discharged and the hull surface area per ton of cargo carried, a consequence of the canal's locks limiting the size of many vessels in the world's cargo fleet.

First, the canal serves as a migration pathway between the Caribbean and Panamic regions for marine species that can tolerate exposure to fresh water as they move through the canal — the first such pathway to be opened in probably 2-3 million years. It also provides a channel that freshwater species on one side of the continental divide can use to migrate to watersheds on the other side.

Table 8. Effects of the canal on voyage distance and voyage duration, and the consequent effect on organism survival. a = Assumes an 8 hour transit time through the canal, and ship speeds of 20 knots elsewhere, b = Assumes an exponential decline in concentration in ballast tanks over time, $N(t)=N(o)e^{-rt}$, with r = 0.3 based on the mean value from studies by Carlton et al. 1982, Wonham et al. 1996, Gollasch et al. 2000a,b, Olenin et al. 2000 and Drake et al. 2002.

Voyage	Voyage Distance (nautical miles)		Voyage Duration (days)[a]		Increase in End-of-Voyage Organism Concentration[b]
	Without Canal	With Canal	Without Canal	With Canal	
London to Auckland	12,743	11,319	27	24	2.5x
New York to Yokohama	16,642	9,698	35	21	67x
New York to San Francisco	13,188	5,261	27	11	122x
Kingston to Guayaquil	9,669	1,418	20	3	164x
Colon to Balboa	10,542	39	22	0.4	652x

Second, by making shipborne trade faster and cheaper between many of the world's ports, the canal has increased the volume of such trade overall. To carry the increased trade, more cargo voyages or bigger cargo vessels are needed, which means more organisms transported in hull fouling or ballast water. Some trading partners, especially those whose distance apart by sea was greatly reduced by the opening of the canal, have seen a much greater increase in trade between them than have other partners. In addition, by shortening the length of voyages between points, the numbers of organisms carried in ballast tanks that survive these voyages has been increased, probably substantially. Studies have generally reported large declines in the number of organisms and the number of species present in ballast tanks over the course of a voyage (reviewed in Cohen 1998 and Gollasch et al. 2000b), probably due to either food depletion or to exposures to environmental stresses. For example, Gollasch et al. (2000a) measured roughly exponential declines in the concentrations of organisms in a ballast tank on a voyage from Singapore to Bremerhaven, with a 90% drop in zooplankton and phytoplankton concentrations after 4 and 9 days, respectively. While there is a lot of variation in the rate of decline,[18] the effect of shorter voyages due to passage through the canal can be estimated using mean values from a number of ballast water sampling studies (Table 8). Thus, a ship traveling between London and Auckland, a voyage shortened by only three days, would have 2.5 times as many living organisms in its ballast tanks at the end of the trip compared to the longer pre-canal voyage; on a voyage between New York and San Francisco, shortened by sixteen days, there would be more than 120 times as many organisms alive at the end of the trip. Because of the variation in the rate of decline, these figures are meant to be illustrative of the possible effect of the canal, rather than predictive.

Besides shortening voyages, the canal also reduces the temperature variation, and therefore the thermal stress, that organisms carried in hull fouling or ballast water are exposed to. For example, prior to 1914, the thermal experience of a ship sailing from New York to San Francisco would include passing from temperate waters to tropical to temperate to subantarctic to temperate to tropical and back to temperate waters again. Both the extremes of temperature and the rate of change of temperatures would stress both the hull fouling organisms, who were exposed to the full force of these changes, and the organisms carried in ballast tanks, since the water in these tanks does heat up and cool down in response to the temperature of the water that the ship passes through (e.g. Carlton

[18] For example, Gollasch et al. (2000) reported that in another tank on a shorter portion of the same voyage there was a much sharper decline of phytoplankton (the density dropping over 80% in 2 days, which the authors suggest resulted from damage during ballast uptake), and a large increase in the density of one copepod species (presumed to be a fast-reproducing, epibenthic detritivore) toward the end of that voyage.

1985, Gollasch et al. 2000a, Olenin et al. 2000). After the canal opened in 1914, the voyage's thermal changes were reduced to passage from temperate to tropical to temperate waters, lowering the stress on organisms and presumably improving their survival.

Another direct effect arises because ships using the canal spend typically around 5-8 hours in the fresh waters of Gatun Lake, crossing between Gatun and Pedro Miguel Locks. Thus, freshwater organisms from other parts of the world that are carried in ballast tanks, sea chests or other seawater system components can be introduced into Panamanian fresh waters. On the other hand, the freshwater passage presumably kills or damages some of the marine organisms that occur as fouling on the hulls of ships, thus reducing the potential for successful introduction of these species into marine waters later in the voyage — though there are many marine fouling species that can survive a few hours of fresh-water exposure.

Perhaps the most significant indirect effect of the canal stems from the size of its locks and the largest vessel that can fit through them, which canal regulations set at 32.3 meters wide and 294.1 meters long (about a meter narrower and ten meters shorter than the lock chamber). Ships built to fit these dimensions are called Panamax ships (Fig. 4), and until 1992 the largest containerships in the world were of the Panamax type. Through the 1980s, the containership industry tried to pack ever-larger numbers of containers onto each ship, in part by stacking the containers higher and higher on the decks, which required correspondingly greater amounts of ballast water to maintain stability. Beamier Post-Panamax containerships are inherently more stable and carry and discharge much less ballast water per voyage - on the order of a few hundred metric tons of ballast water discharged rather than several thousand tons for Panamax ships (Herbert Engineering 1999) - while carrying many more containers. For example, it has been estimated that at the Port of Oakland in California each 3,000 TEU[19] Panamax containership discharges about 4,000 metric tons of ballast water per voyage, compared to 1,000 metric tons of discharge for a 5,000 TEU Post-Panamax containership (Port of Oakland 1999). This works out to around 1.3 metric tons of ballast water discharged per TEU transported in a Panamax vessel, versus 0.2 metric tons of discharge per TEU in a larger Post-Panamax vessel. Similarly, carrying the cargo in a larger number of smaller vessels increases the amount of hull surface area per TEU carried. The same principle holds for other types of vessels: with vessel widths limited by the need to fit through the lock of the Panama Canal, vessels have had to carry more ballast water to maintain stability than they would if they had been built beamier.

[19] A TEU is a "Twenty-foot Equivalent Unit," the standard measure for container cargo.

Fig. 4. A high-stacked, Panamax containership in Gatun Locks, and a Panamax car carrier in Miraflores Locks. Photos courtesy of CanalMuseum.com.

Thus, for many decades the size of the locks in the Panama Canal restricted ship sizes and delayed the production of larger vessels that can transport a given cargo volume while carrying and discharging less ballast water and providing less hull surface for fouling organisms to attach to. These restrictions applied not only to vessels on regular voyages through the canal, but also to vessels throughout the world that might only rarely pass through the canal. The overall impact on species introductions from this indirect effect of the canal has never been calculated, but could be substantial.

About 874,000 vessels have passed through the canal since its opening in 1914, and current traffic runs to about 14,000 vessels each year, carrying 200 million tons of cargo (ACP 2005). Through its combined effects of increasing trade, rearranging global shipping patterns, shortening voyages, and controlling the size and design of a substantial portion of the world's merchant fleet, the Panama Canal has probably had a greater impact on the global scale and pattern of ship-borne species introductions than have any of the world's other great canals.

3 The canal as a biological pathway between the oceans

3.1 Biological studies

The Panama Canal lies in a region of the world where the marine biota is both diverse and relatively poorly known, and there has been remarkably little investigation of the effect that the canal has had on the distribution of that biota. The canal has also affected the spread of organisms from distant regions, though much of the evidence for this is circumstantial and open to different interpretations. Accordingly, the present account will necessarily be both incomplete and somewhat tentative.

In the first edition of *Man and Nature*, published in 1864, George Perkins Marsh speculated that a sea level canal through Central America might divert the flow of the Gulf Stream, depress the temperature of western Europe and eastern North America, and trigger a new ice age resulting in "the extinction of vast multitudes of land sea plants and animals, and a total revolution in the domestic and rural economy of human life in all those countries from which the New World has received its civilized population." Regarding an isthmian canal's potential for translocating organisms, he thought a sea level canal could "produce very interesting revolutions in the animal and vegetable population" in the waters on either side, but that a lock canal would "scarcely possess a geographical character."

Hildebrand provided the first significant description of the canal's effects on the distribution of local fauna, with a 1937 paper on the presence of Atlantic tarpon (*Megalops atlanticus*) in Gatun and Miraflores lakes and a 1939 review of fish and a few invertebrates collected in the canal locks when they were drained for cleaning (Fig. 5). He noted several types of coastal fish whose tolerance of low salinities suggested they might be capable of traversing the canal, including tarpon, snook, guavinas, sardinas and others. His collections from the locks provided evidence that ten Atlantic or Atlantic slope fish and two Pacific slope fish had crossed the continental divide via the canal.

After Hildebrand there was nearly a 30-year hiatus in such studies until Rubinoff and Rubinoff (1968) reported on a Caribbean goby, *Lophogobius cyprinoides*, collected in the Miraflores Third Lock lagoon, as the first record of a fish "passing through the Panama Canal and successfully colonizing the opposite coast." The following year they reported on an eastern Pacific goby, *Gobiosoma nudum*, that they had collected on a reef near the Atlantic entrance to the canal, and suggested that it transited the canal as eggs deposited among the hull fouling of a ship (Rubinoff & Rubinoff 1969).

In a review of trans-isthmian migration, Bayer et al. (1970) noted that the gobies *Gobiosoma* (=*Garmania*) *homochroma* and *Gobiosoma* (=*Garmania*) *hildebrandi* were known only from collections in or near the canal, and their origin was therefore "obscure;" and that *G. hildebrandi,* like *G. nudum*, was established on both coasts. They also reported the Indo-Pacific blenny *Omobranchus punctatus* (as *Omobranchus dasson*), which was collected in Trinidad in the 1930s, as having arrived in the Caribbean after transiting the canal from west to east, traveling among hull-fouling organisms. Later, Springer and Gomon (1975) reported the collection of *O. punctatus* in Limon Bay and Gatun Locks at the Atlantic end of the canal, and argued that it had arrived in the ballast tanks of a ship approaching the canal from the east.

Fig. 5. Sampling the locks when they are drained for maintenance can provide information on the migration or transport of organisms through the canal. Here, Mark Torchin and Ross Robertson seine for fish in Miraflores Locks during draining in 2005. Photo courtesy of Mark Torchin.

Dawson (1970) reported a Caribbean blenny, *Lupinoblennius dispar*, established in the Miraflores Third Lock lagoon, noting the possibility of either adults passing through the canal or of eggs being carried through on hull fouling. Three years later he reported collecting an Indo-West Pacific eleotrid fish, *Butis* (=*Prionobutis*) *koilomatodon*, in Miraflores Upper Locks and suggested that it had been introduced in ballast water discharged into the lock (Dawson 1973).

McCosker and Dawson (1975) reviewed the biota of the Miraflores Third Lock lagoon and the records of exotic fish on either side of the canal. To the examples already published, they added four fish species that had apparently passed through the canal. They argued, as Rubinoff and Rubinoff (1968) had earlier, that the Miraflores Third Lock lagoon provided a kind of "safe haven" on the Pacific Coast for immigrants from the Caribbean. Gunter (1979) reviewed the data and concluded that there were a total of three fish that were "actual migrants" through the canal and one that had been carried through in ballast water. Carlton (1985) argued that six of the fish reported as having transited through or into the canal had possibly or probably been carried in ballast water. Smith et al. (2004) considered the freshwater fish in the watersheds tributary to the canal and concluded, based on their own collections and other data, that 8-11 species of primary or secondary fish (found only or primarily in fresh water and only occasionally in brackish or saline water) and four species of peripheral fish (regularly occurring in brackish or marine as well as fresh water) had become established across the continental divide as a result of migration through the canal.

Besides fish, Hildebrand (1939) listed several invertebrates collected in the locks when they were drained in 1935 and 1937. A few of these may have been transported by ships into or through the canal including a cosmopolitan sponge, *Haliclona permollis*, a European freshwater hydroid, *Cordylophora caspia*, and a Caribbean mussel collected in the Pedro Miguel and Miraflores locks, *Mytilopsis sallei.* Bequaert (1943) reported that two species of Caribbean periwinkles had reached the Pacific through the canal. Abele (1972a) reported a freshwater crab, previously known only from Iraq, that was collected and was apparently abundant in Pedro Miguel Locks; and two mangrove crabs and a freshwater shrimp that he believed had migrated or been transported through the canal (Abele 1972c). McCosker and Dawson (1975) listed a Caribbean crab and two Caribbean seaweeds that had become established in the Miraflores Third Lock lagoon.

Jones and Rützler (1975) reported on invertebrates sampled during the draining of the upper east chamber of the Gatun Locks and the east chamber of Pedro Miguel Locks (Fig. 6), focusing on an abundant freshwater sponge, *Trochospongilla leidii,* that was probably introduced from the eastern United States. Rosewater (1975, 1976) reported on mollusks collected in the drained locks in 1974-75, noting that two out of the 74 species, the snail *Alexania floridana* and the mussel *Mytilopsis sallei,* were collected at both ends of the canal. Jones (1976) reported that 19 out of the 164[20] invertebrate species collected were present in locks at both ends of the canal, and suggested that they had probably been transported between them in hull fouling. Hendrickx (1980) discussed two species of shipworms as having dispersed from the Caribbean to

20 Jones (1976) actually stated that 165 species had been collected in the locks, but the other numbers reported (57 in Atlantic locks, 126 in Pacific locks, with 19 species shared) add up to 164.

the Pacific through the canal. Carlton (1985) listed 17 invertebrates that he believed had possibly or probably been carried through or into the canal in ballast water. Abele and Kim (1989), reporting on decapods collected in the canal, added one species of shrimp and four species of crab to those already reported as having travelled through the canal across the continental divide.

Fig. 6. Barnacles being scraped from the bottom of Miraflores Lock during draining in 2005. Photo courtesy of Mark Torchin.

There has been remarkably little investigation of the mechanism of ship-borne transport of organisms through the canal. In the only published field experiment, Menzies (1968) wrapped a variety of Atlantic and Pacific marine invertebrates in cheesecloth and towed the bundle through the canal alongside a naval vessel, to test the ability of fouling organisms to survive exposure to freshwater during the transit. However, at cruising speed the bundle planed and skipped over the water so the animals were unsubmerged for a significant portion of the trip, and some of the animals were smashed or removed by the force of the water striking them. Most of the rest survived, though it's hard to know what this signifies. Rubinoff and Rubinoff (1969) and McCosker and Dawson (1975) tested the hyposalinity tolerance of five species of gobies and blennies that had migrated or been transported through the canal or to the canal region, in order to assess whether they were capable of migrating through the canal or passing through it as adults or as eggs among hull fouling; and Hay and Gaines (1984) tested the freshwater tolerance of ten common species of seaweeds from Caribbean reefs to assess their potential for transport through the canal as hull fouling. In May 2002, I had the opportunity with a team of taxonomists to examine the hull of a large tuna fishing boat that had spent several months working off the coast of Africa before entering drydock in Balboa near the Pacific end of the canal. The submerged portion of the hull was covered from bow to stern with a layer of fouling that included living seaweeds, serpulid worms, oysters, barnacles and bryozoans (Fig. 7). Among these were several species of campanularid hydroids - delicate, soft-bodied animals - that had apparently survived passage across the Atlantic and through the canal and were still alive on the hull of the ship in drydock on the Pacific coast (D. Calder, A. Cohen unpublished data). Jones (1972) and Glynn (1972) noted the lack of data on hull fouling organisms transported through the canal; there appear to be unutilized opportunities to examine this question, taking up from where Menzies' initial, flawed experiment left off.

Even less investigation has been conducted on the ballast water carried through the canal, which Chesher (1968) first argued could transport large numbers of organisms. In 1967, Rubinoff found no living organisms in samples taken from a few ballast tanks in ships transiting the canal, and stated that "the environment in most ballast tanks is remarkably inhospitable and frequently completely abiotic," though noting that ballast water's role in transporting organisms through the canal had not yet been properly evaluated (Rubinoff 1970). In 1970, Cross (1971) found live crustaceans (probably copepods) in a single vertical tow that he collected from the ballasted cargo tank of an oil tanker transiting the canal en route to Venezuela. The ballast had been loaded from Balboa Harbor 4-8 hours before sampling, and was contaminated with oil that had not been cleaned from the tank. Several researchers noted the need for a program of sampling the ballast water that is carried through the canal (Jones 1972; Dawson 1973; Challinor 1978). Carlton (1985), in a global review of ballast water, listed nine species "probably" introduced through or into the Panama Canal via ballast water and another sixteen species "possibly" introduced via ballast water, including a freshwater sponge, jellyfish, snails, clams, barnacles, crabs and fish. Numerous studies since then in many parts of the world have fully demonstrated the ability of a wide variety of organisms to survive voyages in ballast tanks (e.g. Carlton & Geller 1993; see Cohen 1998 for a review). However, there is little information on current and historic ships' practices regarding the locations and volumes of ballast uptake and discharge relative to the Panama Canal.

Fig. 7. Taxonomists examine a vessel in drydock in Balboa in 2002. Most of the fouling that had covered the hull when the vessel entered drydock has already been hosed and scraped off by the drydock workers. Photo courtesy of Ernesto Campos-González.

In 1965, U.S. President Lyndon Johnson appointed the Atlantic-Pacific Interoceanic Canal Study Commission to assess possible routes for a sea-level canal, including canals created by “nuclear excavation” (part of an effort to find a peaceful use for nuclear explosives). The proposal sparked a heated debate among members of the scientific community, who disagreed about the extent to which the existing canal served as a conduit for the migration or transportation of organisms between the oceans, about the extent to which a new, sea-level canal would serve as a conduit, about the main direction (Atlantic to Pacific, or Pacific to Atlantic) in which invasions were most likely to occur, and about the nature and severity of the consequences that would ensue (e.g. Rubinoff 1968, 1970; Briggs 1968, 1972a,b; Sheffey 1968, 1978; Clarke 1969; Topp 1969). Views about the first question, on the role of the existing canal, ranged from that of Newman (1972), who was strongly opposed to contructing a sea-level canal without a biotic barrier, who wrote:

> *“If it were known that a spectrum of marine organisms had been getting through the existing canal, primarily in ships’*

> *ballast tanks, it would be well that measures be taken to stop the transport, but there is no evidence that successful transports of this kind have been made."*

to that of Bayer et al. (1970), who believed there was no biological reason to oppose a sea-level canal:

> *"Undoubtedly, numerous species and millions of individuals of invertebrates have successfully crossed from one ocean to another on ship bottoms and in ballast tanks through the present canal ever since the first day it went into operation."*

The Commission solicited two reviews of the ecological effects of a sea-level canal, one by Battelle Memorial Institute and one by a committee convened by the National Academy of Sciences. Both reviews found that there was little information available with which to predict impacts, and both reviews and the Commission called for a comprehensive research program including studies of the potential for the region's organisms to pass through a canal, become established on the opposite side of the isthmus, and affect the existing biota (Boffey 1971; Leschine 1981a). A second National Academy of Sciences committee, created in 1978, found that little of this research had been done in the interim (Leschine 1981b). The same could be said today.

3.2 Migration of marine or estuarine species through the canal

The first and best known marine or estuarine species reported to have migrated through the canal is the tarpon, *Megalops atlanticus*. It is native to the western Atlantic from North Carolina to Brazil (with occasional records north or south of this range) and the eastern Atlantic from Senegal to Angola. Hildebrand (1937) reported that tarpon were regularly fished from Gatun Lake, and observed several large specimens breaking the surface in Gatun and Miraflores lakes in 1935. In 1935-37 he and his colleagues collected numerous 1- to 2-meter-long tarpon in the dewatered Gatun Middle and Upper Locks, Pedro Miguel Locks, and Miraflores Upper and Lower Locks (Table 9), including females with roe; and noted that they had been "reliably reported from the sea level end of the Canal below Miraflores Locks" (Hildebrand 1939). He found no evidence of tarpon breeding in Gatun Lake, nor any evidence that they are capable of breeding in fresh water. Several specimens, 1-1.5 meters long, were collected in the drained Miraflores Upper Locks in 1972 (McCosker & Dawson 1975), and there are periodic anecdotal observations of tarpon breaking the water surface in the locks. Heilner (1965) stated that there had been unsuccessful efforts to transplant tarpon to the Pacific Coast, including "attempts to tow them through the Panama Canal alive in slatted, submerged pens," though he gave no dates and I have seen no other references to these efforts.

Bayer et al. (1970) and Voss (1972) reported there was no evidence that tarpon had become established on the Pacific side of the isthmus, Martin et al. (1970; also Voss 1978) stated that tarpon had "failed to establish a breeding colony" on the Pacific side, and McCosker and Dawson (1975) noted that though they often received reports of anglers catching tarpon in Panama Bay, they had not been able to substantiate these reports nor had they seen tarpon for sale in the Panama City fish markets. Heilner (1965), however, reported that tarpon had been caught off the Pacific entrance to the canal, and there have continued to be regular reports of tarpon, including fish weighing over 100 kilograms, being caught in Panamanian waters on the Pacific side of the isthmus in the Gulf of Chiriqui, at Coiba Island, near the canal entrance, in the Bayano River and at Pinas Bay; and on the Pacific coasts of Costa Rica and Colombia (Fig. 8) (Fishbase 2005; Bayano River Sportfishing Charters 2005; Morey 2005; Ruhlow 2005; West 2005).

One charter boat captain reports that he has been catching tarpon on the Pacific side of Panama for at least 30 years, and has regularly seen abundant juveniles and large schools of large adult fish along the Gulf of Panama and in the Bayano River. Until the last few years he observed an annual, gradual migration of large adult fish through the canal, reporting that he would first catch them

> "*on the Atlantic side in August for a couple of months, and then follow them into Lake Gatun for a couple of months then to the Pacific side entrance of the canal around the Miraflores locks, along the Causeway and on out to Taboga. Later, usually in February to March, when Gulf of Panama waters annually turned cold we would catch them in the Bayano River and the Gulf of San Miguel*" (Tony Herndon, pers. comm.).

Despite the lack of collections by biologists, it appears that tarpon are regularly present on the Pacific side of the isthmus. The reports of large numbers of juvenile fish suggest they are reproducing in the Pacific, although it's possible that all reports in the Pacific are of individuals that have migrated through the canal.

Fig. 8. An Atlantic tarpon caught in the Bayano River on the Pacific side of the isthmus. The photograph was taken at the Diablo Spinning Club near Balboa. Photo courtesy of Tony Herndon.

The crested goby *Lophogobius cyprinoides* is native and widespread in Caribbean waters, including records from Porto Bello not far from the canal's Atlantic entrance (Hildebrand 1930; Rubinoff & Rubinoff 1968; Dawson 1972). Hildebrand (1939) found it to be common in the dewatered Gatun Lower Locks in 1935, and it was collected in those locks on several occasions from 1945 to 1974 (Table 9) (Fishbase 2005). It was collected in the Miraflores Lower Locks in 1937 (Dawson 1972), and regularly collected in the Miraflores Third Lock lagoon near the canal's Pacific entrance in 1968-1976, where it had become established as the numerically dominant fish (Rubinoff & Rubinoff 1968, 1969; Dawson 1972; Fishbase 2005). Individuals from the Atlantic and from the Miraflores Third Lock lagoon bred freely in the laboratory (Rubinoff & Rubinoff 1968). Other experiments showed that *L. cyprinoides* tolerates low salinities, with nearly 30% surviving 70-100+ hours of exposure to fresh water (McCosker & Dawson 1975), and there are records from fresh waters in Bermuda (Rubinoff & Rubinoff 1969). These data and records suggest that *L. cyprinoides* may have migrated on its own through the canal (as suggested by Rubinoff & Rubinoff 1969 and Springer & Gomon 1975), although it could also have been transported in association with hull-fouling (suggested by McCosker

& Dawson 1975) or in a ship's ballast tank (suggested by Carlton 1985) or sea chest. While it is well established in the Miraflores Third Lock lagoon, there are no records of it on the Pacific Coast outside of the lagoon. Rubinoff and Rubinoff (1986) and McCosker and Dawson (1975) argue that it may not be able to compete against the native Panama Bay fish fauna outside of the unusual environment in the lagoon.

The Caribbean blenny *Lupinoblennius dispar* ranges from Mexico to Trinidad, and is common in the upper parts of Limon Bay (Dawson 1970; McCosker & Dawson 1975). It was collected in the drained Gatun Lower Locks in 1935 (Hildebrand 1939, as *Blennius* sp.), and in all three levels of the Gatun Locks and on several occasions in 1966-1974 (Table 9) (McCosker & Dawson 1975; Fishbase 2005). In 1967, 1968, 1970 and 1976 it was found in the Miraflores Third Lock lagoon; collections there included adults, juveniles and settled larvae, indicating an established population (Dawson 1970; Springer & Gomon 1975; McCosker & Dawson 1975; Fishbase 2005). *Lupinoblennius dispar* is tolerant of fresh water, with about 60% surviving experimental exposures of 60-100+hours (McCosker & Dawson 1975). As with *Lophogobius cyprinoides*, its repeated collection in the canal system and its tolerance to fresh water indicate that it probably migrated through the canal (as suggested by Springer & Gomon 1975), though transport in hull fouling (as eggs or adults, suggested by Dawson 1970 and McCosker & Dawson 1975), ballast tanks (suggested by Carlton 1985) or sea chests is also possible. *Lupinoblennius dispar* is established in the Pacific only in Miraflores Third Lock lagoon, and like *Lophogobius cyprinoides,* it may not be able to compete with the native fauna outside of the lagoon (McCosker & Dawson 1975).

The anchovy *Anchoa* (=*Anchovia*) *parva* is native to the Caribbean from Cuba and Jamaica south to Venezuela. It occurs mainly in coastal marine and brackish water but sometimes ranges into fresh water. It has been collected on a few occasions from all levels of the Gatun Locks. It was collected in both levels at Miraflores Locks in 1937, but has not become established on the Pacific coast.

The western Atlantic pipefish *Oostethus lineatus*, which is found in both marine and fresh water, was common in brackish and fresh waters up to Gatun Lake prior to the construction of the canal. Males carrying eggs were collected in the lake in 1928 and 1935, suggesting that it was then a permanent resident (Chickering 1929; Hildebrand 1939; McCosker & Dawson 1975). A single specimen was collected in Panama Bay near the Pacific entrance to the canal in 1971, and another in Miraflores Upper Locks in 1972 (McCosker & Dawson 1975).

Table 9. Records in the canal of native Caribbean species that have apparently migrated through to the Pacific drainage. a = Record given as "Gatun Locks" without specifying which level, b = Miraflores Third Lock lagoon, c = Publication date. 1 = Hildebrand 1939, 2 = Hildebrand 1939; Rubinoff & Rubinoff 1968, 1969; McCosker & Dawson 1975; Fishbase 2005, 3 = Hildebrand 1939; Dawson 1970; McCosker & Dawson 1975; Fishbase 2005, 4 = Hildebrand 1937, 1939; McCosker & Dawson 1975, 5 = Chickering 1929; Hildebrand 1939; Zaret & Paine 1973; McCosker & Dawson 1975.

Date	Gatun Lower Locks	Gatun Middle Locks	Gatun Upper Locks	Gatun Lake	Pedro Miguel Locks	Mira-flores Lake	Mira-flores Upper Locks	Mira-flores Lower Locks	Pacific coast of Panama	Ref.
Anchoa parva	1935 1972 1974	1935 1974	1935	–	–	–	1937	1937	–	1
Lophogobius cyprinoides	1935 1945 1974	1966[a] 1972[a] 1974	1974	–	–	–	–	1937	1967-1976[b]	2
Lupinoblennius dispar	1935 1974	1966[a] 1972[a] 1974	1974	–	–	–	–	1937	1967-1976[b]	3
Megalops atlanticus	–	1935	1935	1935	1937	1935	1937 1972	1937	<1939	4
Oostethus lineatus	–	–	–	1910 1928 1933[c] 1935 1972	–	–	1972	–	1971	5

In all, five marine or species are here considered to be probable migrants through the canal, all of them native to Caribbean and Atlantic slope waters (freshwater migrants are considered in Section 3.6 below). Two species, the anchovy *Anchoa parva* and the pipefish *Oostethus lineatus*, are apparently not established in the Pacific. Two others, the goby *Lophogobius cyprinoides* and the blenny *Lupinoblennius dispar*, are established on the Pacific Coast only in the Miraflores Third Lock lagoon. The fifth species, the tarpon *Megalops atlanticus*, is reported by anglers to be common in and around Panama Bay, though we have found no scientific records of their presence in Pacific waters. Reports of large numbers of young fish at sites around the Gulf of Panama suggest that it is breeding there. However, the tarpon is very long lived (with a maximum reported age of 55 years), has been at least seasonally common in Gatun Lake for over 60 years, has been frequently reported from other parts of the canal system, and in recent decades schools of large tarpon are reported to have migrated through the canal into the Pacific in most years. Thus it is conceivable that the large numbers of tarpon reported in Panama Bay and adjoining waters are all fish that were born in the Atlantic and migrated through the canal.

With the possible exception of the pipefish, *Oostethus lineatus*, all of these species appear to have reached the Pacific slope and coast by the migration of individual fish rather than the gradual spread of a population. There is some evidence that *O. lineatus* breeds in Lake Gatun, but it was collected in that area before the lake was created, so that probably can't be counted as dispersal. It was collected twice on the Pacific side, once in Miraflores Locks and once in Panama Bay, but it is probably not established there. *O. lineatus* is a very weak swimmer, and normally would not be a good candidate for colonization over a significant distance via the movement of individual fish; however, since water from Lake Gatun is regularly released down through the locks, individual *O. lineatus* could make it into Panama Bay simply by drifting with the flow.

3.3 Species transport by vessels passing through the canal

Undoubtedly, a large number of species introduced to tropical American waters or other waters of the world have been transported by vessels on voyages that involve a transit of the Panama Canal. The evidence of this for any particular species is circumstantial, and includes consideration of the areas of origin and introduction, the relative amount of ship traffic on different trade routes, the organism's life history and environmental tolerances relative to possible transport mechanisms, and the existence of other non-ship transport mechanisms that could account for the species' distribution.

In this section I describe some representative examples of organisms in different taxonomic groups that appear to have been transported and introduced into regions outside their native range by vessels passing through the canal. In the discussion and tables that follow, the information is organized by taxonomic groups for species transported by vessels transiting the canal from the Atlantic to the Pacific (Table 10) and from the Pacific to the Atlantic (Table 11). The species are also classified by pathway in each direction (Tables 12-17).

Seaweeds

The red seaweed *Polysiphonia denudata* is native to the western Atlantic from southwestern Canada to tropical waters. It was collected in the northeastern Pacific in San Francisco Bay by 1983, and possibly observed there as early as 1963 (Josselyn & West 1985; Cohen & Carlton 1995). *Polysiphonia* species are common in hull fouling and are resistant to many anti-fouling treatments (Cohen & Carlton 1995), suggesting that this species traveled to the Pacific as hull fouling. Transport in ballast water is less likely, as few seaweeds have been reported in ballast water (Wysor 2004). McCosker and Dawson (1975) reported two Caribbean species growing on mangrove roots in the Miraflores Third Lock lagoon, *Polysiphonia atlantica* (as *P. macrocarpa*) and *Cladophora montagneana* (as *C. deliculata*), which presumably would also have been transported in hull fouling or possibly ballast water. However, *C. montagneana* was earlier reported in the Gulf of California, so the Miraflores Third Lock population may not be a trans-isthmian introduction (Wysor 2004; Guiry et al. 2005). Hay and Gaines (1984) noted that *Acanthophora spicifera*, *Centroceras clavulatum* and *Spyridia filamentosa* are found on boat hulls or buoys in the Caribbean, can stand at least six hours of exposure to fresh water, and are abundant on Caribbean reefs but extremely rare on the Pacific coast near Panama, and thus might have been introduced to the Pacific as hull fouling on boats or ships passing through the canal. Although a large number of other seaweed species are reportedly common to both coasts of Panama (see Table 2), it is unclear whether these represent introductions or morphologically similar but genetically distinct species (Wysor 2004).

Table 10. Possible introductions through the canal from the Atlantic to the Pacific. a = Some species listed in Table 9 as migrants through the canal could also have been transported by ship in hull fouling, ballast tanks or sea chests (including Lophogobius cyprinoides and Lupinoblennius dispar as suggested by various authors (Dawson 1970; McCosker & Dawson 1975; Carlton 1985)). They are not included in this table and Table 12 to avoid double-counting, b = Includes introductions of marine or brackish water species to locks on the Pacific side of the isthmus, c = The date refers either to the date the species was first collected in the invaded region, or to the earliest date associated with a report of the species in the invaded region, and is not necessarily the date of introduction which could have been earlier, d = HF = transport as hull fouling or in hull borings; HFE = transport as eggs among hull fouling; BW = transport in ballast water tanks, sea chests or other parts of ships' seawater systems; OA = with oyster aquaculture; AQ = release from an aquarium. Less likely mechanisms are in parentheses, e = References include both collection records and discussions of anthropogenic transport.

Species[a]	Transport: Source Region *to* Destination[b] [Collection Date(s)[c]]	Possible Mechanism[d]	References[e]
Seaweeds			
Polysiphonia atlantica	North Atlantic and Caribbean *to* Miraflores Third Lock lagoon [1975]	HF, (BW)	McCosker & Dawson 1975 (as *P. macrocarpa*)
Polysiphonia denudata	Northwestern Atlantic *to* northeastern Pacific in San Francisco Bay [1978-83]	HF, (BW)	Josselyn & West 1985; Cohen & Carlton 1995
Sponges			
Clathria prolifera	Northwestern Atlantic *to* northeastern Pacific in San Francisco Bay [1945-49]	HF, BW, OA	Carlton 1979; Cohen & Carlton 1995
Geodia gibberosa	Caribbean *to* Pacific Panama [1933]	HF, BW	De Laubenfels 1936
Halichondria bowerbanki	North Atlantic *to* northeastern Pacific in San Francisco Bay [1950-53]	HF, BW, OA	Carlton 1979; Cohen & Carlton 1995

Table 10. continued.

Species[a]	**Transport: Source Region *to* Destination**[b] **[Collection Date(s)**[c]**]**	**Possible Mechanism**[d]	**References**[e]
Haliclona loosanoffi	North Atlantic *to* northeastern Pacific in San Francisco Bay [1950]	HF, BW, OA	Carlton 1979; Cohen & Carlton 1995
Haliclona coerulescens	Caribbean *to* Pacific Panama [1933]	HF, BW	De Laubenfels 1936
Microciona atrasanguinea	Europe *to* Pacific Panama [1933]	HF, BW	De Laubenfels 1936
Prosuberites sp.	Northwestern Atlantic *to* northeastern Pacific in San Francisco Bay [1953, 2004]	HF, OA	Carlton 1979; Cohen & Carlton 1995; J.T. Carlton, A.N. Cohen unpublished data 2004
Hydrozoans			
Blackfordia virginica	Black Sea or Atlantic sites *to* northeastern Pacific in San Francisco Bay [1970] and Coos Bay [1998]	HF, BW	Mills & Sommer 1995; Cohen & Carlton 1995; Mills & Rees 2000
Chelophyes appendiculata	Atlantic *to* Pacific Panama [1962-1969]	BW	Alvariño 1974; Carlton 1985
Maeotias marginata	Black Sea or Atlantic sites *to* northeastern Pacific in San Francisco Bay [1992]	HF, BW	Mills & Sommer 1995; Cohen & Carlton 1995; Mills & Rees 2000
Muggiaea kochi	Atlantic *to* Pacific Panama [1962-1969]	BW	Alvariño 1974; Carlton 1985
Polychaetes			
Boccardiella ligerica	Northwestern Europe *to* California [1935]	BW	Carlton 1979; Cohen & Carlton 1995
Hydroides alateralis	Caribbean *to* Pacific Colombia [1988]	HF, BW	Bastida-Zavala & ten Hove 2003a,b
Hydroides gairacensis	Caribbean *to* Pacific Panama [1933] and Ecuador [1966]	HF, BW	Bastida-Zavala & ten Hove 2003a,b
Hydroides sanctaecrucis	Caribbean *to* Pacific Panama [1972, 2002], Pacific Mexico [2000] and Australia	HF, BW	Bastida-Zavala & ten Hove 2003a,b
Marenzelleria viridis	Northern Atlantic *to* northeastern Pacific in San Francisco Bay [1991]	BW	Cohen & Carlton 1995
Gastropods			
Alexania floridana	Caribbean *to* Miraflores Lower Locks [1974-75] (not established)	HF, BW	Rosewater 1976
Conus mus	Caribbean *to* Pacific Panama at Veracruz [1960] (not established)	BW	Burch 1960
Echinolittorina ziczac	Caribbean *to* Pacific Panama at Panama City [1914, 1933] (not established)	BW, AQ	Bequaert 1943; Carlton 1985

Table 10. continued.

Species[a]	Transport: Source Region *to* Destination[b] [Collection Date(s)[c]]	Possible Mechanism[d]	References[e]
Littoraria angulifera	Caribbean *to* Pacific Panama at Panama City [1933] (not established)	BW, AQ	Bequaert 1943; Carlton 1985
Tenellia adspersa	Europe *to* northeastern Pacific in San Francisco Bay [1953]	HF, BW	Carlton 1979; Cohen & Carlton 1995
Bivalves			
Bankia cieba	Tropical western Atlantic *to* Pacific Panama at Balboa [1946] and the Gulf of California [1971]	HF, BW	Clench & Turner 1946; Turner 1971; Carlton 1985
Bankia destructa	Tropical western Atlantic *to* Pacific Panama at Puerto Armuelles [1946] and to Pacific Mexico near Mazatlan [1978-79]	HF, BW	Clench & Turner 1946; Hendrickx 1980; Carlton 1985
Bankia fimbriatula	Tropical western Atlantic *to* Pacific Panama at Balboa [1943]	HF, BW	U.S. Navy 1951; Carlton 1985
Martesia cuneiformis	Tropical/subtropical western Atlantic *to* Pacific Panama at Balboa [<1955] (not established)	HF, BW	Turner 1955; Carlton 1985
Mytilopsis sallei	Caribbean *to* Pacific Panama [1937] and the Indo-West Pacific (Fiji [1929], India [1967], Japan [1974], Taiwan [1977], Hong Kong [1980], Singapore [1997] and Australia [1998])	HF, BW	Hildebrand 1939; Hertlein & Hanna 1949; Morton 1981, 1987; Rao et al. 1989; Bax 1999; but taxonomy disputed by Marelli & Gray 1985 (see discussion in text)
Teredo bartschi	Western Atlantic *to* Pacific Mexico at La Paz [1971] and near Mazatlan [1978-79], to Los Angeles [2000], and to Hawaii	HF, BW	Hendrickx 1980; Carlton 1992; Benson et al. 2004; A.N. Cohen unpublished data 2000
Barnacles			
Amphibalanus amphitrite	Caribbean *to* Pacific Panama at Balboa [1974]	HF, BW	Spivey 1976
Amphibalanus eburneus	Western Atlantic *to* Pacific Panama at Balboa [1964], possibly also to Pacific Mexico [1959, 1963, 1968]	HF, BW, (OA)	Spivey 1976; Carlton 1985
Amphibalanus reticulatus	Caribbean *to* Pacific Panama at Balboa and Fort Rodman [2000]	HF, BW	F.B. Pitombo & A.N. Cohen unpublished data
Balanus calidus	Western Atlantic *to* Pacific Panama in Miraflores Locks [1974]	HF, BW	Spivey 1976

Table 10. continued.

Species[a]	Transport: Source Region *to* Destination[b] [Collection Date(s)[c]]	Possible Mechanism[d]	References[e]
Balanus trigonus	Caribbean *to* Pacific Panama at Punta Culebra [2000]	HF, BW	F.B. Pitombo & A.N. Cohen unpubl. data
Chthamalus proteus	Caribbean *to* Hawaii [1995]	HF, BW	Southward et al. 1998
Fistulobalanus pallidus	Caribbean *to* Pacific Panama in Miraflores Locks [1972]	HF, BW	Jones & Dawson 1973; Spivey 1976
Isopods			
Pleurocope floridensis	Caribbean *to* southern California [2000]	HF, BW	Fairey et al. 2002
Amphipods			
Gammarus daiberi	Northwestern Atlantic *to* San Francisco Bay [1983]	HF, BW	Cohen & Carlton 1995
Shrimp			
Leander paulensis	Caribbean *to* Miraflores Upper Locks [1972]	HF, BW	Abele & Kim 1989
Crabs			
Callinectes exasperatus	Warmwater/tropical western Atlantic *to* creek draining the Miraflores Third Lock lagoon [1972]	HF, BW	Abele & Kim 1989
Eurypanopeus dissimilis	Caribbean *to* Miraflores Third Lock lagoon [1971]	HF, BW	McCosker & Dawson 1975; Carlton 1985
Goniopsis cruentata	Caribbean *to* creek draining the Miraflores Third Lock lagoon [1972] and Miraflores spillway [1972, 1973]	HF, BW	Abele & Kim 1989
Panopeus rugosus	Caribbean *to* Miraflores Upper Locks [1972, 1974] and Miraflores spillway [1973]	HF, BW	Abele & Kim 1989
Rhithropanopeus harrisii	Northwestern Atlantic *to* San Francisco Bay [1937] (established) and the Pedro Miguel Locks [1969] (not established)	HF, BW, OA	Abele 1972c; Carlton 1985; Abele & Kim 1989; Cohen & Carlton 1995
Bryozoans			
Anguinella palmata	Atlantic *to* Peru, Panama, western Mexico, southern California [1933-42]	HF	Osburn 1953; Cohen & Carlton 1995
Bugula stolonifera	Atlantic *to* Panama Bay [1924, 1969]	HF	Powell 1971
Electra monostachys	Atlantic *to* Panama Bay [1924, 1944-47, 1969]	HF, BW	U.S. Navy 1951; Powell 1971; Carlton 1985
Membranipora annae	West Africa *to* Panama Bay [1924, 1944-47, 1969]	HF, BW	U.S. Navy 1951; Powell 1971
Tunicates			
Botryllus schlosseri	North Atlantic *to* Australia [1928], Pacific Panama at Balboa [1944-47] and California [1945-47]	HF	U.S. Navy 1951; Carlton 1979; Kott 1985; Cohen & Carlton 1995

Table 10. continued.

Species[a]	Transport: Source Region *to* Destination[b] [Collection Date(s)[c]]	Possible Mechanism[d]	References[e]
Fish			
Barbulifer ceuthoecus	Tropical western Atlantic *to* Pacific Panama [1973] (not established)	BW	McCosker & Dawson 1975
Hypleurochilus aequipinnis	Tropical eastern and western Atlantic *to* Miraflores Third Lock lagoon [1971]	HF, HFE, BW	McCosker & Dawson 1975

Plants

Bayer et al. (1970) and Voss (1972, 1978) reported that the common Caribbean turtle grass *Thalassia testudinum* had briefly become established on the Pacific side of the isthmus. Martin et al. (1970) and Earle (1972) described the Pacific records of this species and of another common Caribbean seagrass, *Halodule wrightii*, published in 1935 and 1960, as "questionable" and "doubtful". These records are apparently based on Setchell (1935), who noted *T. testudinum's* "possibly sporadic" occurrence on the Pacific coast, "whether attached or established and whether of recent or ancient occurrence is uncertain," and noted *H. wrightii's* "outlying record" in the Gulf of Panama. If these records are correct, these seagrasses could have been carried as seeds in ballast water, or possibly as plants or seeds snagged in a dredge or other marine equipment; but they could also have been carried across the isthmus by birds (see Martin et al. 1970 at p. 76) as seeds stuck to their feathers or as undigested seeds in their guts. Because of uncertainty regarding the records, and the existence of an alternate, natural transport mechanism, these seagrasses are not included in Tables 10 and 12.

Sponges

De Laubenfels (1936) investigated sponges near the ends of the Panama Canal in 1933. He collected the primarily Caribbean species *Haliclona coerulescens* and *Geodia gibberosa* on both coasts, and the European species *Microciona atrasanguinea* on the Pacific coast. These species probably represent introductions through the canal in hull fouling, or possibly in ballast water. *Laxosuberites zeteki* and *Placospongia intermedia* are common on both Panama coasts; they may have been introduced from one to the other through the canal. *Haliclona permollis*[21], *Halichondria panicea, Adocia cinerea, Tethya aurantia, Tethya diploderma, Chondrilla nucula* and *Oscarella lobularis* are widespread or cosmopolitan species that may have also been introduced to Panama.

---

[21] Also reported by Hildebrand (1939) in Miraflores Upper Locks in 1937.

Four species of north Atlantic sponges, *Clathria prolifera*, *Halichondria bowerbanki*, *Haliclona loosanoffi* and *Prosuberites* sp., were collected in the northeastern Pacific in the 1940s and early 1950s in San Francisco Bay (Carlton 1979; Cohen & Carlton 1995). The first three species are well established in San Francisco Bay and a few other sites on the coast. *Prosuberites* sp., after its initial collection in 1953, was not seen again on the North American Pacific coast until the spring of 2004 (J.T. Carlton, A.N. Cohen unpublished data). All four of these sponges might have initially arrived with shipments (by rail) of Atlantic oysters, *Crassostrea virginica*, planted in San Francisco Bay, though most of these shipments were made in the late 1800s and the first decade of the 1900s and the last, small shipments were in the 1930s. Alternately, they could have arrived as hull fouling or in the ballast tanks of ships traveling through the canal.

Jellyfish

Alvariño (1974) reported on siphonophores collected in 1962, 1963 and 1969 in tropical American waters. He argued that the distributions of five siphonophore species suggested introductions through the canal, either by migration or transport in ships' ballast waters or cooling system waters. *Chelophyes appendiculata* and *Muggiaea kochi* were mainly reported from the tropical Western Atlantic but were also found at a few stations on the Pacific side of the isthmus in the Gulf of Panama. *Chelophyes contorta* was found mainly in the tropical Eastern Pacific off Central America and Mexico, and *Muggiaea atlantica* ranged from the Pacific coastal waters of Central America to Japan, but both species were collected at a few stations in the Caribbean near the entrance to the canal. *Lensia challengeri*, an Indo-West Pacific and Eastern Pacific species, was also found at a station near the Caribbean entrance to the canal. Carlton (1985) listed these species as probable ballast water introductions through the canal.

Two species of brackish-water hydrozoans native to the Black Sea region were discovered in San Francisco Bay in the 1970s-1990s, and were probably transported through the canal in ballast water or hull fouling. *Blackfordia virginica* was first collected in low salinity waters in the upper part of San Francisco Bay in 1970, and was also found in Coos Bay, Oregon in 1998. Since it had also been reported as an exotic species from several Atlantic Ocean sites (Chesapeake Bay, Brazil, France and Portugal) and in India and China, the source of the San Francisco Bay population is uncertain (Mills & Sommer 1995; Cohen & Carlton 1995; Mills & Rees 2000). *Maeotias marginata* has been collected in San Francisco Bay since 1992, with possible records going back to 1959. Outside of the Black and Azov seas, it has been found in the

Baltic Sea, The Netherlands, France, Chesapeake Bay and South Carolina (Mills & Sommer 1995; Cohen & Carlton 1995; Mills & Rees 2000).

The spotted jellyfish, *Phyllorhiza punctata*, is native to the Indo-West Pacific region. It is reported from Australia, the Phillipines and Thailand, and as an introduction in Hawaii (possibly collected there as early as 1933; Devaney & Eldredge 1977), and more recently in the eastern Mediterranean Sea (probably via the Suez Canal) and off southern California. It was collected in the southern Caribbean in 1955, having probably been carried through the Panama Canal in hull fouling (as polyps, the jellyfish's sessile life stage) or possibly in ballast water (as ephyrae, the early phase of its pelagic life stage). It ranges south to Brazilian waters, and in the summer of 2000 became abundant in the northern Gulf of Mexico, where economic losses due to interference with the harvest of white shrimp may have been as high as $10 million (Graham et al. 2003).

Polychaetes

Fauchald (1977) reviewed the intertidal polychaetes that had been collected in Panama. He listed 30 species found on both coasts of Panama, of which 19 are cosmopolitan or circumtropical, and 11 were known only from American coasts; and another seven species found on the Panama coast on one side of the isthmus and in the ocean but not on the Panama coast on the other side of the isthmus. Three of the species found on both Panama coasts are mainly distributed in the western Atlantic, while five species and one subspecies are mainly distributed in the eastern Pacific, and these may well represent introductions through the canal.

Bastida-Zavala & ten Hove (2003a,b) cite records of three Caribbean serpulid worms in the eastern tropical Pacific. *Hydroides gairacensis* was collected in Pacific Panama in 1933 and Ecuador in 1966; *Hydroides sanctaecrucis* was collected in Pacific Panama in 1972 and 2002 and on the Pacific coast of Mexico in 2000; *Hydroides alateralis* was collected on the Pacific coast of Colombia in 1988. These species probably arrived in hull fouling, given the many records of serpulid worms on the hulls of ships (Woods Hole Oceanographic Institution 1952), or possibly in ballast water.

The spionid worm *Boccardiella ligerica* is native to the brackish waters and mudflats of northwestern Europe, with records in France, The Netherlands and Germany. It was collected in Newport Bay in southern California in 1935 and has since spread to several California bays from Mission Bay to San Francisco Bay, where it is now abundant (Carlton 1979; Cohen & Carlton 1995). Another spionid worm, *Marenzelleria viridis* is native to the northwestern Atlantic. It was introduced to Europe by 1983 (probably in ballast water), where it is now common from northwestern Europe to the Baltic Sea, and it has been collected in San Francisco Bay since 1991 (Cohen & Carlton 1995). Both worms were probably transported to California in ballast water carried on a ship that passed through the canal, since spionid larvae are among the most commonly collected types of organisms in ballast water (Carlton & Geller 1993).

Gastropods

Four Caribbean snails have been reported as introductions to Panama Bay, but a lack of recent records suggests they did not become established. Two of the most common periwinkles on Panama's Caribbean coast were collected near Panama City, *Echinolittorina* (=*Littorina*) *ziczac* in 1914 and 1933 and *Littoraria* (=*Littorina*) *angulifera* in 1933 (Bequaert 1943; Rosewater 1980; Carlton 1985). These periwinkles could have been transported as larvae in ballast water as Carlton (1985) suggested, but given the ease with which these snails are collected and their discovery each time near an urban area, they could also have been releases from aquariums. Loftin (1965) noted that hobbyists in the Canal Zone often stocked their aquariums with local species. The mouse cone, *Conus mus*, was collected in 1960 in the low intertidal zone at Veracruz near the Pacific entrance to the canal (Burch 1960). Rosewater (1976) reported that *Alexania floridana,* was collected in the Gatun Lower Locks and Miraflores Lower Locks when they were dewatered in 1974-75. The genus is often found on small anemones that are common in hull fouling, which may have been how it traveled through the canal.

The minute sea slug *Tenellia adspersa* is native to European and Mediterranean waters and was collected in the northeastern Pacific in San Francisco Bay in 1953. It has since spread along that coast from southern Oregon to southern California (Carlton 1979; Cohen & Carlton 1995). It has planktonic larvae, and could have passed through the canal into the Pacific in ballast water, or possibly on hydroids (which it lives on and feeds on) fouling a ship's hull. It has since been reported from Chesapeake Bay, Massachusetts and Brazil (Cohen & Carlton 1995; Benson et al. 2004), probably representing one or more separate introductions from Europe, but possibly representing an introduction back through the Panama Canal from the U.S. Pacific Coast. The northeastern Pacific sea slug *Stiliger fuscovittatus*[22] ranges from Alaska to the Gulf of California, where it feeds on filamentous red seaweeds (Behrens 1991). It was collected in eastern Florida in the Indian River region, perhaps only once, by 1995 (Carlton & Ruckelshaus 1997; Benson et al. 2001b; U.S. Geological Survey 2005),

---

[22] Reported by some authors as *Ercolania fuscovittata*, but this is apparently incorrect (Trowbridge 2005).

where it had probably been introduced in ballast water or in seaweed fouling a boat or ship that had passed through the canal.

Table 11. Possible introductions through the canal from the Pacific to the Atlantic. a = Includes introductions of marine or brackish water species to locks on the Atlantic side of the isthmus, b = The date refers either to the date the species was first collected in the invaded region, or to the earliest date associated with a report of the species in the invaded region, and is not necessarily the date of introduction which could have been earlier, c = HF = transport as hull fouling or in hull borings; HFE = transport as eggs among hull fouling; BW = transport in ballast water tanks, sea chests or other parts of ships' seawater systems; AQ = release from an aquarium, d = References include both collection records and discussions of anthropogenic transport, e = May have been transported through the canal to Pacific Panama after becoming established in the Caribbean.

Species	Transport: Source Region *to* Destination[a] [Collection Date(s)[b]]	Probable Mechanism[c]	References[d]
Jellyfish			
Chelophyes contorta	Pacific Panama *to* Atlantic Panama [1962-1969]	BW	Alvariño 1974; Carlton 1985
Lensia challengeri	Pacific Ocean *to* Atlantic Panama [1962-1969]	BW	Alvariño 1974; Carlton 1985
Muggiaea atlantica	Pacific Panama *to* Atlantic Panama [1962-1969]	BW	Alvariño 1974; Carlton 1985
Phyllorhiza punctata	Indo-West Pacific *to* Caribbean [1955]	HF, BW	Graham et al. 2003
Gastropods			
Stiliger fuscovittatus	Northeastern Pacific *to* Florida [1995] (not established)	HF, BW	Carlton & Ruckelshaus 1997; Benson et al. 2001b; U.S. Geological Survey 2005
Bivalves			
Electroma sp.	Indo-West Pacific *to* Atlantic coast of Colombia & Venezuela [1983]	BW	Borrero & Diaz 1998
Hyotissa hyotis	Indo-Pacific *to* southeastern Florida [2001]	HF, BW	Bieler et al. 2004
Lyrodus medilobatus	Indo-West Pacific *to* Florida [1983]	HF, BW	Mikkelsen et al. 1995; Carlton & Ruckelshaus 1997; Benson et al. 2001b
Pinctada margaritifera	Indo-Pacific *to* southeastern Florida [1990]	HF, BW	Frank 1993; Carlton & Ruckelshaus 1997
Perna viridis	Indo-West Pacific *to* Caribbean (Trinidad, Venezuela, Jamaica and Florida) [1990]	HF, BW	Agard et al. 1992; Benson et al. 2001a
Teredo furcifera	Southwestern Pacific *to* U.S. Atlantic coast at New Jersey [1974]	HF, BW	Benson et al. 2004

Table 11. continued.

Species	Transport: Source Region *to* Destination[a] [Collection Date(s)[b]]	Probable Mechanism[c]	References[d]
Barnacles			
Amphibalanus amphitrite	Indo-West Pacific or eastern Pacific *to* Caribbean, Gulf of Mexico and northwestern Atlantic [1940s][e]	HF, BW	Carlton 1979; Carlton & Ruckelshaus 1997
Amphibalanus reticulatus	Indo-West Pacific *to* Caribbean and Gulf of Mexico [1952][e]	HF, BW	Britton & Morton 1989; Carlton & Ruckelshaus 1997
Balanus trigonus	Indo-West Pacific *to* Atlantic Panama [1966-67], Venezuela [1990s] and Florida [1990s][e]	HF, BW	Bayer et al. 1970; Spivey 1976; Carlton & Ruckelshaus 1997
Tanaids			
Zeuxo kuriliensis	Indo-West Pacific *to* Florida	HF, BW	Sieg & Winn 1981; Carlton & Ruckelshaus 1997
Crabs			
Hemigrapsus sanguineus	Northwestern Pacific *to* northeastern Atlantic [1988]	HF, BW	Lohrer 2001
Petrolisthes robsonae	Tropical western Pacific *to* Gatun Locks [<1960, 1974] (not established?)	BW	Gore & Abele 1976; Abele & Kim 1989
Sesarma rhizophorae	Pacific Panama *to* Caribbean at entrance to canal	HF, BW	Abele 1972c
Sesarma sulcatum	Pacific Panama *to* Caribbean at entrance to canal	HF, BW	Abele 1972c
Bryozoans			
Watersipora subtorquata	Western Pacific *to* Florida [1976]	HF	Mook 1976; Carlton & Ruckelshaus 1997
Tunicates			
Botrylloides perspicuum	Indo-West Pacific *to* Caribbean in Belize [2000]	HF	Goodbody 2000, 2004
Diplosoma virens	Indo-West Pacific *to* Caribbean in Belize [2000]	HF	Goodbody 2000, 2004
Cnemidocarpa irene	Western Pacific *to* Caribbean at Guadeloupe [1984]	HF	Monniot & Monniot 1994
Polyandrocarpa zorritensis	Southeastern Pacific *to* the Mediterranean [1974] and the Caribbean at Puerto Rico [2002]	HF	Lambert & Lambert 1998; G. Lambert, A.N. Cohen unpublished data
Fish			
Gnathodon speciosus	Tropical eastern Pacific *to* Gatun Lower Locks [1968] (not established)	BW, AQ	McCosker & Dawson 1975

Table 11. continued.

Species	Transport: Source Region *to* Destination[a] [Collection Date(s)[b]]	Probable Mechanism[c]	References[d]
Gobiosoma nudum	Tropical western Pacific *to* Atlantic Panama at Galeta Reef [1962] (not established)	HFE, BW	Rubinoff & Rubinoff 1969; McCosker & Dawson 1975

Bivalves

Mytilopsis sallei is native to and distributed throughout the Caribbean region (Marelli & Gray 1983). It or a very similar species have been reported in Fiji (by 1929), in India at Visakhapatnam Harbor (in 1967, where it has become a major fouling pest), Bombay (by 1975) and Kakinada (in 1986), in Japan (in 1974), in Taiwan (in 1977), in Hong Kong (in 1980), in Singapore (by 1997) and in northern Australia (at Darwin in 1998, where it was eradicated with massive applications of bleach and copper sulphate, at a cost of over $AU2.3 million). It was also collected from Gatun Middle and Upper Locks in 1974, Pedro Miguel Locks in 1937 and 1975, and Miraflores Upper Locks in 1937 and 1974 (Hildebrand 1939; Hertlein & Hanna 1949; Jones & Rützler 1975; Jones 1976; Morton 1980, 1981, 1987; Marelli & Gray 1983; Rao et al. 1989; Bax 1999; Hutchings et al. 2002). Marelli and Gray (1985) argue that the introduced Indo-West Pacific populations of *Mytilopsis* are an eastern Pacific species, *M. adamsi* from the Pacific coast of Panama. They suggest that this species was introduced to Fiji (where the genus had been reported in 1898) in the 19th century prior to the construction of the Panama Canal, on British mail steamers that ran between a coaling station in Panama and Australia with apparent stops in Fiji; and that it was later introduced from Fiji to India. However, most authors have considered the introduced Indo-West Pacific populations to be *M. sallei*. For example, Morton (1981) suggested that *M. sallei* had been transported to Fiji through the Panama Canal, possibly on the hull of a ship, aided by its broad salinity tolerance. *M. sallei* has been reported as preferring water of 22-32 ppt, surviving in waters varying between 3 and 35 ppt, living in coastal lagoons with rainy season salinities <1 ppt, and surviving 0-50 ppt (Morton 1981). Hutchings et al. (2002) reported that *M. sallei* could have reached northern Australia on the hull of one of three international cruising yachts that called at Darwin after a voyage through the canal. Morton (1981) noted that *M. africana* of West Africa is similar and possibly identical to *M. sallei*, and if so then *M. sallei* could have arrived in India by way of Africa rather than through the Panama Canal.

The green mussel *Perna viridis* is native to the northeastern Indian Ocean and from southeast Asia to the South China Sea, where it is commercially cultivated. It has been introduced to Tokyo Bay (Siddall 1980; Benson et al.

2001a). It was first found in the Caribbean at Pt. Lisas in Trinidad in 1990 (Agard et al. 1992), and over the next decade spread along the west coast of Trinidad and the Venezuelan coast, and appeared in Kingston Harbor, Jamaica and on the west coast of Florida from Tampa Bay to Charlotte Harbor (Benson et al. 2001a). In the Caribbean it grows abundantly on pilings, seawalls and other artificial structures and on the roots of red mangrove *Rhizophora mangle,* where it may exclude native fouling species. It is generally found in estuarine waters with salinities of 27-33 ppt but can survive salinities at least as low as 16-20 ppt, and its larvae are planktonic for 2-3 weeks (Agard et al. 1992; Benson et al. 2001a). It probably arrived in the Caribbean in ballast water, or possibly as hull fouling, carried through the canal.

Three species of Indo-Pacific oyster that have been collected in the western Atlantic were probably introduced as hull fouling or in ballast water after passing through the canal. An unidentified pearl oyster in the exclusively Indo-West Pacific genus *Electroma* was collected near Santa Marta on the Atlantic coast of Colombia in 1983. It has proliferated and spread to other sites in Colombia and Venezuela (Borrero & Díaz 1998). The black-lipped pearl oyster *Pinctada margaritifera* is found from the Persian Gulf to the South Pacific and Japan, and in the Gulf of California and Panama in the eastern Pacific. At least four specimens were collected on reefs at different sites off southeastern Florida between 1990 and 2003 (Frank 1993; Carlton & Ruckelshaus 1997; Benson et al. 2001b; U.S. Geological Survey 2005). Three specimens of *Hyotissa hyotis*, a graphaeid oyster reported from the Indian Ocean and the western and eastern Pacific, were collected at three sites off southeast Florida in 2001-05 (Bieler et al. 2004; U.S. Geological Survey 2005).

Martesia cuneiformis is a wood-boring clam that is native to the Western Atlantic from North Carolina to Brazil (Turner 1955). It was collected at Balboa near the Pacific entrance to the canal prior to 1955 (Turner 1955),[23] and apparently not since. Turner (1955) suggested that it had arrived via the canal, and Carlton (1985) suggested that it had traveled either in ballast water or bored into a hull.

Shipworms are highly-modified, wood-boring clams, and in the past they were frequently passengers in and major pests of wooden ship hulls. Since many *Bankia* species have a free-swimming larval stage that may last up to a month (Clench & Turner 1946), these species may have traveled in ballast water, especially in recent decades. Collection records suggest that four species of

---

[23] Turner's record may be based on specimens that the U.S. Navy collected in fouling panels at Balboa in 1944-47, in 1948 and in 1949, and identified only as *Martesia* (U.S. Navy 1951).

shipworms have been transported through the canal from the Caribbean to the eastern Pacific and that two species have been transported from the Indo-West Pacific region eastward through the canal, although some species' native ranges may be masked by movement around the globe at an early date in ship hulls.

The western Atlantic shipworm *Bankia fimbriatula* ranges from Florida to Brazil. It was collected at Cristobal near the Caribbean entrance to the canal in 1944 and at Balboa on the Pacific side in 1943 and 1944 (Bartsch 1944 as *B. canalis*; Clench & Turner 1946). *Bankia cieba* is recorded from Cuba to Colombia and was collected at Balboa near the Pacific entrance to the canal by 1946 and in the Gulf of California by 1971 (Clench & Turner 1946; Turner 1971). *Bankia destructa* is recorded from Honduras to Venezuela and was collected at Puerto Armuelles in Pacific Panama by 1946 and near Mazatlan, Mexico in 1978-79 (Clench & Turner 1946; Hendrickx 1980). *Teredo bartschi*, which ranges from South Carolina to the Gulf of Mexico and Caribbean,[24] was collected in Pacific Mexico at La Paz by 1971 and near Mazatlan in 1978-79, in California in Los Angeles Harbor in 2000, and in Hawaii (Hendrickx 1980; Carlton 1992; Benson et al. 2004; A.N. Cohen unpublished data 2000). The southwestern Pacific shipworm *Teredo furcifera* was collected at New Jersey on the U.S. east coast in 1974 (Benson et al. 2004) and in Florida prior to 1997 (Carlton & Ruckelshaus 1997); and the Indo-West Pacific shipworm *Lyrodus medilobatus* was collected in eastern Florida in the Indian River region by 1983 (Mikkelsen et al. 1995; Carlton & Ruckelshaus 1997).

Arthropods

Seven barnacles appear to have been introduced through the canal with shipping. The Indian Ocean barnacle *Amphibalanus* (=*Balanus*) *amphitrite* became established in Hawaii in 1902. It reached the Pacific coast of the Americas at Los Angeles Harbor by 1914. By the late 1940s it was reported in the western Atlantic, where it has been collected at many sites from the Caribbean and Gulf of Mexico and as far north as Massachusetts (Carlton 1979; Cohen & Carlton 1995). Its introduction to the western Atlantic could have occurred either eastward through the canal or westward from England, where it was established at least by 1937, and possibly as early as 1917 (Bishop 1950). In 1974 it was collected near both entrances to the canal (Spivey 1976). Another Indo-West Pacific barnacle, *Amphibalanus* (=*Balanus*) *reticulatus*, was collected in the Gulf of Mexico in 1952, where it has become a dominant fouler on

---

24 Carlton (1992) describes *T. bartschi* as a tropical Atlantic species introduced to western Mexico, but Carlton and Ruckelshaus (1997) list it as possibly a Pacific species introduced to Florida.

offshore drilling platforms from central Louisiana to Eastern Texas (Britton & Morton 1989), and later spread to Florida (Carlton & Ruckelshaus 1997). It was collected at Coco Solo near the Atlantic entrance to the canal in 1974 and at several sites near both entrances in 2000 (Spivey 1976; F.B. Pitombo, A.N. Cohen, unpublished data). A third Indo-West Pacific barnacle, *Balanus trigonus*, was collected in Limon Bay in 1966-67, at Fort Randolph in 1974 and on Galeta Reef by 1975, all near the Atlantic end of the canal; in Venezuela and Florida by the early 1990s, and in Panama near both ends of the canal in 2000 (Bayer et al. 1970; Spivey 1976; Cubit & Williams 1983; Carlton & Ruckelshaus 1997; Benson et al. 2001b; F.B. Pitombo, A.N. Cohen, unpublished data). All three species are common ship and harbor foulers, and could have traveled to the Atlantic via the canal either in hull fouling or in ballast water. Collection records suggest that after they became established in the Caribbean, these barnacles then traveled back through the canal to Panama's Pacific coast.

The Atlantic barnacle *Balanus calidus* is widespread in the Caribbean and the Gulf of Mexico and to North Carolina. It was collected in Miraflores Lower Locks in 1974 (Spivey 1976). *Fistulobalanus* (=*Balanus*) *pallidus* is reported from the west coast of India, West Africa, the Mediterranean and the Caribbean with scattered records south to Argentina. It was abundant in the upper and lower Miraflores Locks in 1972 and 1974 (Jones & Dawson 1973; Spivey 1976). These collection records suggest that these barnacles were transported from the Caribbean or western Atlantic through the canal to the locks on the Pacific side. *Amphibalanus* (=*Balanus*) *eburneus* is also native to the western Atlantic where it has been collected from Massachusetts to Brazil, and was introduced to Atlantic Europe, the Mediterranean, the Black Sea, India, Hawaii and Japan (Carlton 1979). It was reported from Caribbean Panama near the entrance to the canal both as fossil material and as living barnacles collected at various sites in 1966, 1974 and 2000. It was collected in the eastern Pacific in the Gulf of California in 1959, on the mainland west coast of Mexico in 1963 and 1968, at Balboa in Pacific Panama in 1964, and near Los Angeles in southern California in 2000 (Spivey 1976; Carlton 1979; F.B. Pitombo, A.N. Cohen, unpublished data 2002). While it's possible that all these eastern Pacific records derived from a secondary introduction from Hawaii (where *A. eburneus* has been collected since 1929) or from an introduction with imported Atlantic oysters, it seems more likely that some at least were the result of transport through the canal as hull fouling or in ballast water, as suggested by Carlton (1985).

The high intertidal barnacle *Chthamalus proteus* is native to the western Atlantic from eastern Florida to Brazil, including the Caribbean coast of

Panama. It was discovered on Oahu in Hawaii in 1995 (Southward et al. 1998), where it is thought to have become established sometime after the last thorough barnacle survey in 1973, presumably arriving via the Panma Canal in hull fouling or possibly in ballast water. In Hawaii it has been collected on the four largest islands and Midway Atoll. It has also been reported in Apra Harbor on Guam, probably as a secondary introduction from Hawaii (Pauley et al. 2002). The Australasian barnacle *Elminius modestus* was discovered in England in 1944, presumably on vessels from Australia or New Zealand. Bishop (1951) suggested that the war-time practice of vessels traveling in convoys could have enhanced the potential for establishment, by the simultaneous transport of a larger area of hull surfaces and therefore delivery at one time of a larger number of barnacles. Transport from the southwestern Pacific to England could have occurred either through the Panama Canal or the Suez Canal, so this introduction is not included in the tables and tallies in this chapter.

Two peracarid crustaceans that were recently introduced from the western Atlantic into northeastern Pacific waters probably arrived in ballast water or hull fouling via the canal. The amphipod *Gammarus daiberi* is native to north-western Atlantic estuaries, where it occurs in waters of low to nearly fresh salinities from Delaware Bay to South Carolina. It was introduced to the Hudson River in New York in 1975, possibly in ballast water, and collected from San Francisco Bay in 1983 where it is established in low salinity tidal waters (Cohen & Carlton 1995; U.S. Geological Survey 2005). The Caribbean isopod *Pleurocope floridensis* is known from the Turks and Caicos Islands, Belize and off Florida (Kensley & Schotte 1989). It was collected in southern California in Los Angeles/Long Beach harbors in 2000 (Fairey et al. 2002).

The western Atlantic palaemonid shrimp *Leander paulensis* is reported from Florida to Brazil, including collections in Gatun Lower and Middle Locks in 1972-74. Its collection in Miraflores Upper Locks in 1972 suggests transport in ballast water or hull fouling (Abele & Kim 1989).[25]

The collection records of several tropical crabs suggest transport through the canal in hull fouling or ballast water, four in a westward direction and three eastward. The western Atlantic portunid crab *Callinectes exasperatus* ranges from northern Florida to Brazil; nine specimens were collected in the creek draining the Miraflores Third Lock lagoon in 1972 (Abele & Kim 1989). The western Atlantic xanthid crab *Eurypanopeus dissimilis* also ranges from Florida to Brazil, and was frequently collected in the Gatun Lower and Middle Locks in

---

25 In San Francisco Bay we frequently collect the Asian palaemonid shrimp *Palaemon macrodactylus* in fouling (A.N. Cohen unpublished data).

1972 and 1974. It had become established in the Miraflores Third Lock lagoon by 1971 (McCosker & Dawson 1975; Carlton 1985; Abele & Kim 1989). Another Caribbean xanthid, *Panopeus rugosus*, was collected in Miraflores Upper Locks in 1972 and 1974 (Abele & Kim 1989). The Atlantic grapsid crab *Goniopsis cruentata* ranges from Bermuda to Brazil and from Senegal to northern Angola. It was collected in the creek draining the Miraflores Third Lock lagoon in 1972 and in the Miraflores Locks spillway in 1972 and 1973 (Abele & Kim 1989). The eastern Pacific porcelain crab *Petrolisthes robsonae* is recorded from Mexico, El Salvador, Panama and Ecuador, and it is common in both Miraflores Upper and Lower Locks. A single specimen was collected in the Gatun Lower Locks before 1960 (Gore & Abele 1976), and additional specimens were collected there in 1974 (Abele & Kim 1989). The eastern Pacific grapsid crabs *Sesarma rhizophorae* and *Sesarma sulcatum* are common in mangroves on the Pacific coast of Panama but are known in the Caribbean only near the entrance to the canal (Abele 1972b,c).

The brackish water xanthid crab *Rhithropanopeus harrisii* is native to the northwestern Atlantic from New Brunswick, Canada to Veracruz, Mexico (Rathbun 1930), and has also been reported in northeastern Brazil (Abele & Kim 1989).[26] It was introduced into Europe by 1874. It was collected in San Francisco Bay, California in 1937 and subsequently found in a few bays in Oregon (Cohen & Carlton 1995). Five specimens were collected in the Pedro Miguel Locks in 1969 (Abele & Kim 1989). It may have been introduced to San Francisco Bay with shipments of Atlantic oysters, *Crassostrea gigas*, although most of these shipments were made in the late 1800s and the first decade of the 1900s. Alternately, it could have been transported by way of the canal in ships' ballast water or possibly hull fouling (Carlton 1985). It could have arrived in Pedro Miguel Locks from either direction, transported in ballast water or hull fouling (Carlton 1985).

The grapsid crab *Hemigrapsus sanguineus* is native to the Asia coast from southern Russia to Hong Kong, and was collected in the northwestern Atlantic at New Jersey in 1988. It has since spread along that coast from Maine to North Carolina (Lohrer 2001; Benson et al. 2001b, 2004). *H. sanguineus* has a planktonic larval stage of at least 25 days and is usually considered a ballast water introduction, though Lohrer (2001), argues that its habit of nestling in dock

---

[26] Although Abele (1976c at page 4) earlier reported this species to be "widely distributed throughout the American tropics," that appears to be an error, since other authors including Abele and Kim (1989) report the more restricted tropical distribution listed here (in Brazil and north of Veracruz in the Atlantic, and a single collection in Pedro Miguel Locks in the tropical Pacific).

fouling and a report of a congeneric crab found on the outside of vessel hulls indicates that it could also have traveled in hull fouling.

Bryozoans

Powell (1971) reported that the Atlantic bryozoans *Electra monostachys*, recorded from Europe and the western Atlantic from Canada to Brazil, *Membranipora annae*, recorded from West Africa, and *Bugula stolonifera*, recorded from Europe, West Africa and from Massachusetts to Brazil, had been collected in the Pacific Ocean only in Panama Bay in 1924 and 1969. *E. monostachys* and *M. annae* (as *Acanthodesia serrata*) were also collected on test panels at Balboa Harbor near the Pacific entrance to the canal in 1944-47 (U.S. Navy 1951); and *B. stolonifera* became established in Hawaii on Oahu, where it was first collected in 1940 (U.S. Geological Survey 2005), and in California in Los Angeles Harbor and San Francisco Bay, where it was collected by 1978 (Cohen & Carlton 1995). Powell considered the distribution of these three species to be "compelling evidence" that they had been introduced through the canal in hull fouling. Carlton (1985), noting the long-lived planktonic larvae of *Membranipora* and *Electra* species, listed *Electra monostachys* as a possible ballast water introduction through the canal, and suggested that *Membranipora savartii*, which is present at both ends of the canal as well as elsewhere in tropical waters, may be another bryozoan that has been introduced through the canal. *Electra bengalensis*, known from West Africa, the Indian Ocean and Australia and collected in the eastern Pacific only in Panama Bay in 1950 and 1969, may be another case of transport through the canal in either hull fouling or ballast water, or perhaps it was released from a ship heading eastward before it passed through the canal, as suggested by Martin et al. (1970).

The Atlantic ctenostome bryozoan *Anguinella palmata* is recorded from northern Europe, West Africa and the western Atlantic from Massachusetts to the Gulf of Mexico, Puerto Rico and Brazil. It was collected in the Pacific Ocean in Peru, at Panama City, in Mexico and in southern California in 1933-1942 (Osburn 1953), and was still present in Panama near the Pacific entrance to the canal in 2002 (A.N. Cohen unpublished data). It has been collected in San Francisco Bay, California since 1993 (Cohen & Carlton 1995), and has also been reported from Australia and New Zealand (Allen 1953; Gordon 1967). It is a brackish water species that has been collected in salinities of 13-32 ppt (Cohen & Carlton 1995). It was most likely transported through the canal in hull fouling. The bryozoan *Watersipora subtorquata* is native to the western Pacific. It was established in the Indian River Lagoon in southeastern Florida by 1976, probably having been introduced in hull fouling (Mook 1976; Carlton & Ruckelshaus 1997).

Tunicates

The fouling tunicates *Botrylloides perspicuum* and *Diplosoma virens* are native to the Indo-West Pacific region and have been introduced in the Caribbean to the Belize barrier reefs (Goodbody 2000, 2004). The common western Pacific tunicate *Cnemidocarpa irene* was introduced to the island of Guadeloupe in the Caribbean by 1984 (Monniot & Monniot 1994). The southeastern Pacific tunicate *Polyandrocarpa zorritensis* was described from the northern end of Peru in the eastern Pacific in 1931. It was collected in Italy in 1974, where it is established (Brunetti & Mastrototaro 2004), and was collected in the northwestern Atlantic by the 1990s and in Puerto Rico in 2002 (Lambert & Lambert 1998; G. Lambert, A.N. Cohen unpublished data). It was also reported from southern California in 1994, Oahu in Hawaii in 1997, Baja California in 2000, San Francisco Bay by 2005, and Japan since the 1990s (Lambert & Lambert 1998, 2003; Ruiz et al. 2005). Another fouling tunicate, *Botryllus schlosseri*, is native to the North Atlantic and appeared in Australia by 1928 and in Pacific Panama at Balboa Harbor and in California in San Francisco Bay by 1944-47 (U.S. Navy 1951; Carlton 1979; Kott 1985; Cohen & Carlton 1995).

Fish

Gobiosoma nudum is a burrow-dwelling marine goby native to the eastern tropical Pacific with records from Mexico to Peru (Rubinoff & Rubinoff 1969; Fishbase 2005). In 1962 six specimens were collected at Galeta Reef, near the Caribbean end of the canal, though none have been reported from the Atlantic since then (Rubinoff & Rubinoff 1969; McCosker & Dawson 1975; Fishbase 2005). Fish that were placed in fresh water became moribund or died within 2-3 hours (Rubinoff & Rubinoff 1969; McCosker & Dawson 1975), so they probably didn't move through the canal on their own or in hull fouling. However, they could have traveled as eggs laid in fouling, since eggs hatched successfully though at slightly reduced rates after eight hours experimental exposure to fresh water (Rubinoff & Rubinoff 1969), or traveled in ballast water tanks (suggested as a possibility by Carlton 1985), sea chests or other components of ships' seawater systems. One of the same mechanisms may have transported the bearded goby *Barbulifer ceuthoecus* through the canal in the other direction. This stenohaline species is native to the Caribbean from the Bahamas to Brazil and is moderately abundant in lower Limon Bay and other Caribbean coastal waters of Panama (McCosker & Dawson 1975). A single specimen was collected on Panama Reef near the Pacific entrance to the canal in 1973 (McCosker & Dawson 1975). McCosker & Dawson (1975) and Carlton (1985) suggested that it had probably been transported in ballast water.

The oyster blenny *Hypleurochilus aequipinnis*, whose range includes both the eastern and western tropical Atlantic, was collected in Gatun Locks in 1972 and

1974, and in the Miraflores Third Lock lagoon in 1971 and 1976, where it is established (McCosker & Dawson 1975; Fishbase 2005). It is more tolerant of freshwater than *Gobiosoma nudum*, with over 65% survival after 10 hours of exposure and about 8% survival after 30 hours of exposure (McCosker & Dawson 1975), so that it could have been transported as either adults or eggs in hull fouling (as suggested by McCosker & Dawson 1975), or in a ballast tank (suggested by Carlton 1985) or some other part of a ship's seawater system. Spring and Gomon (1975) suggested that *Hypleurochilus* migrated through the canal, but with only a couple of records of this species in the canal system, and 30 hours exposure to fresh water causing over 90% mortality, migration seems much less likely than transport on a ship.

A single 16-centimeter specimen of the yellow jack *Gnathodon speciosus*, a tropical Indo-Pacific and eastern Pacific species, was collected in Gatun Lower Locks in 1968 (McCosker & Dawson 1975). It might have been transported in ballast water, but since this species is often kept in aquariums (Fishbase 2005), it might instead have been released from one into Limon Bay.

3.4 Transport by pathways toward, from and through the canal

The species discussed in this chapter probably represent only a small fraction of the species introduced by voyages that transit the canal. Introductions between the Caribbean and Panamic regions probably account for some of the numerous species reported to occur on both sides of the isthmus (see Table 2). Many cosmopolitan or circumtropical species found near one or both entrances to the canal, including sponges, hydroids, serpulid worms, bryozoa, tunicates and other species (e.g. De Laubenfels 1936; A.N. Cohen, unpublished data 2002), may represent introductions between these waters and distant regions by ships that passed through the canal. Given the substantial role of the canal in global commerce, some significant portion of the ship-borne introductions between distant regions may also involve voyages through the canal.

Of the introductions discussed in this chapter, 25 or 26 species appear to have been introduced by ships passing westward through the canal from the Caribbean to the Pacific, and six species by ships passing eastward (Table 12). Many of these introductions are represented by only one or two records, and may not be established, although limited sampling could also account for the sparse records of some species. Despite the limitations of these data, they do suggest that on the Near-Canal-Near Pathway either organism transport or the survival of organisms after release is greater from the Caribbean to the Pacific than in the opposite direction.

Table 12. Possible introductions via the Near-Canal-Near Pathway (from the Caribbean through the canal to the Panamic region, or the reverse). a = If *Mytilopsis zeteki* or *M. adamsi* are junior synonyms of *M. sallei*, b = After introduction to the Caribbean from the Indo-West Pacific.

Caribbean to Panamic Region	Panamic to Caribbean Region
Polysiphonia atlantica	*Chelophyes contorta*
Geodia gibberosa	*Muggiaea atlantica*
Haliclona coerulescens	*Petrolisthes robsonae*
Chelophyes appendiculata	*Sesarma rhizophorae*
Muggiaea kochi	*Sesarma sulcatum*
Alexania floridana	*Gobiosoma nudum*
Conus mus	
Echinolittorina ziczac	
Littoraria angulifera	
Martesia cuneiformis	
Bankia cieba	
Bankia destructa	
Bankia fimbriatula	
Mytilopsis sallei[a]	
Amphibalanus amphitrite[b]	
Amphibalanus reticulatus[b]	
Balanus trigonus[b]	
Leander paulensis	
Callinectes exasperatus	
Eurypanopeus dissimilis	
Goniopsis cruentata	
Panopeus rugosus	
Barbulifer ceuthoecus	
Hypleurochilus aequipinnis	
Lophogobius cyprinoides	
Lupinoblennius dispar	

Seven of the Caribbean species introduced through the canal have been collected on the Pacific Coast only or mainly in the Miraflores Third Lock Lagoon and its drainage channel (Table 13). In some ways, the environment in the lagoon resembles certain Caribbean habitats more than any found on Panama's Pacific coast. The small tidal range in the lagoon is closer to the tidal range in the Caribbean than in the eastern Pacific. The subtidal oyster-encrusted rocks of the lagoon walls are also an unusual habitat on the Pacific coast due to the great tidal range there, and are more typical of the Caribbean. In addition, the red mangrove (*Rhizophora mangle*) that grows in the lagoon's shallow areas also grows in the Caribbean. These habitat similarities, combined with an initial absence of biota in the lagoon when it was created in the 1940s and a scarcity of Pacific species adapted to these conditions, may have made it easier for estuarine oyster-reef or mangrove-associated species from the Caribbean to become established in the lagoon (Rubinoff & Rubinoff 1968). Dependence on these habitat conditions and an inability to compete with the organisms of

Panama Bay in their native habitat may keep the Caribbean species restricted to the lagoon (McCosker & Dawson 1975).

Table 13. Exotic species collected in the Miraflores Third Lock Lagoon or its drainage channel.

Species	Collection Date(s)	Native Range	Other Eastern Pacific Records	References
Seaweeds				
Polysiphonia atlantica	by 1975	Atlantic	none	McCosker & Dawson 1975 (as *P. macrocarpa*)
Crabs				
Callinectes exasperatus	1972	Northern Florida to Brazil	none	Abele & Kim 1989
Eurypanopeus dissimilis	1971	Western Florida to Brazil	none	McCosker & Dawson 1975
Goniopsis cruentata	1972	Bermuda to Brazil, and Senegal to Angola	Miraflores Locks spillway in 1972 and 1973	Abele & Kim 1989
Fish				
Hypleurochilus aequipinnis	1971	Eastern and western tropical Atlantic	none	McCosker & Dawson 1975
Lophogobius cyprinoides	1967, 1968, 1970, 1971	Caribbean	Miraflores Locks in 1937	Rubinoff & Rubinoff 1968; Dawson 1970; McCosker & Dawson 1975
Lupinoblennius dispar	1967, 1968, 1971	Caribbean	none	Dawson 1970; McCosker & Dawson 1975
Vieja maculicauda	by 1975	Atlantic slope	Miraflores Lake since 1921, Pedro Miguel and Miraflores Locks in 1937, Rio Grande in 2002	Hildebrand 1939; McCosker & Dawson 1975; Smith et al. 2004; Fishbase 2005

Twenty species are discussed in this chapter as examples of introductions from distant regions through the canal into the Caribbean or Panamic regions, on the Distant-Canal-Near Pathway (Table 14). Eastward introductions dominate with twelve species, compared to eight species transported westward. Two of the species listed here as westward transported, the bryozoans *Bugula stolonifera* and *Electra monostachys*, have a broad range in the Atlantic that includes Caribbean waters, and thus may have traveled instead on a Near-Canal-Near Pathway; and one of the westward species, the crab *Rhithropanopeus harrisii* which was collected in Pedro Miguel Locks, could in fact have been introduced on the Distant-Near-Canal pathway while traveling eastward from its introduced range on the U.S. Pacific Coast.

Species can also be introduced from the Caribbean or Panamic region through the canal to distant regions, the Near-Canal-Distant Pathway, but only three examples of this are reported here, the introductions of the Caribbean barnacle *Chthamalus proteus* into Hawaii, the Caribbean isopod *Pleurocope floridensis* into southern California, and the eastern Pacific tunicate *Polyandrocarpa zorritensis* into the Mediterranean (Table 15). The much smaller number of species described for the Near-Canal-Distant Pathway compared to the Distant-Canal-Near Pathway (3 versus 20) is partly due to issues of evidence. Eight of the twenty introductions into the Caribbean or Panamic region were first collected in the canal or close to one of the canal entrances, and these collection data are part of the reason for concluding that these species were introduced by ships passing through the canal. For an organism introduced from the Caribbean or Panamic region to a distant region, the site of its initial collection in that distant region provides no information on the likelihood of passage through the canal.

Table 14. Possible introductions via the Distant-Canal-Near Pathway (from a distant region through the canal to nearby waters).

Westward to Panamic Region	Eastward to Caribbean Region
Microciona atrasanguinea	*Lensia challengeri*
Teredo bartschi	*Electroma* sp.
Amphibalanus eburneus	*Hyotissa hyotis*
Rhithropanopeus harrisii	*Lyrodus medilobatus*
Bugula stolonifera	*Perna viridis*
Electra monostachys	*Pinctada margaritifera*
Membranipora annae	*Amphibalanus amphitrite*
Botryllus schlosseri	*Amphibalanus reticulatus*
	Balanus trigonus
	Botrylloides perspicuum
	Cnemidocarpa irene
	Diplosoma virens

Nineteen species are discussed as examples of introductions between distant regions, with fourteen traveling westward through the canal and five traveling eastward (Table 16). The difference may be explained by the large number of invasions known from San Francisco Bay (Cohen & Carlton 1995), including eleven of the westward travelers listed here. Some of these, such as the four sponge species, might have been introduced with shipments of Atlantic oysters, rather than with shipping through the canal. One of the listed species, the bryozoan *Anguinella palmata*, had a wide initial range in the Atlantic including Caribbean waters, and was collected in the Pacific at several sites at about the same time, including sites both within and outside of the Panamic region, so that it might have alternately or additionally traveled on a variety of pathways.

For most ship-borne introductions between the Atlantic and the Eastern Pacific, and between the Pacific and the Western Atlantic, passage through the Panama

Canal is substantially shorter and therefore usually likelier than other routes. For introductions between other distant regions, other routes provide similar or shorter connections—for example, for many locations in the Eastern Atlantic and Western Pacific, voyages through the Suez Canal or around Cape Horn are roughly comparable in distance to a voyage through the canal, and in these cases distance provides no basis for determining whether the introduction occurred through the Panama Canal.

Table 15. Possible introductions via the Near-Canal-Distant Pathway (from nearby waters through the canal to a distant region).

Westward from Caribbean Region	**Eastward from Panamic Region**
Chthamalus proteus *Pleurocope floridensis*	*Polyandrocarpa zorritensis*

Table 16. Possible introductions via the Distant-Canal-Distant Pathway (between distant regions

Westward	**Eastward**
Polysiphonia denudata *Clathria prolifera* *Halichondria bowerbanki* *Haliclona loosanoffi* *Prosuberites* sp. *Blackfordia virginica* *Maeotias marginata* *Boccardiella ligerica* *Marenzelleria viridis* *Tenellia adspersa* *Gammarus daiberi* *Rhithropanopeus harrisii* *Anguinella palmata* *Botryllus schlosseri*	*Stiliger fuscovittatus* *Teredo furcifera* *Zeuxo kuriliensis* *Hemigrapsus sanguineus* *Watersipora subtorquata*

Springer and Gomon (1975) discussed an example of the Distant-Near-Canal pathway (Table 17), in which a vessel introduces an organism into waters near the canal before passing through the canal. The blenny *Omobranchus punctatus* is native to the Indo-West Pacific, ranging from the Persian Gulf and eastern Indian Ocean to eastern Australia and Japan. There is also an outlying population at Delagoa Bay in the western Indian Ocean and a questionable record from Fiji. It appeared in Trinidad in 1930 and was abundant around docks there in the 1960s. It spread to the nearby Venezuela coast by 1961, and was collected in Limon Bay and in the Gatun Lower and Middle Locks starting in 1966 (Bayer et al. 1970; Springer & Gomon 1975; McCosker & Dawson 1975; Fishbase 2005). Although Bayer et al. (1970) stated that *O. punctatus* was introduced to Trinidad in hull fouling on a ship that had passed through the canal from the Pacific, Springer and Gomon argued on morphological and historical grounds that it was transported from India to Trinidad in the ballast tanks of ships carrying indentured laborers from India prior to the opening of the canal, and later transported from Trinidad to Panama by a ship on its way through the canal.

Table 17. Possible introductions via the Distant-Near-Canal Pathway (from a distant region to nearby waters before a ship transits the canal).

Westward to Caribbean Region	Eastward to Panamic Region
Omobranchus punctatus	*Electra bengalensis*
	Butis koilomatodon

The cheilostome bryozoan *Electra bengalensis*, known from West Africa and India and collected in Panama Bay in 1950, is another possible case of a Distant-Near-Canal introduction. Martin et al. (1970) argued for this pathway, suggesting that transport could have been on a Japanese ship, since Japanese vessels account for a significant portion of the traffic through the canal. A third possible case is the collection of a single specimen of the Indo-West Pacific eleotrid fish *Butis* (=*Prionobutis*) *koilomatodon* in Miraflores Upper Locks in 1972. This fish's native range is from Mozambique through the Indian Ocean to Singapore, China, Taiwan and the Philippines. Dawson (1973) argued that the direction of transport was probably from Asia across the Pacific to Panama, based on dominant trade routes and information from Panama Canal Company employees that ships often discharge some ballast water either before entering the canal or (illegally) within the locks (see Section 2.1 regarding ballast water regulations and practices in and near the canal). Alternately, either *E. bengalensis* or *B. koilomatodon* could have been transported from their native Eastern Atlantic or Indian Ocean range across the Atlantic and through the canal, in which case they should be added to Tables 10 and 14 as examples of the Distant-Canal-Near Pathway.

Another possible pathway is the Canal-Near-Distant Pathway, in which a vessel passes through the canal, then acquires an organism from nearby waters which it carries and releases into a distant region. No examples of this pathway have been reported. The small number of records for these last two pathways probably results from problems of evidence. All three of the cases described for the Distant-Near-Canal Pathway involve organisms collected in the canal or near the canal entrances, suggesting that they were introduced from a ship transiting or about to transit the canal. Had these organisms been released from such a ship earlier, so that they were collected at sites within the Caribbean or Panamic regions that were more distant from the canal, there would be no strong reason to think that their introduction involved a ship heading for a canal transit. For an organism traveling the Canal-Near-Distant Pathway, the site of initial collection in the distant, introduced region can give no indication of whether the ship had transited the canal before picking up the organism.

Species that are known only from the canal or its near vicinity are likely to be introductions from some other, distant region where they are not yet described. If introduced by a ship on its way through the canal, these would be additional examples of the Distant-Canal-Near Pathway (Table 14) or the Distant-Near-Canal Pathway (Table 17), depending on their origins and the points where they were initially introduced into the canal region. Five examples are listed in Table 18.

Table 18. Species of unknown origin collected only from the Panama Canal and vicinity.

Species	Records	References
Bivalves		
Bankia zeteki	On the Pacific side in Miraflores? Locks by 1921, Balboa in 1943, Puerto Armuelles by 1946 and Mazatlan in 1978-79; on the Atlantic side at Cristobal in 1944 and Coco Solo by 1946	Bartsch 1921, 1944, Clench & Turner 1946, Hendrickx 1980
Shrimp		
Palaemonetes schmitti	Described from Miraflores Upper Locks in 1950; by 1989 additionally reported only from tide pools in the Miraflores Spillway in 1973, Miraflores Lower Locks in 1974 and 1976, and 2 nearby beaches, Venado Beach and San Francisco Beach	Abele & Kim 1989
Fish		
Gobiosoma hildebrandi	Gatun Locks since 1935; Pedro Miguel Locks since 1937; Miraflores Lake and Miraflores Locks in the 1960s-70s; near Atlantic entrance since 1971; below Miraflores spillway in 1972-73; Barbados in 1983	Hildebrand 1939; McCosker & Dawson 1975; Fishbase 2005
Gobiosoma homochroma	Pedro Miguel Locks in 1937 and 1960s-70s; Miraflores Lake and Miraflores Locks in the 1960s-70s; near the Pacific entrance in 1970	Hildebrand 1939; McCosker & Dawson 1975; Fishbase 2005
Guavina micropus	Miraflores Locks and stream draining the Miraflores Third Lock lagoon by 1975; Pacific Costa Rica in 1970; Pacific Guatemala in 1971	McCosker & Dawson 1975; Fishbase 2005

The shipworm *Bankia zeteki* was described in 1921 from specimens collected in "timber of the Canal Locks at Balboa" (Bartsch 1921), presumably the Miraflores Locks. It was collected in fouling panels at Balboa in 1943 and 1944, and at Cristobal near the Atlantic entrance to the canal in 1944 (Bartsch 1944), with additional records at Puerto Armuelles in Pacific Panama, Coco Solo in Atlantic Panama (Clench & Turner 1946), and near Mazatlan in Pacific Mexico in 1978-79 (Hendrickx 1980). It was probably introduced to the canal region in hull fouling, although Clench and Turner (1946) interpreted its distribution as indicating its spread "with commerce" from the eastern Pacific to the Caribbean, while Hendrickx (1980) described it as having dispersed through the canal from the Caribbean to the eastern Pacific. The shrimp *Palaemonetes schmitti* was described in 1950 from Miraflores Locks, and by 40 years later had been found only in the locks, the lock spillway and two nearby beaches.

McCosker and Dawson (1975) reported on three fish collected in or near the canal whose native region is unknown. The estuarine goby *Gobiosoma* (= *Garmannia*) *homochroma* was collected in the Pedro Miguel Locks in 1937. It was found there, in Miraflores Lake and Locks, and in Panama Bay near the canal entrance in the 1960s-70s (Hildebrand 1939; McCosker & Dawson 1975; Fishbase 2005). *Gobiosoma* (=*Garmannia*) *hildebrandi* was collected and described from the Gatun and Pedro Miguel Locks in 1935-37, and collected in these sites plus Miraflores Lake and Locks in the 1960s-70s. Adults and juveniles have been collected on the Pacific side from the pool and drainage channel below the Miraflores Spillway, and on the Caribbean coast from several sites near the canal entrance. There is also a 1983 record from tidepools on Barbados (Hildebrand 1939; McCosker & Dawson 1975; Fishbase 2005). *Guavina micropus* was collected from Miraflores Locks and the stream draining the Miraflores Third Lock lagoon (McCosker & Dawson 1975), and from the Pacific coasts of Costa Rica and Guatemala (Fishbase 2005). The paucity of records outside of the canal area in either the Caribbean or Panamic regions suggests that these fish are native to other parts of the world and were introduced to the canal region by a ship passing through the canal, most likely via ballast tanks, sea chests or other components of ships' seawater systems.

3.5 Introductions of freshwater species into the canal

Jones and Rützler (1975) reported that a freshwater sponge from the eastern United States, *Trochospongilla leidii,* was abundant in the Gatun Upper Locks and the Pedro Miguel Locks (Table 19). They suggested that it could have been introduced either in ships' ballast water or as a release from an aquarium. Despite the popularity of aquarium-keeping as a hobby in the Canal Zone (Loftin 1965), this seems an unlikely introduction mechanism for a freshwater sponge.

Cordylophora caspia is a freshwater hydroid that is native to the Caspian and Black Sea regions, but has long been widely distributed around the world. It was abundant in Gatun Locks in 1935 and 1974, and was collected in Gatun Lake in 1974 and Pedro Miguel Locks in 1975 (Hildebrand 1939; Jones & Rützler 1975; Jones 1976). As it can tolerate some salinity, it could have been introduced either in ship fouling or in ballast water.

The tanaid *Sinelobus stanfordi* was collected in Gatun Middle and Upper Locks in 1974, in Gatun Lake in 1974 and 2002, and in Pedro Miguel Locks and Miraflores Upper Locks in 1974-75 (Jones & Rützler 1975; Jones 1976; A.N. Cohen unpublished data). This species has been reported from a remarkable geographic and habitat range, including fresh, brackish, marine and hypersaline water, in the Arctic, northern temperate and eastern tropical Pacific, northern and southern temperate and tropical Atlantic, and tropical Indo-West Pacific regions (Sieg 1986). Locations include Brazil, the West Indies, England, the Mediterranean, Senegal, South Africa, the Tuamotu Archipelago, the Kurile Islands, Hawaii, a few sites on the Pacific coast of North America from San Francisco Bay to British Columbia, and the Galapagos Islands (Cohen & Carlton 1995). Its broad habitat distribution suggests that a species complex is involved, and its wide, disjunct geographic distribution suggests that human-associated transport by ships or other mechanisms has occurred. The primarily freshwater form reported in the canal system is thus probably an introduction from some other part of the world. It could have been introduced in ballast water or possibly in hull fouling, since it dwells in tiny mud tubes that are often constructed in fouling and may have some tolerance of salinity.

The isopod *Uromunna* (=*Munna*) *reynoldsi* is believed to be native to the southeastern United States where, however, it has been reported only from collections on oak leaves from the surface of a salt marsh and on oak leaves from the bottom of a tidal creek in Georgia (suggesting that it was probably washed in from fresher water) in 1964 and 1965, and from brackish water (2-12 parts per thousand at the time of collection) in Lake Ponchartrain, Louisiana in 1973 and 1975 (Frankenberg & Menzies 1966; Schultz 1969, 1979; Kensley & Schotte 1989). It was collected in the Gatun Upper Locks and Pedro Miguel Locks in 1974-75 (as *Munna* sp. in Jones & Rützler 1975; Jones 1976) and from Miraflores Locks by 1979 (Schultz 1979). Schultz (1979) described it as "apparently established" in the canal, based on the collection in each lock of males and females including gravid females, the total examined from the canal being 17 males and 30 females, including nine gravid. Carlton (1985) listed this species as possibly introduced to the canal in ballast water or hull fouling.

Abele (1972a) reported collecting 23 small (carapace length 2-6 millimeters) specimens, including eight gravid females, of the freshwater crab *Elamenopsis kempi* (=*Neorhynchoplax kempi* of Abele 1972a, =*Neorhynchoplax alcocki* of Carlton 1985), taken from the east Pedro Miguel Locks when they were drained for cleaning in February 1969. Abele and Kim (1989) reported collecting an additional 23 species in Pedro Miguel Locks in 1975, and 10 specimens in Miraflores Upper Locks in 1974. *E. kempi* had previously been reported only from the Shat-al-Arab marsh at the confluence of the Tigris and Euphrates rivers in Iraq, about 70 miles from the sea. The most common plant fragment in the lock, washed in from Gatun Lake, was a Eurasian aquatic plant, *Hydrilla verticillata*. This is a common aquarium plant that was thought to have been introduced to these waters between 1940 and 1960 (Abele 1972a). Loftin (1965) reported that keeping aquarium fish was a common hobby in the Canal Zone. Abele argued that *E. kempi* had been introduced since 1937, since Hildebrand (1939) had not collected it in his survey of the canal,[27] and speculated that it could have been introduced with *Hydrilla*. Dawson (1973) and Carlton (1985) suggested that it might have been introduced in ballast water.

Table 19. Freshwater species introduced into the Panama Canal. a = The date given refers either to the date the species was first collected in the canal, or to the earliest date associated with a report of the species in the canal, and is not necessarily the date of introduction which could have been earlier, b = HF = transport as hull fouling or in hull borings; BW = in ballast water tanks or other parts of ships' seawater systems; AQ = release from an aquarium, c = References include both collection records and discussions of anthropogenic transport.

Species	Transport: Source Region *to* Destination [Collection Date(s)[a]]	Probable Mechanism[b]	References[c]
Sponges			
Trochospongilla leidii	Eastern North America *to* Gatun Locks [1974] and Pedro Miguel Locks [1975]	BW	Jones & Rützler 1975; Carlton 1985
Hydroids			
Cordylophora caspia	Unknown source *to* all levels of Gatun Locks [1935, abundant in the middle & upper locks; 1974, upper locks], Gatun Lake [1974] and Pedro Miguel Locks [1975]	HF, BW	Hildebrand 1939; Jones & Rützler 1975; Jones 1976
Tanaids			
Sinelobus stanfordi	Unknown source *to* Gatun Middle and Upper Locks [1974], Gatun Lake [1974, 2002], Pedro Miguel Locks [1975] and Miraflores Upper Locks [1974]	HF, BW	Jones & Rützler 1975; Jones 1976; A.N. Cohen unpublished data 2002
Isopods			
Uromunna reynoldsi	Southeastern U.S. *to* Gatun Upper Locks [1974] and Pedro Miguel Locks [1975]	HF?, BW	Jones & Rützler 1975; Jones 1976; Carlton 1985
Crabs			
Elamenopsis kempi	Iraq *to* Pedro Miguel Locks [1969, 1975] and Miraflores Upper Locks [1974]	BW?, AQ?	Abele 1972; Dawson 1973; Carlton 1985; Abele & Kim 1989
Bryozoans			
Asajirella gelatinosa	Indo-West Pacific *to* Rio Chagres at Gatun Lake [1992] and Madden Lake [1998]	BW, AQ	Wood & Okamura 1999

[27] I find this argument unconvincing. Hildebrand's survey focused on fish, reported on only a few species of invertebrates in the locks, and made no collections of invertebrates in Gatun or Miraflores lakes or tributary waters. Clearly, Hildebrand didn't list all of the invertebrates in these waters, and he and his collectors (who were not invertebrate biologists) could have easily missed a half-centimeter-long crab even if it was present in the locks.

The Asian freshwater bryozoan *Asajirella gelatinosa* was collected in the lower reaches of the Chagres River near Gatun Lake in 1992, and further upstream in Madden Lake in 1998 (Wood & Okamura 1999). In its native region it is known mainly from Japan and Korea, with a few additional records in Taiwan, Indonesia and countries bordering the Indian Ocean as far west as India. Some of the *A. gelatinosa* in the Chagres River was growing on leaves and stems of *Hydrilla verticillata*, and Wood and Okamura (1999) suggested that the two might have been introduced together. It could also have been transported in ballast water from a freshwater port in Asia.

3.6 Migration of freshwater species across the divide

In addition to the marine or estuarine species discussed in Section 3.2, freshwater organisms can use the canal to migrate from watersheds on one side of the continental divide through the Culebra Cut to the other side of the divide. One freshwater shrimp and seventeen freshwater fish species have been collected across the continental divide from their native watersheds, including ten species of primary[28] or secondary[29] fish and seven species of peripheral fish[30] (Tables 20 and 21).

One shrimp and ten fish have migrated from the Atlantic slope to the Pacific slope. The Caribbean freshwater shrimp *Palaemon pandaliformis* has been collected on the Atlantic slope of Panama including Gatun Lake. Four juvenile specimens were collected in Pedro Miguel Locks in 1969, but it is not known if it is established on the Pacific slope (Abele 1972c; Abele & Kim 1989). Hildebrand (1939) reported on seven Atlantic slope fish collected in the locks or Miraflores Lake on the Pacific slope, including three primary or secondary fish and four peripheral fish (Table 20). *Eleotris pisonis*, one of the peripheral fish, was at that time reported on the Pacific slope only as apparent hybrids with the Pacific species *Eleotris picta*, but in 1972 it was reported as common in and near the Miraflores Upper Locks (Dawson 1973). Smith et al. (2004), based on a survey conducted in 2002 and a review of the neotropical freshwater fish collection at the Smithsonian Tropical Research Institute, reported on two additional primary or secondary fish and one additional peripheral fish from the Atlantic slope that were collected in the Rio Grande drainage on the Pacific slope. Two of these Atlantic slope species have also been reported in large rivers in Panama that drain to the Pacific south of the canal region: *Brycon*

---

28 Freshwater fish that have little or no tolerance of increased salinities.

29 Freshwater fish that have some tolerance of increased or variable salinities.

30 Fish of marine ancestry that can reside in fresh water.

chagrensis in the Bayano River and *Ancistrus chagresi* in the Bayano and Tuira rivers (Fishbase 2005). The silverside *Atherinella chagresi* (=*Menidia chagresi* of Hildebrand 1939, =*Melaniris chagresi* of McCosker & Dawson 1975) was collected in the creek draining the Miraflores Third Lock lagoon in 1968 (McCosker & Dawson 1975).

Table 20. Migration of Atlantic Slope freshwater species through the canal to the Pacific Slope.

Species	Records on Pacific Slope
Shrimp	
Palaemon pandaliformis	Pedro Miguel Locks in 1969 (Abele & Kim 1989)
Primary and Secondary Fish	
Ancistrus chagresi	Tuira River in 1965 and Bayano River in 1981 (Fishbase 2005); Rio Grande in 2002 (Smith et al. 2004)
Brycon chagrensis	Pedro Miguel Locks and established in Miraflores Lake in 1935-37 (Hildebrand 1939; Fishbase 2005); Bayano River in 1972 (Fishbase 2005); Rio Grande in 2002 (Smith et al. 2004)
Brycon petrosus	Cocoli River above Miraflores Lake in 1937 and 1962 (Hildebrand 1939; Fishbase 2005); Rio Grande in 2002 (Smith et al. 2004)
Roeboides guatemalensis	Common in Gatun Lake but none west of the divide in 1937 (Hildebrand 1939); Rio Grande in 2002 (Smith et al. 2004)
Vieja maculicauda	Miraflores Lake in 1921 (Fishbase 2005); Pedro Miguel Locks, Miraflores Upper Locks and established in Miraflores Lake in 1937 (Hildebrand 1939); Miraflores Third Lock lagoon in 1970, 1971? and 1976 (McCosker & Dawson 1975; Fishbase 2005); Rio Grande in 2002 (Smith et al. 2004)
Peripheral Fish	
Atherinella chagresi	Pedro Miguel Locks and Miraflores Lower Locks in 1937 (Hildebrand 1939); in the creek draining the Miraflores Third Lock lagoon in 1968 (McCosker & Dawson 1975)
Dormitator maculatus	Miraflores Lower Locks in 1937 (Hildebrand 1939); Rio Grande in 2002 (Smith et al. 2004)
Eleotris pisonis	Apparent hybrids with the Pacific species *Eleotris picta* in Miraflores Lower Locks in 1937 (Hildebrand 1939); in Miraflores Upper Locks and "locally common" in 1972 (Dawson 1973)
Leptophilypnus fluviatilis	Abundant in Pedro Miguel Locks, Miraflores Lake and Miraflores Upper Locks in 1935-37 (Hildebrand 1939)
Sicydium altum	Rio Grande in 2002 (Smith et al. 2004)

Seven fish have migrated from the Pacific slope to the Atlantic slope. Hildebrand (1939) reported two Pacific slope fish (*Astyanax aeneus* (= *Astyanax fasciatus*), a secondary fish; and the sleeper goby *Gobiomorus maculatus*, a peripheral fish) in Gatun Lake on the Atlantic slope (Table 21). Smith et al. (2004) reported on four additional primary or secondary fish and one additional peripheral fish from the Pacific slope that were collected in the Chagres River system on the Atlantic slope.

Smith et al. (2002) argued that some of the fish populations that had crossed the continental divide through the canal were now probably isolated from their native source populations, due to the invasion of Gatun Lake by a piscivorous gamefish from the Amazon basin. About 100 fingerlings of *Cichla ocellaris*, known as the peacock bass, were placed in an impoundment in the Chagres River watershed in 1965. *Cichla ocellaris* subsequently entered the river system in rainy season overflows, reached Gatun Lake at Gamboa by 1970, and spread through the lake over the next few years by 1972 it had nearly eliminated several species of previously abundant fish from a large part of the lake (Zaret & Payne 1973). *Cichla ocellaris* may thus be acting as a barrier to migration through the canal for some fish species.

Table 21. Migration of Pacific Slope freshwater species through the canal to the Atlantic Slope. a = Note the potential for confusing records of this Pacific species *Gobiomorus maculatus* (Günther, 1859) with records of the Atlantic species *Dormitator maculatus* (Bloch, 1792) (= *Gobiomorus maculatus* (Bloch, 1792)).

Species	Records on Atlantic Slope
Primary and Secondary Fish	
Astyanax aeneus	Gatun Lake (Hildebrand 1939); Alajuela Reservoir (Maturell 1986); Chagres River in 2002 (Smith et al. 2004)
Cyphocharax magdalenae	Chagres River in 2002 (Smith et al. 2004)
Imparales panamensis	Chagres River in 1972 and 2002 (Smith et al. 2004)
Roeboides occidentalis	In Miraflores Locks and Lake but none east of the divide in 1935-37 (Hildebrand 1939); Chagres River in 2002 (Smith et al. 2004)
Vieja tuyrensis	Chagres River in 2002 (Smith et al. 2004)
Peripheral Fish	
Eleotris picta	Chagres River in 2002 (Smith et al. 2004)
Gobiomorus maculatus[a]	Common in Gatun Lake (Hildebrand 1939); Chagres River in 2002 (Smith et al. 2004)

Hildebrand (1939) discussed 21 fish species as potential or actual migrants across the continental divide through the canal, including both freshwater and coastal euryhaline species. Two of these fish, the mojarra species *Eucinostomus californiensis*[31] and *Gerris cinereus*, are known from both coasts of Panama, but it is unclear whether these represent introductions through the canal or natural distributions of morphologically similar populations. The others are fourteen Atlantic species (Table 22) and five Pacific species (Table 23) whose distributions, habits and salinity tolerances suggested they were likely to make the passage across the divide. Ten of the Atlantic and two of the Pacific species had by then been collected in Panamanian waters across the continental divide from their native ranges.[32] Of these, the Atlantic tarpon *Megalops atlanticus* had

---

31 Hildebrand (1939, footnote 3 on p. 20) noted that *E. californiensis* might consist of two species; McCosker and Dawson (1975) note that Atlantic and Pacific forms are separable by differences in gill raker counts and are probably different species.

32 McCosker and Dawson (1975) reported this as Hildebrand's 14 "probable migrants," apparently including the two mojarra species.

been reported in Pacific Ocean waters, but the other eleven species had been collected across the divide only in the waters of the canal system: the Atlantic species in Pedro Miguel Locks, Miraflores Lake[33] or Miraflores Locks, and the Pacific species in Gatun Lake.

Table 22. Hildebrand's likeliest migrants from the Atlantic to the Pacific. a = Collections on the Pacific side of the continental divide.

Common Name	Species	Status in 1939	Status since 1939
Tarpon	*Megalops atlanticus*	Reported and possibly established in Pacific waters.[a]	Reported and probably established in Pacific waters.[a]
Guavina	*Dormitator maculatus*	Collected in Miraflores Lower Locks.[a]	Collected in the Rio Grande basin.[a]
Guavina	*Leptophilypnus fluviatilis*	Abundant in Pedro Miguel, Miraflores Lake and Miraflores Upper Locks.[a]	
Guavina	*Gobiomorus dormitor*	Collected in Gatun Lake.	
Guavina	*Eleotris pisonis*	Collected as a hybrid in Miraflores Lower Locks.[a]	Collected in Miraflores Upper Locks.[a]
Snook	*Centropomus parallelus*	Collected in Miraflores Lake.[a]	
Rancon	*Pomadasys crocro*	Possibly collected in Pedro Miguel Locks or Miraflores Lake.	
Anchovy	*Anchoa parva*	Collected in both Miraflores Locks.[a]	
Silverside	*Atherinella chagresi*	Collected in Pedro Miguel Locks, Miraflores Lower Locks, and the stream draining the Miraflores Third Lock lagoon.[a]	Collected in Miraflores Third Lock lagoon.[a]
Chogorro	*Vieja maculicauda*	Collected in Pedro Miguel Locks, Miraflores Lake and Miraflores Upper Locks.[a]	
Sabalo pipon	*Brycon chagrensis*	Collected in Pedro Miguel Locks and Miraflores Lake.[a]	Collected in the Rio Grande and Bayano River basins.[a]
Sabalo pipon	*Brycon petrosus*	Collected in a tributary to Miraflores Lake.[a]	Collected in the Rio Grande basin.[a]
Sardina	*Roeboides guatemalensis*	Common in Gatun Lake.	Collected in the Rio Grande basin.[a]
Pipefish	*Oostethus lineatus*	Collected in Gatun Lake.	Collected in Miraflores Locks and in Panama Bay near the entrance to the canal.[a]

Since 1939, two additional Atlantic and two additional Pacific species from Hildebrand's list of potential migrants have been collected across the divide from their native ranges, for a total of sixteen migrants. Leaving aside the two mojarras, 84 percent of the fish species that Hildebrand believed might migrate through the canal have now been collected on the other side of the divide. Four

[33] Including, for *Brycon petrosus*, collection in a tributary just upstream from Miraflores Lake.

of the Atlantic species have been found in Pacific Ocean waters or in waters that connect to the Pacific beyond the Miraflores Lower Locks: the tarpon, the silverside *Atherinella chagresi* (in Miraflores Third Lock lagoon), the pipefish *Oostethus lineatus* (one specimen collected in Panama Bay near the Pacific entrance to the canal) and the sabalo pipon *Brycon chagrensis* (collected in the Bayano River basin).

Table 23. Hildebrand's likeliest migrants from the Pacific to the Atlantic. a = Collections on the Atlantic side of the continental divide.

Common Name	Species	Status in 1939	Status since 1939
Guavina	*Gobiomorus maculatus*	Collected in Gatun Lake.[a]	Collected in Chagres River basin.[a]
Guavina	*Eleotris picta*	Collected in Pedro Miguel Locks.	Collected in Chagres River basin.[a]
Rancon	*Pomadasys bayanus*	Possibly collected in Gatun Lake.	
Sardina	*Astyanax aeneus*	Collected in Gatun Lake.[a]	Collected in Chagres River basin.[a]
Sardina	*Roeboides occidentalis*	Collected in Miraflores Locks and Lake.	Collected in Chagres River basin.[a]

3.7 Conclusions

The construction of the Panama Canal altered maritime commerce by transforming trade routes, reducing the length of voyages and limiting the maximum size of a significant portion of the world's cargo fleet. Though the canal's impact on the translocation of marine organisms has been comparably diverse and and far-reaching, and possibly as significant, it has received remarkably little scientific attention. No study of its potential effect on biota was made prior to construction, and subsequent investigations have been sporadic and modest in scope.

Hildebrand's (1939) and Smith et al.'s (2004) examinations of fish migration through the canal are among the more comprehensive efforts. Studies in the 1960s and 1970s, inspired by the prospect of blasting a sea-level canal through the isthmus with nuclear explosions, documented a few additional species that had migrated or been transported through the canal, primarily on the Near-Canal-Near pathway. Carlton (1985) discussed a few species that may have been transported through in ballast water, including transport between distant regions. These and other data have been assembled and augmented in this chapter, along with an initial effort to assess how the canal's effects on voyage duration and vessel design has affected species introductions. There are, however, many other exotic organisms, both near and far from the canal, whose introductions may have been directly or indirectly influenced by the canal.

In late Paleozoic and early Mesozoic time, the global environment consisted of Pangaea, a single, compact landmass, and Panthalassa, the vast, unbroken, surrounding sea. Subsequent shifting of tectonic plates has resulted in a handful of more or less isolated continents and partially divided and isolated oceans, which the passage of time has endowed with distinct biological communities. In the most recent topographic alteration, the rise of the Central American isthmus separated the Atlantic Ocean from the Pacific.

As inscribed on a plaque in the decaying, old quarter of Panama City, "le génie humain réunit les océans" through the construction of the canal (Fig. 9). The reunion was intended for commerce, but a partial reunion of the far-flung descendents of the Panthalassan biota has also been effected. The challenge remains for human ingenuity to understand, and manage, the biological implications of connecting the oceans.

Fig. 9. (A) A plaque on the French Canal Workers' monument in Casco Viejo, Panama City. (B) Closeup of inscription. Andrew Cohen photos.

Acknowledgements

This chapter draws in part on work conducted in Panama in 2002. I want especially to thank the Smithsonian Tropical Research Institute for hosting me during my stay in Panama; the staff at Naos and Galeta for their assistance and generosity; the participants in the 2002 Panama Exotics Survey (Don Cadien, Dale Calder, Ernesto Campos-Gonzalez, John Chapman, Leslie Harris, Charlie Lambert, Gretchen Lambert, Fabio Pitombo, Sergio Salazar-Vallejo and Brian Wysor) for sharing their knowledge and expertise; and the Pew Fellowship Program for providing financial support. I am also grateful to Jim Carlton for generously sharing his extensive files on the Panama Canal.

4 References

Abele, L.G. 1972a. Introductions of two freshwater decapod crustaceans (Hymenosomatidae and Atyidae) into Central and North America. Crustaceana 23: 209-218.

Abele, L.G. 1972b. Comparative habitat diversity and faunal relationships between the Pacific and Caribbean Panamanian decapod crustacea: a preliminary report, with some remarks on the crustacean fauna of Panama. Bull. Biol. Soc. Wash. 2: 125-138.

Abele, L.G. 1972c. Panamanian Decapod Crustacea: fauna of the Panama Canal, comparative community structure and remarks on the social and economic importance of some species. Proceedings of the International Congress of Zoology 17(3): 1-18.

Abele, L.G. and W. Kim. 1989. The Decapod Crustaceans of the Panama Canal. Smithsonian Contributions to Zoology 482:1-50.

ACP. 2005. Panama Canal Website, Authoridad del Canal de Panama, Balboa-Ancon, Republic of Panama, at http://www.pancanal.com/index.html (accessed in December 2005).

Agard, J., R. Kishore and B. Bayne. 1992. *Perna viridis* (Linnaeus, 1758): first record of the Indo-Pacific green mussel (Mollusca: Bivalvia) in the Caribbean. Carib. Mar. Stud. (3): 59-60.

Agassiz, A. 1869. Preliminary report on the Echini and starfishes dredged in deep water between Cuba and the Florida Reef by L. F. De Pourtalès. Bulletin of the Museum of Comparative Anatomy, Harvard University 1(9): 253-308.

Allen, F.E. 1953. Distribution of marine invertebrates by ships. *Austral. J. Mar. Freshw. Res.* 4: 307-316.

Alvariño, A. 1974. Distribution of the siphonophores in the regions adjacent to the Suez and Panama Canals. Fishery Bulletin 72 (2): 527-546.

Bainbridge, R. 1960. Speed and stamina in three fish.. J. Exp. Biol. 37(1): 129-153.

Baker, P., S.M. Baker and J. Fajans. 2004. Nonindigenous Marine Species in the Greater Tampa Bay Ecosystem - Literature Review and Field Survey of Tampa Bay for Nonindigenous Marine and Estuarine Species. Tampa Bay Estuary Program Tchnical Publication #02-04, Tampa, FL.

Bartsch, P. 1944. A new shipworm from the Panama Canal. Smithsonian Miscellaneous Collections 104(8): 1-3.

Bastida-Zavala, J.R. and H.A. ten Hove. 2003a. Revision of *Hydroides* Gunnerus, 1768 (Polychaeta: Serpulidae) from the Eastern Pacific Region and Hawaii. Beaufortia 53(4): 67-110.

Bastida-Zavala, J.R. and H.A. ten Hove. 2003b. Revision of *Hydroides* Gunnerus, 1768 (Polychaeta: Serpulidae) from the western Atlantic Region. Beaufortia 52: 103-178.

Bax, N.J. 1999. Eradicating a Dreissenid from Australia. Dreissena! 10(3): 1-5.

Bayano River Sportfishing Charters. 2005. Website at http://www.panamafishingandcatching.com (accessed in October 2005).

Bayer, F.M., G.L. Voss and C.R. Robins. 1970. Report on the Marine Fauna and Benthic Shelf-Slope Communities of the Isthmian Region. Bioenvironmental and Radiological-Safety Feasibility Studies, Atlantic-Pacific Interoceanic Canal, BMI-171-38, Battelle Memorial Institute, Columbus Laboratories, Columbus, OH.

Behrens, D.W. 1991. Pacific Coast Nudibranchs. Sea Challengers, Monterey, CA.

Benson, A.J., D.C. Marelli, M.E. Frischer, J.M. Danforth and J.D. Williams. 2001a. Establishment of the Green Mussel, *Perna viridis* (Linnaeus 1758) (Mollusca: Mytilidae) on the west coast of Florida. J. Shellfish. Res. 20(1): 21-29.

Benson, A.J., P.L. Fuller and C.C. Jacono. 2001b. Summary Report of Nonindigenous Aquatic Species in U.S. Fish and Wildlife Service Region 4. U.S. Fish and Wildlife Service, Arlington, VA and U.S. Geological Survey.

Benson, A.J., C.C. Jacono, P.L. Fuller, E.R. McKercher and M.M. Richerson. 2004. Summary Report of Nonindigenous Aquatic Species in U.S. Fish and Wildlife Service Region 5. U.S. Fish and Wildlife Service, Arlington, VA and U.S. Geological Survey.

Bequaert, J.C. 1943. The genus *Littorina* in the western Atlantic. Johnsonia 7: 1-27.

Bermingham, E., and H.A. Lessios. 1993. Rate variation of protein and mitochondrial DNA evolution as revealed by sea urchins separated by the Isthmus of Panama. Proc. Natl. Acad. Sci. 90: 2734-2738.

Bieler, R., P.M. Mikkelsen, T. Lee and D.O. Foighil. 2004. Discovery of the Indo-Pacific oyster *Hyotissa hyotis* (Linnaeus, 1758) in the Florida Keys (Bivalvia: Gryphaeidae). Molluscan Research 24: 149-159.

Bishop, M.W.H. 1950. Distribution of *Balanus amphitrite* Darwin var. *denticulata* Broch. Nature 165(4193): 409-410.

Bishop, M.W.H. 1951. Distribution of barnacles by ships. Nature 167(4248): 531.

Boffey, P.M. 1971. Sea-level canal: how the Academy's voice was muted. Science 171: 355-358.

Borrero, F.J. and J.M. Díaz. 1998. Introduction of the Indo-Pacific pteriid bivalve *Electroma* sp to the tropical western Atlantic. Bull. Mar. Sci. 62(1): 269-274.

Briggs, J.C. 1968. Panama's sea-level canal. Science 162: 511-513.

Briggs, J.C. 1972a. Aquatic ecosystems. Science 176: 581-582.

Briggs, J.C. 1972b. Marine biological effects of a sea level Panama Canal. Proc. International Cong. Zool. 17(3): 1-4.

Britton, J.C. and B. Morton. 1989. Shore Ecology of the Gulf of Mexico. University of Texas Press, Austin, TX.

Brunetti, R. and F. Mastrototaro. 2004. The non-indigenous stolidobranch ascidian *Polyandrocarpa zorritensis* in the Mediterranean: description, larval morphology and pattern of vascular budding. Zootaxa 528: 1-8.

Burch, J.Q. 1960. Notes and News: *Conus mus*. Nautilus 74(2): 81-82.

Carlton, J.T. 1979. History, Biogeography, and Ecology of the Introduced Marine and Estuarine Invertebrates of the Pacific Coast of North America. Ph.D. thesis, University of California, Davis, CA.

Carlton, J.T. 1985. Transoceanic and interoceanic dispersal of coastal marine organisms: the biology of ballast water. Oceanogr. Mar. Biol. Ann. Rev. 23: 313-371.

Carlton, J.T. 1992. Introduced marine and estuarine mollusks of North America: an end-of-the-20th-Century perspective. Journal of Shellfish Research 11(2): 489-505.

Carlton, J.T. and J.B. Geller. 1993. Ecological roulette: the global transport of nonindigenous marine organisms. Science 261: 78-82.

Carlton, J.T., A. Navarret and R. Mann. 1982. Biology of Ships' Ballast Water: The Role of Ballast Water in the Transoceanic Dispersal of Marine Organisms. Final Project Report for the National Science Foundation, Woods Hole Oceanographic Institute, Woods Hole, MA.

Carlton, J.T. and M.H. Ruckelshaus. 1997. Non-indigenous marine invertebrates and algae. Pages 187-201 in: Strangers in Paradise: Impact and Management of Nonindigenous Species in Florida, D. Simberloff, D.C. Schmitz and T.C. Brown (eds.), Island Press, Washington, DC.

Carlton, J.T. and J.A. Scanlon. 1985. Progression and dispersal of an introduced alga: *Codium fragile* ssp. *tomentosoides* (Chlorophyta) on the Atlantic coast of North America. Botanica Marina 28: 155-165.

Challinor, D. 1972. Background for a new sea level Panama canal. Bull. Biol. Soc. Wash. 2: 7-12.

Challinor, D. 1978. Testimony. Page 248 in: Sea-Level Canal Studies. Hearings before the Subcommittee on the Panama Canal of the Committee on Merchant Marine and Fisheries, House of Representatives, 95th Congress, June 21, 27 and 28, 1978, Serial No. 95-51, U.S. Government Printing Office, Washington, DC.

Chesher, R.H. 1968. Transport of marine plankton through the Panama Canal. Limnol. Oceanogr. 13: 387-388.

Chesher, R.H. 1972. The status of knowledge of Panamanian echinoids, 1971, with comments on other echinoderms. Bull. Biol. Soc. Wash. 2: 139-158.

Chickering, A.M. 1929. An Atlantic pipefish caught in transit through the Panama Canal. Copeia 173: 85-86.

Clarke, A.H. 1969. Some possible consequences of a sea-level Panama Canal. *Amer. Malacol. Union Annual Report* 1969: 51-52.

Clench, W.J. and R.D. Turner. 1946. The genus *Bankia* in the western Atlantic. Johnsonia 2(19): 1-28.

Coan, E. 1984. The Recent Crassatellinae of the eastern Pacific, with some notes on *Crassinella*. Veliger 26(3): 153-169.

Cohen, A.N. and J.T. Carlton. 1995. Nonindigenous Aquatic Species in a United States Estuary: A Case Study of the Biological Invasions of the San Francisco Bay and Delta. U.S. Fish and Wildlife Service, Washington, DC.

Cohen, A.N. 1998. Ships' Ballast Water and the Introduction of Exotic Organisms into the San Francisco Estuary: Current Status of the Problem and Options for Management. A report for CALFED and the California Urban Water Agencies. San Francisco Estuary Institute, Richmond, CA.

Coles, S.L., R.C. DeFelice and L.S. Godwin. 1999. The impact of nonindigenous marine species transported to Pearl Harbor on the hull of the U.S.S. Missouri. Hawaii Biological Survey Contribution No. 1999-014, Bernice Pauahi Bishop Museum, Honolulu, HI.

Constant, T.M. 1978. Testimony on behalf of the Panama Canal Company. Pages 189-198 in: Sea-Level Canal Studies. Hearings before the Subcommittee on the Panama Canal of the Committee on Merchant Marine and Fisheries, House of Representatives, 95th Congress, June 21, 27 and 28, 1978, Serial No. 95-51, U.S. Government Printing Office, Washington, DC.

Coutts, A.D.M. 2002. A Biosecurity Investigation of a Barge in the Marlborough Sounds. Cawthron Report No. 744, Cawthron Institute, Nelson, New Zealand.

Cross, T.P. 1967. A study of ballast water; a medium for the translocation of marine species across the Isthmus of Panama. in: George, C.J. Should Oceans Meet? Proceedings of a Conference at Union College, Schenectady, NY.

Cubit, J. and S. Williams. 1983. The Invertebrates of Galeta Reef (Caribbean Panama): A Species List and Bibliography. Atoll Research Bulletin No. 269.

Dawson, C.E. 1970. The Caribbean blenny *Lupinoblennius dispar* (Tribe: Blenniini), with observations on a Pacific population. Proc. Biol. Soc. Wash. 83: 273-286.

Dawson, C.E. 1972. A redescription of *Lophogobius cristulatus* Ginsburg (Pisces: Gobiidae) with notes on *L. cyprinoides* (Pallas). Proc. Biol. Soc. Wash. 84(44): 371-384.

Dawson, C.E. 1973. Occurrence of an exotic eleotrid fish in Panama with discussion of probable origin and mode of introduction. Copeia 1:141-4.

D'Croz, L. and D.R. Robertson. 1997. Coastal oceanographic conditions affecting coral reefs on both sides of the Isthmus of Panama. Proc. 8th Int. Coral Reef Sym. 2: 2053-2058.

DeFelice, R.C. 1999. Fuling invertebrates on the floating dry dock U.S.S. Machinist in Pearl Harbor prior to its move to Apra Harbor, Guam. Hawaii Biological Survey, Bernice Pauahi Bishop Museum, Honolulu, HI.

De Laubenfels, M.W. 1936. A comparison of the shallow water sponges near the Pacific end of the Panama Canal with those at the Caribbean end. Proc. U.S. Nat. Mus. 83: 441-466.

Devaney, D.M. and L.G. Eldredge. 1977. Class Scyphozoa. Pages 108-118 in: Reef and Shore Fauna of Hawaii. Section I: Protozoa through Ctenophora. Devaney, D.M. and L.G. Eldredge (eds.), Special Publication 64(1), Berenice P. Bishop Museum , Honolulu, HI.

Dickerson, R.E. 1917. Ancient Panama Canals. Proc. Calif. Acad. Sci. 7(8): 197-205.

Drake, L.A., G.M. Ruiz, B.S. Galil, T.L. Mullady, D.O. Friedmann and F.C. Dobbs. 2002. Microbial ecology of ballast water during a transoceanic voyage and the effects of open-ocean exchange. Marine Ecology Progress Series 233: 13-20.

Earle, S.A. 1972. A review of the marine plants of Panama. *Bull. Biol. Soc. Wash.* 2:69-87.

Evermann, B.W. and O.P. Jenkins. 1891. Report on a collection of fishes made at Guaymas, Sonora, Mexico, with descriptions of new species. Proceedings of the United States National Museum 14(846): 121-165.

Fairey, R., R. Dunn, C. Roberts, M. Sigala and J. Oliver. 2002. Introduced aquatic species in California Coastal Waters, Final Report. Appendix A in: Report to the Legislature: A Survey of Non-Indigenous Aquatic Species in the Coastal and Estuarine Waters of California, Ashe, M.E. (ed.), California Department of Fish and Game, Office of Oil Spill Prevention and Response, Sacramento, CA.

Fauchald, K. 1977. Polychaetes from intertidal areas in Panama, with a review of previous shallow-water records. Smithsonian Contributions to Zoology 221.

Fishbase. 2005. A database designed by R. Froese and D. Pauly, at http://filaman.uni-kiel.de/home.htm (accessed in October 2005).

Frank, W. 1993. Expect the unexpected. Shell-O-Gram (Newsletter of the Jacksonville Shell Club) Jul-Aug 1993, at http://www.jaxshells.org/selected.htm (accessed in December 2005).

Frankenberg, D. and R.J. Menzies. 1966. A new species of Asellote marine isopod, *Munna* (*Uromunna*) *reynoldsi* (Crustacea: Isopoda). Bulletin of Marine Science 16(2): 200-208.

Glynn, P.W. 1972. Observations on the ecology of the Caribbean and Pacific coasts of Panamá. Bull. Biol. Soc. Wash. 2:13-30.

Glynn, P.W. 1982. Coral communities and their modification relative to past and prospective Central American seaways. Adv. Mar. Biol. 19: 91-132.

Gollasch, S., J. Lenz, M. Dammer, and H.G. Andres. 2000. Survival of tropical ballast water organisms during a cruise from the Indian Ocean to the North Sea. Journal of Plankton Research 22(5): 923-937.

Gollasch, S., H. Rosenthal, J. Botnen, H. Hamer, I. Laing, E. Leppåkoski, E. Macdonald, D. Minchin, M. Nauke, S. Olenin, S. Utting, M. Voigt and I. Wallentinus. 2000. Survival rates of species in ballast water during international voyages: results of the first workshops. Pages 296-305 in: Proceedings of the First National Conference on Marine Bioinvasions, Pederson, J. (ed.), MIT Sea Grant Program, Cambridge, MA.

Goodbody, I. 2000. Diversity and distribution of Ascidians (Tunicata) in the Pelican Cays, Belize. Atoll Research Bulletin 480: 301-326.

Goodbody, I. 2004. Diversity and distribution of ascidians (Tunicata) at Twin Cays, Belize. Atoll Research Bulletin 524: 1-20.

Gordon, D.P. 1967. A report on the ectoproct bryozoa of some Auckland shores. *Tane (J. Auckland Univ. Field Club)* 13: 43-76.

Gore, R.H. and L.G. Abele. 1976. Shallow water porcelain crabs from the Pacific Coast of Panama and adjacent Caribbean waters (Crustacea: Anomura: Porcellanidae). Smithsonian Contributions to Zoology 237: 1-30.

Graham, J.B. 1971. Temperature tolerances of some closely related tropical Atlantic and Pacific fish species. Science 172: 861-863.

Graham, W.M., D.L. Martin, D.L. Felder, V.L. Asper and H.M. Perry. 2003. Ecological and economic implications of a tropical jellyfish invader in the Gulf of Mexico. Biological Invasions 5: 53-69.

Guiry, M.D., F. Rindi and G.M. Guiry. 2005. AlgaeBase version 4.0, a World-wide electronic publication, National University of Ireland, Galway, at http://www.algaebase.org (accessed in December 2005).

Gunter, G. 1979. Marine fishes of Panama as related to the Canal. Gulf Research Reports 6(3): 267-273.

Hastings, A.B. 1930. Cheilostomatous polyzoa from the vicinity of the Panama Canal collected by Dr. C. Crossland on the cruise of the S.Y. St. George. Proc. Zool. Soc. London 1929: 697-740.

Hay, M.E. and S.D. Gaines. 1984. Geographic differences in herbivore impact: do Pacific herbivores prevent Caribbean seaweeds from colonizing via the Panama Canal? Biotropica 16(1): 24-30.

Heilner, V.C. 1965. The lordly tarpon — angler's delight. Pages 96-101 in: The Wondrous World of Fishes, Grosvenor, M.B. (ed.), National Geographic Society, Washington, DC.

Hendrickx, M.E. 1980. Range extensions of three species of Teredinidae (Mollusca: Bivalvia) along the Pacific Coast of America. Veliger 23(1): 93-94.

Herbert Engineering. 1999. Ballast Water Management for Containerships: Implications for the Port of Oakland. Herbert Engineering Corp., San Francisco, CA for the Port of Oakland, Oakland, CA.

Hertlein, L.G. and G.D. Hanna. 1949. Two new species of *Mytilopsis* from Panama and Fiji. Bull. So. Cal. Acad. Sci. 48(1): 13-18.

Hildebrand, S.F. 1939. The Panama Canal as a passageway for fishes, with lists and remarks on the fishes and invertebrates observed. *Zoologica* 24: 15-45.

Hutchings, P.A., R.W. Hilliard and S.L. Coles. 2002. Species introductions and potential for marine pest invasions into tropical marine communities, with special reference to the Indo-Pacific. Pacific Science 56(2): 223-233.

Jones, M.L. 1972. Introduction. Bull. Biol. Soc. Wash. 2: vi-viii.

Jones, M.L. 1976. The invertebrate fauna of the locks of the Panama Canal. Proceedings of the Association of Island Marine Laboratories of the Caribbean 11: 17-18.

Jones, M.L. and C.E. Dawson. 1973. Salinity-temperature profiles in the Panama Canal Locks. Mar. Biol. 21: 86-90.

Jones, M.L. and K. Rützler. 1975. Invertebrates of the upper chamber, Gatún locks, Panama Canal, with emphasis on *Trochospongilla leidii* (Porifera). Marine Biology 33: 57-66.

Jordan, D.S. 1885. A list of the fishes known from the Pacific coast of tropical America, from the Tropic of Cancer to Panama. Proc. U.S. Nat. Mus. 8: 361-394.

Jordan, D.S. 1908. The law of geminate species. Amer. Natur. 42: 73-80.

Josselyn, M.N. and J.A. West. 1985. The distribution and temporal dynamics of the estuarine macroalgal community of San Francisco Bay. Hydrobiologia 129: 139-152.

Kensley, B. and M. Schotte. 1989. Guide to the Marine Isopod Crustaceans of the Caribbean. Smithsonian Intitution Press, Washington, DC.

Knowlton, N. 1993. Sibling species in the sea. Annual Review of Ecology and Systematics 24: 189-216.

Knowlton, N., and L.A. Weigt. 1998. New dates and new rates for divergence across the Isthmus of Panama. *Proc. R. Soc. Lond.* 265: 2257-2263.

Knowlton, N., L.A. Weigt, L.A. Solórzano, D.K. Mills, and E. Bermingham. 1993. Divergence in proteins, mitochondrial DNA, and reproductive compatibility across the Isthmus of Panama. Science 260:1629-1632.

Kott, P. 1985. The Australian Ascidiacea. Part I, Phlebobranchia and Stolidobranchia. Mem. Queensland Mus. 23: 1-440.

Lambert, C.C. and G. Lambert. 1998. Non-indigenous ascidians in southern California harbors and marinas. *Marine Biology* 130: 675-688.

Lambert, C.C. and G. Lambert. 2003. Persistence and differential distribution of nonindigenous ascidians in harbors of the southern California Bight. Marine Ecology Progress Series 259: 145-161.

Leschine, T.M. 1981a. The Panamanian sea-level canal. Oceanus 24(2): 20-30.

Leschine, T.M. 1981b. The Panamanian sea-level canal: problems and prospects from a policy perspective. In: Oceanus 81, Proceedings of the IEEE/MTS Annual Meeting, Boston, MA, Sept. 1981.

Lessios, H.A. 1979. Use of Panamanian sea urchins to test the molecular clock. Nature 280:599-601.

Lessios, H.A. 1981. Divergence in allopatry: molecular and morphological differentiation between sea urchins separated by the Isthmus of Panama. Evolution 35(4): 618-634.

Lessios, H.A. 1984. Possible prezygotic reproductive isolation in sea urchins separated by the Isthmus of Panama. Evolution 38(5): 1144-1148.

Lessios, H.A. 1998. The first stage of speciation as seen in organisms separated by the Isthmus of Panama. Pages 186-201 in: D. Howard and S. Berlocher (eds.), Endless Forms: Species and Speciation. Oxford University Press, Oxford.

Lessios, H.A. and J.R. Weinberg. 1994. Genetic and morphological divergence among morphotypes of the isopod *Excirolana* on the two sides of the isthmus of Panama. Evol. 48:530-48.

Littlewood, D.T.J. 1991. Pest and predators of cultivated Mangrove Oysters. Pages 109-146 in: Oyster Culture in the Caribbean, Proceedings of the Oyster Culture in the Caribbean Workshop, Nov. 19-22, 1990, Kingston, Jamaica, Newkirk, G.F. and B.A. Field (eds.). Mollusc Culture Network, Dalhousie University, Halifax, Nova Scotia, Canada.

Loftin, H. 1965. On unusual distribution records from the Panama Canal Zone region. Carib. J. Sci. 5(1-2): 83-86.

Lohrer, A.M. 2001. The invasion by *Hemigrapsus sanguineus* in eastern North America: A Review. Aquatic Invaders 12(3): 1-11.

Marelli, D.C. and S. Gray. 1983. Conchological redescriptions of *Mytilopsis sallei* and *Mytilopsis leucophaeta* of the brackish Western Atlantic. Veliger 25(3): 185-193.

Marelli, D.C. and S. Gray. 1985. Comments on the status of recent members of the genus *Mytilopsis* (Bivalvia: Dreissenidae). Malacological Review 18: 117-121.

Martin, W.E., J.A. Duke, S.G. Bloom and J.T. McGinnis. 1970. Possible Effects of a Sea-Level Canal on the Marine ecology of the Americam Isthmian Region. Bioenvironmental and Radiological-Safety Feasibility Studies, Atlantic-Pacific Interoceanic Canal, U.S. Atomic Energy Commission Prime Contract No. AT(26-1)-171, Battelle Memorial Institute, Columbus Laboratories, Columbus, OH.

Marsh, G.P. 1864. Man and Nature; or, Physical Geography as Modified by Human Action. Scribner, NY.

Maturell, J.C. 1986. Recopilacion sobre las caracteristicas generales y potencial pesquero de los embalses Gatun, Alajuela y Bayano en la Republica de Panama. *In:* Vila, I. and E. Fagetti (eds.). 1986. Trabajos presentados al Taller internacional sobre ecología y manejo de peces en lagos y embalses. COPESCAL Technical Report 4, U.N. Food and Agriculture Organization, Rome.

Mayr, E. 1954. Geographic speciation in tropical echinoids. Evolution 8: 1-18.

McCosker, J.E. and C.E. Dawson. 1975. Biotic passage through the Panama Canal, with particular reference to fishes. Mar. Biol. 30: 343-351.

McCullough, D. 1977. The Path Between the Seas: The Creation of the Panama Canal 1870-1914. Touchstone Books, Simon & Schuster, New York.McMahon, R. 1996. The physiological ecology of the zebra mussel, *Dreissena polymorpha*, in North America and Europe. American Zoologist 36: 339-363.

Menzies, R.J. 1968. Transport of marine life between oceans through the Panama Canal. Nature 220: 802-3.

Mikkelsen, P.M., P.S. Mikkelsen, D.J. Karlen. 1995. Molluscan diversity in the Indian River lagoon, Florida. Bull. Mar. Sci. 57(1): 94-127

Miller, M.A. 1968. Isopoda and Tanaidacea from buoys in coastal waters of the continental United States, Hawaii, and the Bahamas (Crustacea). Proc. U.S. Nat. Mus. 125: 1-53.

Mills, C.E. and J.T. Rees. 2000. New observations and corrections concerning the trio of invasive hydromedusae *Maeotias marginata* (=*M. inexspectata*), *Blackfordia virginica*, and *Moerisia* sp. in the San Francisco Estuary. Scientia Marina 64 (Supl. 1): 151-155.

Mills, C.E. and F. Sommer. 1995. Invertebrate introductions in marine habitats: two species of hydromedusae (Cnidaria) native to the Black Sea, *Maeotias inexspectata* and *Blackfordia virginica*, invade San Francisco Bay. Marine Biology 122: 279-288.

Morey, S. 2005. Tarpon biological profile. Ichthyology at the Florida Museum of Natural History, at http://www.flmnh.ufl.edu/fish/Gallery/Descript/Tarpon/Tarpon.html (accessed October 2005).

Morton, B. 1980. *Mytilopsis sallei* (Recluz) (Bivalvia: Dreissenacea) recorded from Hongkong: an introduction by Vietnamese refugees? Malacological Review 13: 90-92.

Morton, B. 1981. The biology and functional morphology of *Mytilopsis sallei* (Recluz) (Bivalvia: Dreissenacea) fouling Visakhapatnam Harbour, Andhra Pradesh, India. Journal of Molluscan Studies 47: 25-42.

Morton, B. 1987. Recent marine introductions into Hong Kong. Bulletin of Marine Science 41(2): 503-513.

Olenin, S., S. Gollasch, S. Jonusas and I. Rimkute. 2000. En-route investigations of plankton in ballast water on a ship's voyage from the Baltic Sea to the open Atlantic coast of Europe. Internat. Rev. Hydrobiol. 85(5-6): 577-596.

Olsson, A.A. 1972. Origin of the existing Panamic molluscan biotas in terms of their geologic history and their separation by the isthmus land barrier. Bull. Biol. Soc. Wash. 2: 117-123.

Osburn, R.C. 1953. Bryozoa of the Pacific coast of America: Part 3, Cyclostomata, Ctenostomata, Entoprocta, and Addenda. *Allan Hancock Pac. Exped.* 14(3): 613-841.

Pauley, G., L. Kirkendale, G. Lambert and C. Meyer. 2002. Anthropogenic biotic interchange in a coral reef ecosystem: a case study from Guam. Pacific Science 56(4): 403-422.

Port of Oakland. 1999. Oakland Harbor Navigation Improvement (-50 Foot) Project, Volume V: Final Environmental Impact Statement/Environmental Impact Report, Appendix X: Responses to Comments. Port of Oakland, Oakland, CA and U.S. Army Corps of Engineers, San Francisco District, San Francisco, CA.

Porter, J.W. 1972. Ecology and species diversity of coral reefs on opposite sides of the isthmus of Panama. Bull. Biol. Soc. Wash. 2: 89-116.

Powell, N.A. 1971. The marine bryozoa near the Panama Canal. Bull. Mar. Sci. 21: 766-778.

Rao, K.S., V.V. Srinivasan and M. Balaji. 1989. Success and spread of the exotic fouling bivalve *Mytilopsis sallei* (Recluz) in Indian waters. Pages 125-127 in: Exotic Aquatic Species in India; Proceedings of the Workshop 25-26 April 1988, M.M. Joseph (ed.), Spec. Publ. No. 1, Asian Fisheries Society, Indian Branch.

Rathbun, M.J. 1930. The Cancroid crabs of America of the families Euryalidae, Portunidae, Atelcyclidae, Cancridae and Xanthidae. *Bulletin of the United States National Museum* 152.

Rosewater, J. 1975. Mollusks of Gatun Locks, Panama Canal. Bull. Am. Malacological Union 1974: 42-43.

Rosewater, J. 1976. Some results of the National Museum of Natural History–Smithsonian Tropical Research Institute survey of Panama. Bull. Am. Malacological Union 1975: 48-50.

Rosewater, J. 1980. Subspecies of the gastropod, *Littorina scabra.* Nautilus 94(4): 158-162.

Rubinoff, I. 1968. Central American sea-level canal: possible biological effects. Science 161: 857-861.

Rubinoff, I. 1970. The sea-level canal controversy. Biological Conservation 3(1): 33-36.

Rubinoff, I. 1972. A sea level canal in Panama. Proc. International Congress of Zoology 17(3):1-13.

Rubinoff, R.W. and I. Rubinoff. 1968. Interoceanic colonization of a marine goby through the Panama Canal. *Nature* 217: 476-478.

Rubinoff, R.W. and I. Rubinoff. 1969. Observations on the migration of a marine goby through the Panama Canal. Copeia 2: 395-397.

Ruhlow, J. 2005. Pacific tarpon catch confirmed, Costa Rica fishing report for August 16, 2005, at http://www.costaricaoutdoors.com/fishing-report.htm (accessed in October 2005).

Ruiz, G.M., K. Larson, T. Huber, B. Steves, A. Hines and G. Lambert. 2005. Spatial analysis of ascidian assemblages in North America: comparison of standardized field measures across bays, coasts and latitudes. Abstract for International Invasive Sea Squirt Conference, April 21-22, 2005, Woods Hole Oceanographic Institution, Woods Hole, MA.

Sambilay, V.C., Jr. 1990. Interrelationships between swimming speed, caudal fin aspect ratio and body length of fishes. Fishbyte 8(3): 16-20.

Schormann, J., Carlton, J.T. and M.R. Dochoda. 1990. The ship as a vector in biotic invasions. Trans. Inst. Mar. Engineers (C) 102: 147-152.

Schultz, G.A. 1969. How to Know the Marine Isopod Crustaceans, W.C. Brown Co., Dubuque, IA.

Schultz, G.A. 1979. Louisiana and Panama Canal locations and ecology of *Munna* (*Pangamunna* nov. subgen.) *reynoldsi* Frenkenberg & Menzies (Isopoda: Asellota). Proceedings of the Biological Society of Washington 92(3): 577-579.

Setchell, W.A. 1935. Geographic elements of the marine flora of the North Pacific Ocean. Amer. Nat. 69: 560-577.

Sheffey, J.P. 1968. When Caribbean and Pacific waters mix. Science 162: 1329.

Sheffey, J.P. 1972. Physical characteristics of the proposed sea-level isthmian canal. Bull. Biol. Soc. Wash. 2: 31-39.

Sheffey, J.P. 1978. Testimony. Pages 302-324 in: Sea-Level Canal Studies. Hearings before the Subcommittee on the Panama Canal of the Committee on Merchant Marine and Fisheries, House of Representatives, 95th Congress, June 21, 27 and 28, 1978, Serial No. 95-51, U.S. Government Printing Office, Washington, DC.

Shelton, D.N., R.K. Wallace and H.M. Perry. 2002. An infestation of marine mussels in the firemain system of a Navy vessel. Aquatic Invaders 13(4): 21-23.

Siddall, S.E. 1980. A clarification of the genus *Perna* (Mytilidae). Bull. Mar. Sci. 30(4): 858-870.

Sieg, J. 1986. Distribution of the Tanaidacea: synopsis of the known data and suggestions on possible distribution patterns. Pages 165-194 in: Gore, R.H. and K.L. Heck (eds.), Crustacean Biogeography. Crustacean Issues 4, A.A. Balkema, Rotterdam & Boston.

Sieg, J. and R.N. Winn. 1981. The Tanaidae (Crustacea: Tanaidacea) of California, with a key to the world genera. Proc. Biol. Soc. Wash. 94(2): 315-343.

Smith, S.A., G. Bell and E. Bermingham. 2004. Cross-cordillera exchange mediated by the Panama Canal increased the species richness of local freshwater fish assemblages. Proceedings of the Royal Society: Biological Sciences 271(1551): 1889-1896.

Southward, A.J., R.S. Burton, S.L. Coles, P.R. Dando, R.C. DeFelice, J. Hoover, P.E. Parnell, T. Yamaguchi, and W.A. Newman. 1998. Invasion of Hawaiian shores by an Atlantic barnacle. Mar. Ecol. Prog. Ser. 165: 119-126.

Spivey, H.R. 1976. The cirripeds of the Panama Canal. Corrosion Marine-Fouling 1(1):43-50.

Springer, V.G., and M.F. Gomon. 1975. Revision of the Blenniid fish genus *Omobranchus* with descriptions of three new species and notes on other species of the tribe Omobranchini. Smithsonian Contributions to Zoology 177: 1-135.

Topp, R.W. 1969. Interoceanic sea-level canal: effects on the fish faunas. *Science* 165: 1324-7.

Trowbridge, C.D. 2005. Re: Information on *Ercolania fuscovittata.* Message in Sea Slug Forum on April 12, 2005, Australian Museum, Sydney, at http://www.seaslugforum.net/find.cfm?id=13436 (accessed in December 2005).

Turner, R.D. 1971. Identification of wood-boring molluscs. Pages 17-64 in: Marine Borers, Fungi and Fouling Organisms of Wood, Jones, E.B.G. and S.K. Eltringham (eds.), Organization for Economic Cooperation and Development, Paris.

U.S. Geological Survey. 2005. Nonindigenous Aquatic Species Database, Gainesville, FL, at http://nas.er.usgs .gov (accessed in December 2005).

U.S. Navy. 1951. Report on marine borers and fouling organisms in 56 important harbors and tabular summaries of marine borer data from 160 widespread locations. Navdocks TP-Re-1, Bureau of Yards and Docks, Department of the Navy, Washington, DC.

Verrill, A.E. 1866. On the polyps and corals of Panama, with descriptions of new species. Proc. Boston Soc. Nat. Hist. 10: 325-357.

Voss, G.L. 1967. Phase I-Final Report: Marine Resources and Ecology. Bioenvironmental and Radiological-Safety Feasibility Studies, Atlantic-Pacific Interoceanic Canal. BMI-171-5, Battelle Memorial Institute, Columbus Laboratories, Columbus, OH. 143 pp.

Voss, G.L. 1972. Zoogeographical considerations of the impact of a sea-level canal across Panama. Proc. International Congress of Zoology. 17(3): 1-13.

Voss, G.L. 1978. Panama Sea-Level Canal-II. Biological catastrophe or grand experiment? *Sea Frontiers* 24: 206-13.

West, D. 2005. World records and Pacific tarpon, at http://www.worldseafishing.com/international/other/ pacific_tarpon.shtml (accessed in October 2005).

Wonham, M.J., W.C. Walton, A.M. Frese and G.M. Ruiz. 1996. Transoceanic transport of ballast water: biological and physical dynamics of ballasted communities and the effectiveness of mid-ocean exchange. Report for the U. S. Fish and Wildlife Service, Washington, DC.

Woods Hole Oceanographic Institution. 1952. Marine Fouling and its Prevention. United States Naval Institute, Annapolis, MD.

Wysor, B. 2004. An annotated list of marine Chlorophyta from the Pacific Coast of the Republic of Panama with a comparison to Caribbean Panama species. Nova Hedwigia 78: 209-241.

Zaret, T.M. and R.T. Payne. 1973. Species introduction in a tropical lake. Science 182: 449-455.

The Suez Canal

The Marine Caravan – The Suez Canal and the Erythrean Invasion

BELLA S. GALIL

National Institute of Oceanography, Israel Oceanographic and Limnological Research, P.O.B. 8030, Haifa 31081, Israel

"Le Canal de Suez, pomme de discorde dès sa creation"
(Faouzi 1951).

1 Ancient isthmian canals

The rulers of ancient Egypt appreciated the economic and strategic importance of a navigable waterway connecting the prosperous civilization of the Nile Valley with the Red Sea, East Africa and the Indian Ocean. Schooled early in drainage and irrigation engineering, and trained in colossal construction projects, the nilotic civilization was equal to the challenge.

Herodotus [5th century BCE], the Greek geographer and historian, visited Egypt and supplied us with the earliest reference to a trans-isthmian canal joining the easternmost arm of the Nile with the northern Red Sea: "Psammetichus had a son called Necos [reigned 609-594 BCE], who became king of Egypt. It was he who began the making of the canal into the Red Sea, which was finished by Darius the Persian [521-486 BCE]. This is four day's voyage in length, and it was dug wide enough for two triremes to move in it rowed abreast. It is fed by the Nile, and is carried from a little above Bubastis, by the Arabian town of Patumus; it issues into the Red Sea. The beginning of the digging was in the part of the Egyptian plain which is nearest to Arabia; the mountains that extend to Memphis (in which mountains are the stone quarries) come close to the plain;

the canal is led along the lower slope of these mountains in a long reach from west to east; passing then into a ravine it bears southward out of the hill country towards the Arabian Gulf. Now the shortest and most direct passage from the northern [Mediterranean Sea] to the southern or Red Sea is from the Casian promontory, which is the boundary between Egypt and Syria, to the Arabian Gulf, and this is a distance of one thousand furlongs [200 km], neither more nor less; this is the most direct way, but the canal is by much longer, inasmuch as it is more crooked. In Necos' reign a hundred and twenty thousand Egyptians perished in the digging of it." (in Godley 1975, II: 158).

Diodorus of Sicily [1st century BCE] too visited Egypt, and while partly repeating Herodotus' account, he related the conviction that the level of the waters in the Red Sea is higher than in the Mediterranean: "From the Pelusiac mouth [of the Nile] there is an artificial canal to the Arabian Gulf and the Red Sea. The first to undertake the construction of this was Necho the son of Psammetichus, and after him Darius the Persian made progress with the work a time but finally left it unfinished; for he was informed by certain persons that if he dug through the neck of land he would be responsible for the submergence of Egypt, for they pointed out to him that the Red Sea was higher than Egypt. At a later time the second Ptolemy [Ptolemy Philadelphus, 285-246 BCE] completed it and in the most suitable spot constructed an ingenious kind of lock. This he opened, whenever he wishes to pass through, and quickly closed again, a contrivance which usage proved to be highly successful. The river which flows through this canal is named Ptolemy, after the builder of it, and has at its mouth the city called Arsinoë." (in Oldfather 1968, I: 34).

The Greek geographer and historian Strabo [1st century CE] traveled in Egypt and Ethiopia and lived for several years in Alexandria. He described the canal in the last volume of his *Geographia*, antedating its origin to a Pharaoh of the 19th Dynasty, and rebutting the elevation conundrum: "There is another canal which empties into the Red Sea and the Arabian Gulf near the city of Arsinoë, a city which some call Cleopatris. It flows also through the Bitter Lakes, as they are called, which were indeed bitter in earlier times, but when the above-mentioned canal was cut they underwent a change because of the mixing with the river, and now are well supplied with fish and full also of aquatic birds. The canal was first by Sesostris before the Trojan War – though some say by the son of Psammetichus, who only began the work and then died- and later by Dareius the First, who succeeded to the next work done upon it. But he, too, having been persuaded by a false notion, abandoned the work when it was already near completion; for he was persuaded that the Red Sea was higher than Aegypt, and that if the intervening isthmus were cut all the way through, Egypt would be inundated by the Sea.

The Ptolemaic kings, however, cut through it and made a strait a closed passage, so when they wished they could sail out without hindrance into the outer sea and sail in again. … The canal which empties into the Red Sea… has a breadth of one hundred cubits and a depth sufficient for very large merchant vessels..." (in Jones 1967, 17.1.25,26). Strabo insisted that the seas are level "the sea outside the Pillars [of Hercules, the Atlantic Ocean], the Red Sea, and the Mediterranean Sea too…. all have the same level." (in Jones, 1.3.13).

Pliny the elder [23-79 CE] had no first hand knowledge of Egypt, but compiled his encyclopedic *Natural History* from many sources, reiterated the prevalent conviction concerning the level of the waters in the Red Sea, though with some reservation: " .. the Harbour of the Daneoi, from which there was a project to carry a ship-canal through to the Nile at the place where it flows into what is called the Delta, over a space of 62½ miles, which is the distance between the river and the Red Sea; this project was originally conceived by Sesostris King of Egypt, and later by the Persian King Darius and then again by Ptolemy the Second, who did actually carry a trench 100 ft broad and 30 ft. deep for a distance of 34½ miles, as far as the Bitter Springs. He was deterred from carrying it further by fear of causing a flood, as it was ascertained that the level of the Red Sea is 4½ ft. above that of the land of Egypt. Some persons do not adduce this reason for the abandonment of the project, but say that it was due to the fear lest making an inlet from the sea would pollute the water of the Nile, which affords to Egypt its only supply of drinking water." (in Rackham 1969, VI: 33).

All classical authors were in agreement concerning Darius' enterprise, and gratifyingly, while digging the isthmian sands near the 150 km point, an ancient stela was unearthed bearing Darius' haughty words "I am a Persian. From Persia I captured Egypt. I commanded this canal to be built from the Nile, which flows in Egypt, to the sea which comes from Persia. So was this canal built, as I had commanded, and ships passed from Egypt through the canal to Persia, as was my purpose." Darius' canal did exist, and the king commemorated his achievement by erecting stele along its route, of which fragments of five were discovered (Rogers 1929). Darius' canal ran from Pelusium (Tel el-Farama) which was situated at the mouth of the easternmost [Pelusiac] branch of the Nile, now extinct, to Bubastis, and through the east-west depression of Wadi Tumilat to Lake Timsah, the Bitter Lakes, and to the Gulf of Suez (Fig. 1).

Evidence for the earlier pharaonic canals was sought in Egyptian and Hebrew paleography, where references to Egypt's eastern border indicate the presence of an inland navigable waterway (Gardiner 1920).

Fig. 1. Geographic map of the pre canal isthmus (Source: Fontane, 1869).

This is possibly the waterway depicted on the wall of the great temple of Ammon at Karnak commemorating the campaigns of Seti I [1306-1290 BCE] in Syria and Palestine. The pharaoh, grand in his war chariot decorated with the decapitated heads of the vanquished kings, drives files of bound captives towards a well-fortified bridge to enter Egypt (Fig. 2).

Fig. 2. Triumphal procession of Seti I depicted on the walls of Ammon's Temple at Karnak, Egypt. Crocodiles fill the reed-lined canal that forms Egypt eastern border. After Sneh et al. 1975.

The crocodiles inhabiting the waterway and the reeds fringing its shores signify a fresh or brackish water channel (Lake Timsah, crocodile in Hebrew, may preserve the memory of the ancient crocodile-infested waterbody). Sneh et al. (1975) proposed that the artificial waterway served as the eastern border barrier known as "Shur of Egypt" in the ancient texts, and may have existed as early as 2000 BCE. An aerial survey of the northeastern Delta revealed the presence of a previously unrecorded ancient canal southeast of the defunct Pelusaic branch, about 70 m wide between the embankments, and extending from Tell el-Farama to Qantara, just north of El Ballah sabkha (Sneh et al. 1975). The authors suggest that the canal may have been part of the ancient barrier moat, joining the Nile through the east-west gorge of Wadi Tumilat, and was later enlarged and used for transport as part of Necos' trans-isthmian waterway. Indeed, a 12 km long ancient canal extending between the northern shore of Lake Timsah and El Ballah depression was identified by the French engineer Linant de Bellefonds (1872-1873) as part of Necos' canal. Aly Bey Shafei (1946), an irrigation engineer who designed major irrigation projects in the eastern Delta, suggested that Necos' canal was designed to link the Bitter Lakes and Lake Timsah to the Pelusiac branch and the sea, but owing to its alignment failed to provide access to the Nile water, whereas Darius' canal linked Lake Timsah with the Nile through Wadi Tumilat. The canal, traversing marshy lands and desert, required constant maintenance to keep it from being silted up by the Nile floods or smothered in desert sands. The Ptolemies re-excavated the canal, and built a new port city, Arsinoë, at its Red Sea terminus. The gradual drying up of the Pelusiac branch made navigation through the old canal unreliable, so under the Roman rule, possibly during the reign of Trajan [98 CE], the canal was reconditioned and lengthened to meet the Nile at Babylon, near present day Cairo. Papyrological documents discussing repairs made on the canal in 287 and 332 CE confirm that it remained at least in intermittent service through the fifth century. The construction and repeated reconditioning of the "Amnis Trajanus" attest to the importance of the direct trade with Arabia and India (Sidebotham 1986, Mayerson 1996).

Following the Arab conquest of Egypt, the Caliph Umar ibn al Khattab ordered the Governor of Egypt, Amr ibn-al-Asi, to rehabilitate/rebuild Trajan's Canal and ready the waterway for navigation in 642/3. The canal became the main route for transporting Egypt's grain and produce to Medina and Mecca, and garnered the Caliph and his descendents the appellation "Emir al Muminin" [Prince of the Faithful]. But Caliph Umar forbade extending the canal from Lake Timsah to the Mediterranean for fear it would allow the Christian fleets direct access to Arabia (Butler 1978). The canal was ordered blocked in 775 by the Abbasid Caliph, Abu Jafar al Mansur, to quell a revolt in Arabia by withholding grain shipments. Inundated during the annual Nile floods, segments of the canal persisted till Mehemet Ali [1811] ordered it filled in, and even then it continued its millennia-long existence: sections were refurbished in the 19th century and utilized as part of the fresh water canal leading from Cairo to Suez.

The Ottoman conquest of Egypt raised the prospect of a trans-isthmian canal: the Grand Vizier of Sultan Selim II, Sokullu Mehmed Pasha [1568], sought to furnish the powerful Ottoman navy with a direct route to Arabia, East Africa, and the Persian Gulf, to protect the pilgrim route to Mecca and the eastern trade. He charged the Governor of Egypt with preparation of a report on the possible route and dimensions of the canal, and its military advantages (Sarioglu 2002). This quest, as well as the discussions with the Venetians worried over the decline in the Republic's fortunes with the ebbing of Mediterranean trade, came to naught because of the immense expenditure of the project. The idea of a trans-isthmian canal "cut from Suez to Cairo, such as was effected under the ancient Egyptian kings" fit in with the general optimism of 17th century Europe (Schonfield 1969). In the "Age of Enlightment", when philosophers corresponded with kings, the German philosopher and statesman Leibnitz advised Louis XIV to seize Egypt: "There is in Egypt the most important isthmus in the world, that separating its great seas, the Ocean and the Mediterranean: a place that cannot be avoided without circling all the sinuosities of Africa; the connecting point, the obstacle, the key, the only possible door between two areas of the world, Asia and Africa; the meeting-point and marketing-place of India on one side, and Europe on the other" (Thompson 1988). The French Secretary of state for foreign affairs, Arnaud de Pomponne, replied wryly that Holy Wars "ont cessé d'être à la mode depuis Saint Louis" (Roux 1901). But the dream of joining the two seas gained force with the increasingly profitable maritime trade with the East. Savary, a distinguished merchant and lawyer, advanced the idea of cutting a canal through the isthmus, and judiciously examined the engineering and fiscal limitations of the project in his book entitled "Le parfait négociant ou Instruction générale pour ce qui regarde le commerce des merchandises de France et des pays étrangers". Comte de Volney, whose widely read account of travels in the Levant, "Voyage en Syrie et en Égypte", was published with the "Approbation & Privilége du Roi", devoted a chapter "De l'Isthme de Suez, & de la jonction de la Mer rouge à la Méditerranée". Volney mentioned the pharaonic and caliphate canal [as Anc. Canal de Qulzoum], and identified and mapped its course: "Ce canal est le même qui, de nos jours, passe au Kaire, & qui va se perdre dans la campagne au nord-est de Berket-el-Hadj" (Volnay 1787:195).

2 Forty centuries are watching us

On the eve of Napoleon's departure to Egypt he was handed instructions ostensibly formulated by the Republic's Executive Directory, but in effect drafted by himself "..to seize Egypt;... chase the English from all their possessions in the Orient;.... Destroy all their settlements on the Red Sea. .... then cut the Isthmus of Suez ... for the French Republic" (Silvera 1975). In December 1798, but five months after the Battle of the Pyramids that sought to secure French domination of Egypt, Napoleon, attended by Generals Berthier and Caffarelli, a few scientists and an armed escort, rode from Cairo to Suez to investigate the possibility of a trans-isthmian canal. The company was tormented by hunger and cold, marauding Bedu tribesmen, and a hazardous crossing of the Gulf of Suez that nearly cost Caffarelli his life (Herold 1962). Undeterred, Napoleon rode over the desolate land and tracked the remains of the ancient canal from Suez to the Bitter Lakes, and onwards to the Nile.

The excursion reinforced Napoleon's resolve to excavate a canal through the Isthmus. He entrusted the topographical survey to Le Père, his chief engineer: "Le vif intérêt que le general Bonaparte montrait dans ces diverses reconnaissances, était un témoignage de son désir d'avoir des resultants plus précis" (Le Père 1822: 63). Napoleon's zeal and urgency must have been contagious for on January 16, 1799, ten days after Napoleon's return from Suez,

Le Père's hastily assembled team of surveyors and engineers left Cairo on their mission to survey the land and determine its elevation in comparison with the bordering seas and the Nile. In one year Le Père and an ever changing crew of surveyors traveled over the deserts and marshes of the Isthmus. Beset by Bedu tribesmen and Mamelouk brigands, afflicted by hunger and disease, and hampered by the pillage of their instruments during the revolt in Cairo, the exhausted team carried three successive expeditions. The results of the first (January-February, 1799) confirmed the possibility of cutting a canal from the Gulf to the Bitter Lakes. After a six months' hiatus caused by the deployment of their escort in Syria, Le Père mounted the second expedition (September-October, 1799) to study the eastern part of Wadi Tumilat and the Bitter Lakes, ending it "extrêmement fatigué" after a forced march in the scorching heat (Le Père 1822: 75). Desperately pressed for time, Le Père divided his team in two for the third expedition (November-December, 1799). One party under his leadership traveled to Lake Menzaleh and the Mediterranean, the other party, under Févre was to measure the level of the Nile in the vicinity of Cairo. In the following year the Nile flood was extraordinarily high and Le Père mounted another expedition (October 1800) to study the extent of the inundation. He found the submerged Wadi Tumilat "avait l'aspect d'une mer" (Le Père 1822: 83), the Nile waters reaching as far as the ruins of Serapeum by the Bitter Lakes. The preliminary report was sent to the "Citoyen premier consul", by then back in Paris, on 6 December 1800 : "Plus j'y ai réfléchi, citoyen premier consul, (et j'apprécie toutes les consequences d'une opinion hasardée), plus je me suis convaincu que le rétablissement du canal ne présente aucune difficulté majeure: au moyen d'écluses, ouvrages d'invention moderne, on pourra profiter plus avantageusement des eaux du Nil pendant toute la durée des crues, quell que soit le niveau variable de ces eaux par rapport à celui de la mer Rouge, également variable par l'effect des marées" (Le Père 1822: 95). Despite the care taken over the measurements, a mistake in the calculations that placed the level of the Red Sea waters at high tide 10 meters above the Mediterranean sea in low tide, and the impressions left by the 1800 inundation, moved Le Père to recommend to Napoleon a canal that followed the route of the previous canals – through the Nile! At the time, Laplace and Fourier protested that the calculations are wrong, but to no avail, like Strabo before them, they were defeated by a myth. The full report (Mémoire sur la communication de la mer des Indes a la Méditerranée par la mer Rouge et l'Isthme de Soueys) was published long after "le génie créateur du général Bonaparte" and "la gloire du chef de l'expédition d'Égypte" faded somewhat, in the 11th volume (Dédiée au Roi) of the encyclopaedic "Description de L'Égypte ou Recueil des observations et des recherches qui ont été faites en Égypte pendant l'Expédition de l'Armée Francaise". On reading the report, and considering the weight given to the ancient texts (all translated and introduced in an Appendix), and the efforts expended in endeavoring to reconcile them with the geographical and archeological evidence, one can not escape the suspicion that Le Père, who was doubtlessly aware of the survey's shortcomings, chose to err on the side of the classical authors who asserted time and again that the waters of the Red Sea would flood the land.

In 1805 Mehemet Ali seized control of Egypt and introduced sweeping reforms. He invested in modernizing the irrigation and transport infrastructure and enlisted the services of two French engineers, Linant-Bey [Linant de Bellefonds] and Mougel Bey. They built the Mahmoudieh Canal, from Alexandria to the Nile, allowing navigation upstream to Cairo, that followed, unbeknown to Mehemet Ali and his engineers, the yet unpublished Le Père plan. Between 1822 and 1833 Linant surveyed the Isthmus five times, studying and mapping in detail its topography and geology. He was confident of the feasibility of a direct isthmian canal traversing Lake Timsah and the Bitter Lakes, and discreetly worked with Mougel on its plan. Linant communicated his plans to the French consul, Mimaut, and his vice-consul, de Lesseps, then newly arrived in Egypt. In 1833 a group of French intellectual-utopians led by Enfantin, the Saint-Simonians, traveled to Egypt to promote their idea of a "Canal de jonction des deux mers". They too met with Linant. Failing to elicit the consent of Mehemet Ali, the Saint-Simonians returned to France and formed in 1846 the 'Société d'études pour le Canal de Suez' The next year the Société sponsored a joint team of French-Austrian-British engineers, headed by Bourdalou who specialized in geodesy, to confirm Linant's measurements. Interest in a direct trans-isthmian canal revived when the survey proved there is no significant difference in the levels of the Mediterranean and the Red Sea (Roux 1901). The members of the Société, though realizing a direct trans-isthmian canal is viable, approved nevertheless a plan, put forward by Talabot, for a canal running from Alexandria through Cairo to Suez. The one dissenting voice belonged to the engineer Negrelli, who reiterated Linant's plan and proposed an alternative route: from Pelusium, through Menzaleh, Ballah, and the Bitter Lakes.

3 A man, a plan, and two canals: Ferdinand de Lesseps [1805-1894]

Ferdinand de Lesseps (Fig. 3) was born to an influential and well-connected family: his uncle, the distinguished diplomat Barthélemy de Lesseps, was ennobled by Louis XVI; his father, Mathieu de Lesseps, also a career diplomat, was made count by Napoleon, and his mother, Catherine de Grivégnée y Gallegos, was related to the Countess of Montijo. Born in Versailles, de Lesseps spent his childhood on the family estate in Pisa, before moving to Paris to attend the college of Henry IV. Following a brief stint in the army, he was posted in junior consular positions in Lisbon, and Tunis, where his father was consul-general, before arriving in Alexandria in 1832 as the vice-consul. De Lesseps enjoyed Muhammad Ali's favor and befriended his son, Said. In 1837 he returned to France to wed Agathe Delamalle. He was appointed consul in Barcelona in 1842 and spent that decade in Spain, where he distinguished himself during the insurrection in Catalonia. In 1848 he was posted as minister plenipotentiary in Madrid, where he met his young cousin Eugénie de Montijo. When Louis-Napoleon was elected to the presidency, he appointed his brother the ambassador to Madrid, and de Lesseps was dispatched to Rome where the forces of the Republic assailed the Pope while themselves besieged by the Austrians and the French. De Lesseps managed to negotiate a provisional agreement, but political events in Paris put his efforts to naught, the National Assembly nullified his agreement, and he was censured by the Council of State.

Fig. 3. Ferdinand de Lesseps (Source: "Homenaje a Ferdinand de Lesseps" at the Museo de Arte Contemporaneo, Panama City, November 1994).

Disgraced and disheartened, de Lesseps resigned his post and retired to the family estate. In 1853 he lost his wife and a child to scarlet fever. Throughout the enforced retirement, he studied the patterns of international trade and in 1852 wrote a memorandum on the feasibility of a trans-isthmian canal at Suez, and a year later sent it to Abbas, then ruler of Egypt. In 1854 he received an invitation from his one-time protégé, Mohammed Said, who acceded to the viceroyalty, to visit him. Before leaving France for Alexandria, de Lesseps acquired the notes of the Société d'Études pour le Canal de Suez from Enfantin and Talabot, including the detailed maps of the Isthmus prepared by Linant de Bellefonds and Talabot. De Lesseps presented to Said, on November 15, 1854, a plan for a direct trans-isthmian canal, adopting the layout advocated by Negrelli. By the end of November, 1854 the initial concession to build a canal, with a port at each end (Fig. 4), was approved and signed.

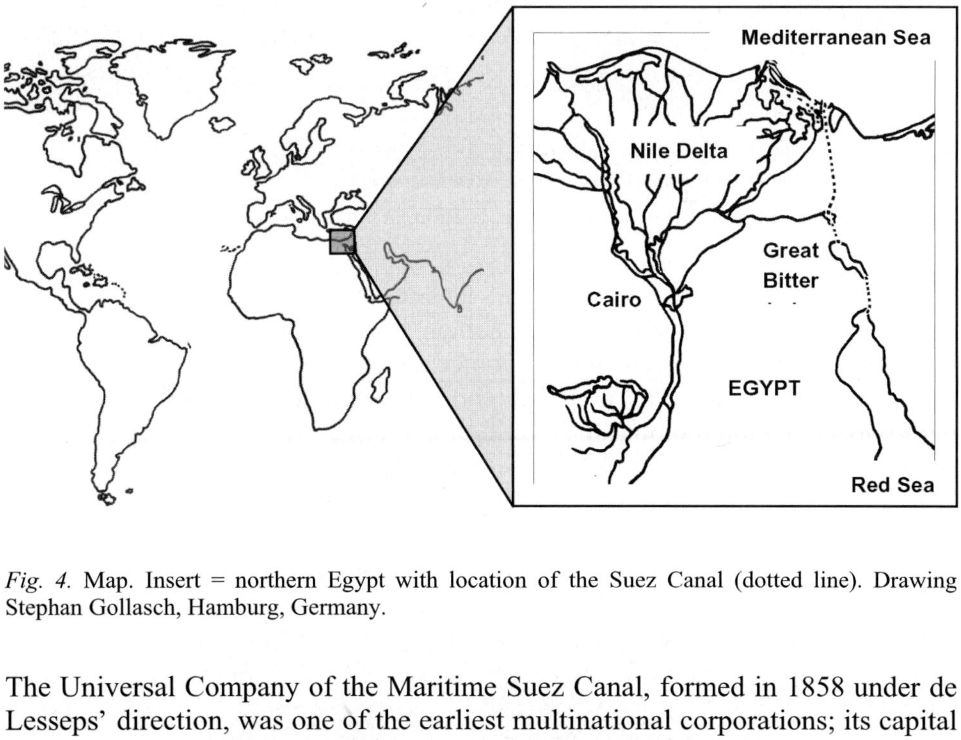

Fig. 4. Map. Insert = northern Egypt with location of the Suez Canal (dotted line). Drawing Stephan Gollasch, Hamburg, Germany.

The Universal Company of the Maritime Suez Canal, formed in 1858 under de Lesseps' direction, was one of the earliest multinational corporations; its capital was set at 200 million gold francs divided into 400,000 shares. De Lesseps raised, by popular subscription in France, over half the capital needed, and much of the rest was invested by Said himself. With only half of the funds and lacking the permission of the Sultan of Turkey, the construction began April 25, 1859, on the site of the future Port Said. Twenty thousand conscripted *fellahin* and prisoners formed the bulk of the laborers digging largely by hand through the waterless sandy wastes. They were later replaced by steam-powered bucket dredges (Fig. 5).

Fig. 5. Steam-powered bucket dredges (Source: E. Riou, in Fontane, 1869).

Said died in 1863, and his successor, Ismail, backed the project. Crises and delays abounded, many provoked by Britain's opposition to the project, but de Lesseps, undaunted, tirelessly promoted and prodded the enterprise to completion on August 15, 1869. The construction of the Suez Canal involved the excavation and dredging of 74 million m^3 of sediments. The canal was 8 m deep, 58-90 m wide and 160 km long, and along its banks three new cities were built: Suez, Ismailia, and Port Said (Fig. 6).

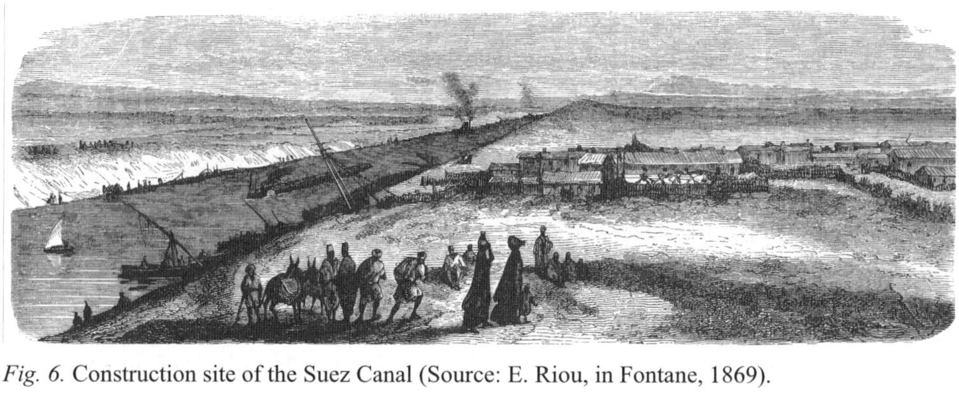

Fig. 6. Construction site of the Suez Canal (Source: E. Riou, in Fontane, 1869).

The inauguration of the Suez Canal was attended by European royalty, including Empress Eugénie of France, Arab notables, and thousands of foreign visitors. Empress Eugénie awarded de Lesseps the Grand Cross of the Legion of Honor, Queen Victoria bestowed on him the Grand Cross of the order of the Star of India, and similar honorifics were received from the Khedive and the emperor of Austria.

His labors over, de Lesseps married in Ismailia the 20 years old Louise-Hélène Autard de Bragard, who was to bear him 12 children. Between 1875-1881 he published five volumes of letters and records documenting the history of the Suez Canal, his memoirs "Souvenirs de quarante années" (1878), and "Origines du canal du Suez" (1890). In 1884 de Lesseps presented the United States, in the name of the 'Committee of the Franco-American Union', the statue that graces New York Harbor - "Liberty Enlightening the World".

In 1878 he assumed the presidency of the 'Compagnie Universelle du Canal Interocéanique de Panama' to construct a salt water, sea-level canal without locks. Work began in 1881, but the difficulties of digging through the continental divide at the Culebra Cut and the Chagres River, combined with tropical diseases and labor disputes, impeded the canal's progress. Financial mismanagement, stock failure and adverse publicity contributed to the failure of the company, and forced it into bankruptcy. In 1889 the construction was halted. De Lesseps and his eldest son were found guilty by a French court of mismanagement and misappropriation of funds, fined and sentenced to prison. Disgraced, de Lesseps retired again to the family estate, and died five years later. But the project was too far advanced to scuttle, and a few months before de Lesseps' death, the 'Compagnie Nouvelle du Canal de Panama' was formed in order to continue the work.

He was honored posthumously (1899) with a colossal statue at the entry to the Suez Canal, the 'Monument de Lesseps'. The statue was destroyed in 1956 by the Egyptians in the aftermath of the Suez Crisis.

4 Ever deeper, ever wider – the expansion of the Suez Canal

From Port Said in the Mediterranean to Port Taufiq on the Red Sea, cutting through the sandy desert, the Suez Canal is about 162.5 km long. When first opened the Canal was not quite 8 m deep, 22 m wide at the bottom, and 60-90 m wide at the surface. To allow ships to pass each other, passing bays were built along its banks, but the narrowness of the channel resulted in some 3,000 groundings of ships between 1870 and 1884. By 1872 the permissible draft increased from 6.7 m to 7.2 m, and the Assémbleé Générale approved an annual budget for dredging (Nourse 1884). The *Compagnie Universelle du Canal Maritime de Suez* embarked on major improvements including the widening and deepening of the channel in 1876, and by 1880 the number of ships transiting the Canal was 2026, compared with 486 in 1870. By 1882 the increased volume of shipping resulted in a revival of Talabot's plans and a proposal of a "fresh-water maritime canal" including five locks (London Times, 13 October 1882). In 1884 an 'International Consultative Commission' was convened to discuss the enlargement of the Canal as an alternative to the excavation of a new waterway. It was agreed that the widening of the canal was preferable to the construction of a second waterway for it would be less costly in time, effort, and expense (Wilson 1933). The British delegation to the Commission argued in favor of increasing the channel depth to 9 m. Further expansion was initiated in 1908 and completed in 1914. Under this program the channel bottom was excavated to 11 m, and widened to 45 m over its whole length, facilitating faster transit time. Yet, even as the improvements were carried out, it was recognized that the canal was again inadequate for coping with the increasing volume of traffic and size of vessels. Another expansion program was adopted in 1912, before the previous one was completed, to both deepen the Canal and channel the roadsteads of Suez and Port Said to allow the passage of vessels with 10 m draft. By 1939 the Canal was 12 m deep and 130 m wide at the surface (Ghazzawi 1939). In 1948 a by-pass canal, seven and a half miles long, was excavated, to speed up convoys in the northern section of the Canal, and the channel deepened by half a meter over its entire length. In 1955 14,666 ships traversed the canal, and the Suez Canal Company announced a plan to excavate over 80 million tons of sediments to allow passage of vessels of 11 m draft. The events of the summer and autumn of 1956 – the nationalization of the Suez Canal Company, followed by the Anglo-French invasion of the Canal Zone, the Arab-Israeli war, and the blockage of the Canal by the Egyptians – derailed those plans. During the months of closure and blockage two feet of sand had accumulated, and when traffic finally resumed on March 1956, only vessels with maximum draft of 10 m could pass. In 1958 work began on the 1955 improvement plan adopted with slight modifications by the Egyptian Suez Canal Authority (SCA). It was designed to widen the Canal by almost 30 m, and to increase the maximum draft to 11.3 m (bottom depth 13.5-14 m). In 1959 the World Bank extended a loan of 56.5 million dollars for an ambitious expansion project, the so-called 'Nasser plan'. The first stage entailed doubling the canal's width along its entire length and deepening it to handle laden tankers of 70,000 tons with a maximum draft of 14.6 m. In 1966, a six-year program, the second stage of the 'Nasser Plan', was launched. Its object was to allow the navigation of 110,000 ton loaded tankers and 125,000 tons partially loaded vessels by 1972. The Six Days War forced the closing of the Suez Canal in June 1967, blocking it for the second time in ten years.

The canal remained inoperative until June 1975, when maintenance work was recommenced to clear the sand that filled the channel bed. By that time the Canal was incapable of handling half of the world's tanker fleet, with Very Large Crude Carriers (VLCC) [200,000-300,000 dead weight tons (DWT)] plying alternate sea routes (Schonfield 1969). Once again the SCA undertook a dredging program. In the late 1970s the waterway was deepened to 16.2 m, which meant a laden 150,000 DWT tanker could transit northbound, and a vessel of over 300,000 DWT in ballast southbound. A new category of tanker, the "Suezmax" was designed expressly for this route. Subsequent dredging increased the Canal draft to 17.7 m in 1999 (www.sis.gov.eg/public).

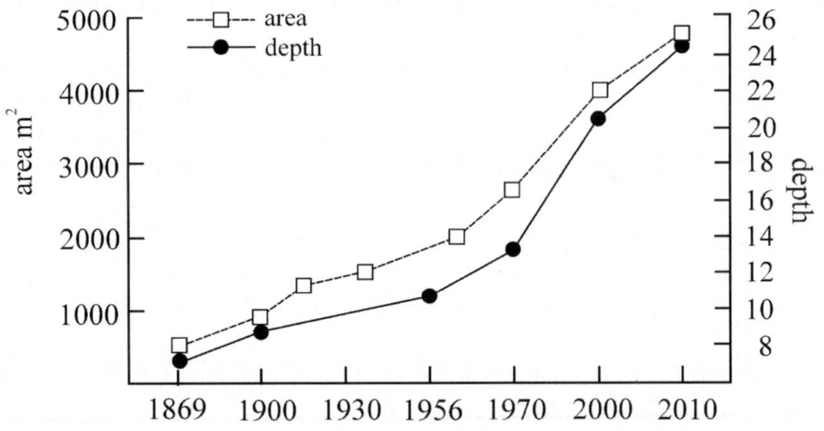

Fig. 7. Depth and area of typical cross-section of the Suez Canal 1869-2010.

The canal is at present between 300 and 365 m wide and its maximum permissible draught is 17.7 m (www.imsalex.com/suez_canal). The Canal has been doubled in five parts (Port Said, Ballah, Timsah, Deversoir, Kabret) and this allows the transit of ships in both directions. The banks of the Canal are protected against the erosion generated by the transit of ships by revetments of stone and steel piles. The SCA board chairman, Admiral Fadel, announced in 1999 plans to enlarge the Canal to permit passage of 'Ultra-Large-Crude-Carrier' (ULCC) class tankers with oil cargos of up to 350,000 DWT, in order to maintain the Canal's market share against the inexorable increase in the size of ships. The expansion will increase the canal width to 400 m and its depth to 25 m by 2010, as well as extending the approach channel to Port Said to almost 25 km into the Mediterranean, compared with 10 km at present (Al-Ahram

Weekly, no. 440, 29 July 1999). The typical cross-sectional area was 304 m² in 1869, 1200 m² in 1956, 1800 m² in 1970s, 3600 m² in 2000, but is being increased to 4600 m² (Hassan & El-Sabh 1975, Halim 1990, Abril & Abdel – Aal 2000) (Fig. 7).

5 Shipping

The sea level, lockless Suez Canal is one of the world's great short cuts and one of the most dense shipping lanes. The opening of the Suez Canal had an immediate effect on shipping and trade, shifting some of the world shipping routes. The Canal shortened dramatically voyages from Europe to Asia: the distance between London and Kuwait was reduced from 11,300 nm via the Cape to 6500 nm via Suez. Whereas in 1875 the Canal share of the world sea borne trade was just about 1.5%, increasing to 10% in 1948, by the 1960s some 15% of the world sea borne trade used the Canal (Farnie 1969). In 1870 the number of ships transiting the Canal was 486 (436,609 net tonnage NT). The Canal handled 2026 transits (3,057,422 NT) in 1880, 3,389 (6,890,094 NT) in 1890, 4,533 (16,581,898 NT) in 1910, 5,761 (31,668,759 NT) in 1930, 11,751 (81,795,523 NT) in 1950, and 18,734 (185,322,000) in 1960 (Schonfield 1969) (Fig. 8).

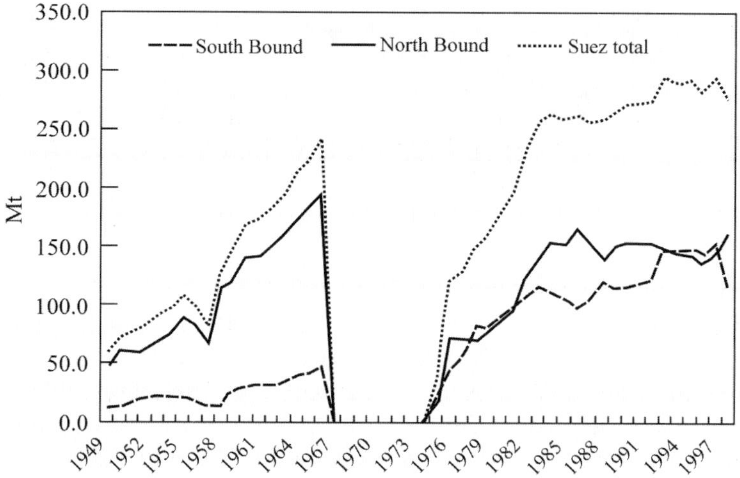

Fig. 8. Suez Canal traffic, 1949-1998.

Since that time, Canal traffic, although maintaining its growth, underwent a change. The number of passenger ships, which had been of major importance on the shipping routes from Europe to the South East Asia, declined with the advent of commercial air travel. At the same time the Canal benefited from the development of the Middle-Eastern oil fields, being closely associated with the oil trade from the Gulf – oil shipments increasing from nearly 20% of the total volume in 1938, to over 72% of the traffic in 1966 (Quéguiner 1978). The closure of the Canal in 1967-1975 launched a rapid increase in tanker sizes and the emergence of the VLCC and ULCC vessels specifically designed for long-haul routes, so that in the 1990s north bound laden tanker transits declined (Fig. 8) (INTERTANKO 2004). In order to maintain its market share, the SCA has to balance the cost of engineering and the income from larger vessels which it hopes will be attracted to use the waterway. According to statistics from SCA, in 2003 about 2,800 oil tankers passed through the Suez Canal, of which 1224 were north bound laden vessels, including 111 VLCCs, transiting after offloading part of their cargo to the SUMED pipeline. 1369 vessels transited the canal in January 2004, as compared with 1225 in January 2003, with receipts rising from 190.8 to 243.9 M US$ (www.economic.idsc.gov.eg/suez). Annual Receipts from the Canal are expected to top 2 billion US$ in 2004.

6 Shipping impacts and pollution in the Suez Canal

Due to its strategic location, the Suez Canal is a main artery of transport – about 6% of total world seaborne trade passes through it. Oil historically has represented about 25% of Suez Canal revenues. Though tanker traffic has been competing with the SUMED pipeline and the alternate route around the Cape of Good Hope, thousands of laden and partly-laden oil tankers (see above) transit the Canal annually transporting about 1.3 million bbl/d (www.eia.doe.gov/emeu/cabs). At present the Canal accommodates Suezmax class tankers with 200,000 dead-weight-ton maximum cargos, but by the end of the decade will accommodate ULCC with oil cargos of up to 350,000 dead-weight-tons. The movement of so many large vessels is the source of turbulence, anti-fouling biocides, and hydrocarbon pollution.

The sediment resuspension and turbulence caused by the ever increasing volume of shipping, the continuous dredging and the recurrent widening and deepening of the Canal make for a greatly disturbed environment. Fox (1929:12) wrote: "The bottom of the Canal is periodically turned over by dredgers. … The water of the Canal proper is almost invariably turbid". A survey of the changes in mineral assemblages along the Canal revealed a marked disparity between sediment provinces and hydrographic conditions suggesting that some net dispersal of sediment occurs in the direction of the prevalent traffic, particularly in areas of weak currents, contributing to southward displacement of resuspended bottom sediments (Stanley et al. 1982).

The large number of vessels transiting the Canal make the waterway vulnerable to organotin pollution. Since the early 1960s' Tributyltin (TBT), and other organotin based compounds, have been utilized widely in anti-fouling marine paints. A decade later it was discovered that the organotin compounds have severe toxicological effects: imposex in prosobranch molluscs – the phenomenon whereby male sex characters are superimposed on females – has proved to be the most sensitive parameter to assess TBT pollution (Gibbs et al. 1987, Alzieu et al. 1991). The sudden decline in the populations of the gastropod *Murex forskoehli* Röding, 1798 (described as *M. tribulus*) in the Bitter Lakes was attributed to "the failure of reproduction as a result of imposex induction due to TBT contamination" (Hanafy 1996: 137). A study of the populations of *Thais lacera* (Born 1778) (described as *Th. carnifera*) at two sites along the Canal (north of Lake Timsah - Fig. 9), and on the Mediterranean coast (west of Port Said), indicated that all three populations suffered. But whereas all female specimens collected from the Canal exhibited the late stages of imposex (large penis, blockage of genital pore, sterility), 40% of the Mediterranean female specimens were normal, and the rest manifested only early stages of imposex. The Canal's populations of *Th. Lacera* "very abundant before" have declined precipitously, and the absence of juveniles indicated recruitment failure due to TBT pollution (Hanafy 1996: 142).

Fig. 9. View of Lake Timsah (Source: E. Riou, in Fontane, 1869).

The expansion of the Middle-Eastern oil production resulted in an increase in oil shipments from the Gulf through the Canal. Already Gruvel complained that the area south of the Canal had been "literalment empoisonnée par pétrole, le mazout ou leurs déchêts de fabrication" (1934), and later ".. les usines à mazout installées sur la partie voisine de la Baie [Suez] et le lavage des tanks des navires à moteur, de plus en plus nombreaux, chassent les animaux de la Baie quand ils sont suffisamment mobiles et les tuent sur place quand ils ne peuvent échapper à cet empoisonnement general de la Baie de Suez" (Gruvel 1936: 223). Tar and oil mark the banks of the Canal: "oil pollution of canal banks from ships' tanks was noticeable. In certain places, some banks were covered to a height of about 50 cm above the water level by a thin layer of tar" (Aleem 1984: 904). In 1981 water and sediment samples were collected in the Canal for detection of hydrocarbons (El Samra et al. 1983). Total hydrocarbons in water (aromatic fraction) reached 13.75 $\mu g\ l^{-1}$ at Toussoum, north of the GBL, and accumulated hydrocarbons in the sediments were highest (2.45 mg/kg dry wt) in Port Said. El Samra et al. (1983:100) suggested that the net northward water transport in November allows "the petroleum hydrocarbons originating in the Gulf of Suez (El Zaitia harbor) to merge with the Suez Canal water and form a mixed-water mass extending horizontally to distances of about 75 kilometers", whereas the summer current reversal prevents the hydrocarbon-contaminated waters of El Zaitia from entering the Canal. Moreover "The continuous and vigorous excitation of bottom sediments caused by tankers crossing the canal, together with the northward high-speed current, transported the suspended sediment particles with their high content of pollutants into the Mediterranean. The result is an accumulation of petroleum hydrocarbons in the sediments of the north end of the Suez Canal. " According to El Samra et al. (1983:100) their results point to a need "to evaluate the Suez Canal as a new source of pollution to the Eastern Mediterranean". In the 1990s Barakat et al. (1996) again analysed sediments collected from nine points along the Canal for oil pollution using gas chromatography. They concluded that "all the studied locations are suffering from pollution of oil that is spilled while shipping petroleum through the Suez Canal." (there, 755), reaching levels of 1067 and 1235 ppm dry weight between Lake Timsah and the Great Bitter Lakes. The presence of iron oxide in the samples was considered as footprint of oil pollution from tanker ballast washings, as tankers' sludge is often mixed with rust from the tanks' walls.

In the past half century the population of Egypt has grown rapidly: the combined population of 'Greater' Cairo and Alexandria was estimated at 5.8 million in 1965, 11.6 million in 1985, and 16.4 million by 1995. Following the construction of the Aswan Dam, Egypt moved from the age-old flood irrigation to agriculture based on constant irrigation and synthetic fertilization. Extension of urban water supplies and sewage collection systems, agricultural drainage and fertilizer use expanded dramatically. Municipal, agricultural and industrial wastewaters discharge into Lakes Menzalah and Timsah. Recent complaints of pollution in Lake Timsah, affecting both the Lake's fishery and local tourist industry, resulted in a study of the levels of chlorinated hydrocarbons, polycyclic aromatic hydrocarbons and polychlorinated biphenyls in the Lake's biota (Ahmed et al. 2001). Residues of endosulfan, DDE, Dieldrin, heptachlor epoxide and Dicofol were detected in the tissues of the grey mullet, *Mugil cephalus*, the Indo-Pacific swimming crab *Portunus pelagicus* and *Ruditapes decussata*, with concentrations of endosulfan reaching 124.8 mu g kg^{-1} in *M. cephalus* muscles. The concentration of polycyclic aromatic hydrocarbons was highest in *R. decussata* 48.9 mu g kg^{-1}, with Flourine and anthracene the most commonly detected compounds.

7 Hydrography and hydrodynamics

The Suez Canal is a hydrographicaly-complex body of water, passing five man-made lakes of widely diverse salinity – Menzaleh, Ballah, Timsah, and the two Bitter Lakes – on its course from Port Said to Suez. The perpetual enlargement (see above), and the region's other major engineering feat, the Aswan Dam, had profound impact on the Canal's hydrography and hydrodynamics.

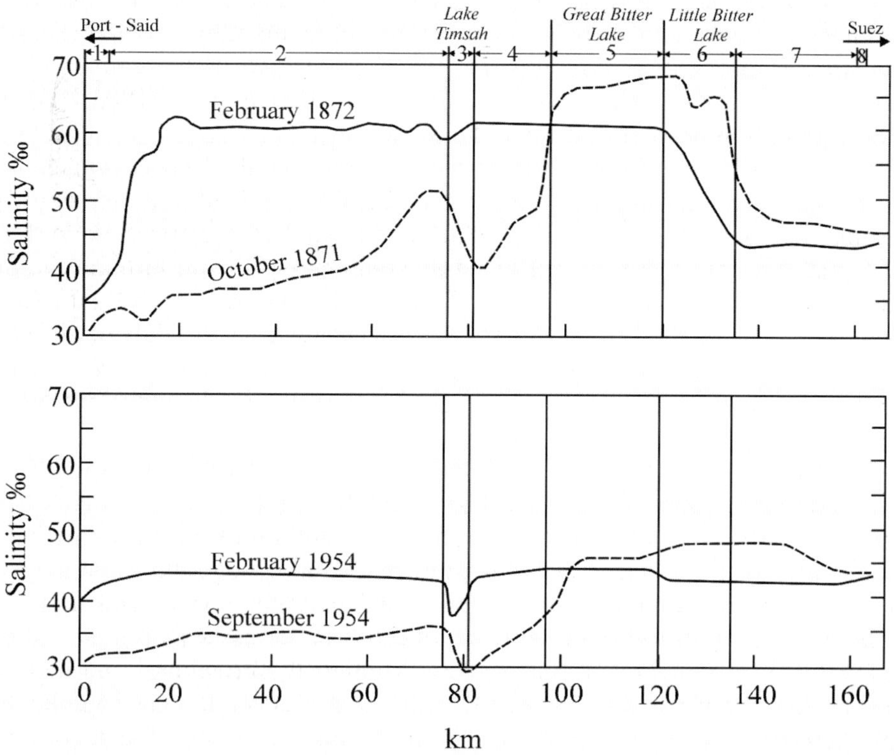

Fig. 10. Distribution of surface salinity along the Suez Canal (according to Morcos 1980).

The strategic importance of the Canal was such that the Director of Engineering and Architectural Works of the British Admiralty and a Navy hydrographer visited it three months after it had been opened for navigation on November 1869 (Richards & Clarke 1870). The first set of water temperatures and density measurements were taken at the behest of the British Admiralty in May, 1870 at 14 sites along the Canal. Tissot, an engineer in the service of the Egyptian Government, replicated the measurements at the same sites in February 1872, and published both sets (Tissot 1872, *fide* Morcos 1972, 1980). Unfortunately, the data proved useless due to difficulties in determining its accuracy (see Morcos 1972: 456).

Analyses made between 1872-1874 by the laboratories of the 'École des Ponts-et-Chaussées de Paris' determined that the salinity of the Mediterranean at Port Said ranged from 26 in October at the peak of the Nile flood to 37-38 in June; the salinity of the Bitter Lakes was 68, whereas in the Gulf of Suez it was 43 (Roux 1901(II):230). Lt. J. C. Richards of the H.M. Indian troopship *Malabar* is credited with the "most systematic and best recorded observations" (Morcos 1980: 296), collected during five cruises conducted between March 1871 and April 1872. The accuracy of the measurements is uncertain because the higher salinities encountered were beyond the upper limit of the hydrometer's scale, nonetheless they constitute the first evidence for the great variation in the salinity of the water along the Canal, and its extreme seasonal fluctuation (Fig. 10).

Richards' measurements taken in March, 1871, and February and April, 1872, indicate that the salinity of Canal waters from about 20 km southwards of Port Said and through the Great Bitter Lake (GBL) was 60-64 ppt, then dropping precipitously through the Little Bitter Lake and the southern end of the Canal to 42-44 ppt, thus matching the salinity of the water in the Bay of Suez. After the enlargement of the Canal in the late 1970s, the salinity in the Bay of Suez decreased, and does not exceed 42.7 ppt (Soliman 1995a). The measurements taken on October 1871 were radically different: the salinity of the water in the northernmost part of the Canal was less than the Levantine seawater (32 ppt) due to the incursion of the Nile waters, though the salinity in the GBL remaining high (maximum value 68 ppt) – effecting a salinity gradient within the Canal of 36 ppt. The seasonal change in the salinity in the northern part of the Canal was noted also by Captain G. S. Nares of the H.M.S. *Newport* : ".... in October, when from the overflowing of the Nile, the Lake [Menzalah] is at its highest, and the water in it is the freshest: the water in the Canal is considerably fresher than saltwater; and in April, when the Lake is low and salt[y], the water in the Canal is much salt[i]er than sea-water; and even equals that of the Bitter Lakes" (Nares 1871).

Evidently the salinity of the Bitter Lakes plays a significant role in determining the hydrography of the Canal. The salt bed at the bottom of the dry basin – composed of halite interbedded with calcium sulfate and alluvial sediments – that was later flooded and renamed the 'Great Bitter Lake', was 13 km long, 5 km wide and up to 13.2 m thick, and estimated at 97 million tons (Aillaud 1868, Lesseps 1874). The high salt contents of the Lake's water were attributed to the ancient salt bed and the high evaporation from the Lake's surface. Richards (*fide* Morcos 1980: 302), and Krukenberg (1888a) ascribed the formation of the high salinity waters mainly to the surface evaporation, whereas Keller (1883) considered that both evaporation and leaching of the salt deposits were producing the high salinity.

Measurements of the increasing depth and decreasing salinity of the waters of the GBL made by Voisin Bey, M. Levasseur, L. Durand-Claye and K. Natterer were used by of Fox (1929) to calculate that all the salt will have been dissolved before the end of the 20th century: "At the close of this century all of the salt will have gone into solution" (Fox 1929:843). Wüst (1934: 447, footnote) used those same measurements to predict: "eine lineare Extrapolation würde bei der Annahme einer Salzlagermächtigkeit von 13 m aus der bisherigen Tiefenzunahme von 4 m ungefähr das Jahr 2030 ergeben; würde man jedoch die Salzgehaltkurve linear fortsetzen, so käme man auf das Jahr 1970, wo der ozeanische Betrag von 42% erreicht wäre". Wüst (1951) argued that salinity had been decreasing asymptotically since the rate of salinity decline in the 1930s was smaller than in the late 19th century. He has also shown a linear correlation between the bottom salinity and the depth of the GBL based on data collected between 1868 and 1924. He attributed the sinking of the Lake's bottom, from 7,65 m in 1869 to 11,7 m in 1921, to the dissolution of the salt bed, though the continuous expansion of the Canal (see above) may have contributed in no small measure to that result. Krauss (1958), while using Wüst's data (1934 1935, 1951), refuted his assumptions and calculated that the salt bed has already been dissolved, and the salinity of the Lake's water stabilized. Core and grab samples from the bottom of the GBL obtained by the RV *Atlantis II* in 1965 were composed of mud and gypsum crystals, suggesting that "the salt deposits of the Great Bitter Lake seem to be exhausted and that the maximum salinity is probably due to evaporation" (Miller & Munns 1974: 295).

The incursion of Nile flood waters into the northern Canal was considered a major determinant of the seasonal salinity fluctuation. The Nile poured into the Mediterranean through the Damietta channel during the flood months (Aug. – Nov.) of 1955, 143 million m^3/day (Morcos 1960). Though the Damietta opens 60 km west of Port Said, the Nile waters moved with the prevailing counter clockwise coastal current, and effected nearly a 10 ppt decrease in surface water salinity at the entry of the Canal (Morcos 1960, table 1). Additionally, 14 million m³/day Nile waters were discharged into Lake Menzaleh (during the flood months of 1955, *fide* Morcos 1960). Lake Menzaleh is a shallow marshy lagoon, rarely exceeding 1m in depth, opening into the Canal 3.5 km south of Port Said through the El-Raswah Canal. The lake's main source of seawater is the El-Gamil outlet; the main source of freshwater before the construction of the Aswan High Dam was the annual Nile flood and groundwater (Shaheen & Yousef 1978), whereas after the Dam's completion freshwater are derived from a series of large drains (El-Wakeel & Wahby 1970). A recent study of surficial samples and cores from the Lake has shown that the brackish water (4-13 ppt) biota that existed before the closure of the Dam was replaced by freshwater (1-2 ppt) taxa (Reinhardt et al. 2001). The Nile is also connected with Lake Timsah in a canal, dug in 1863, that supplies the region with freshwater, and opening at the Ismailia lock. A small decrease in the Lake's surface water salinity was already discerned in the measurements taken in November 1871 (Morcos 1980), but the enlargement of the freshwater canal, to meet the needs of the growing population and irrigation demands, and the construction of the Mahsamah and Abu Gamous drains (Morcos 1975), meant a greater drop in surface salinity (Ghazzawi 1939, Morcos 1960, 1975, Miller & Munns 1974). The shallow (< 3m deep), much embayed Lake Timsah is distinguished by large surface salinity variation (El-Serehy 1992) and striking seasonal salinity stratification, with up to 20 ppt difference between surface and 12 m depth (Morcos 1960), and as much as 5 ppt difference within the top 5 m layer in October and November 1966, after the completion of the Dam (Miller & Munns 1974).

Salinity measurements taken at stations along the entire length of the Canal, from February 1924 to January 1925 by the "Compagnie du canal maritime" showed the surface salinity in the Lake was no higher than 53.5 ppt (Fox 1929). A decade later, monthly measurements of salinity (Jan. – Sept.) in the GBL, at 5 m depth, ranged from 44.4 ppt to 47.5 ppt, but the surface salinity gradient in the Canal in October 1934 was still a steep 25 ppt (Ghazzawi 1939). The distribution of salinity at the surface and at depths of 6 and 12 m, was studied from hydrographical data collected during thirteen monthly cruises conducted between November 1953 and November 1955 (Morcos 1960). The salinity in the GBL ranged from 44 ppt in April to 46.6 ppt in September, when it was exceeded by the salinity of the Little Bitter Lake (47.9 ppt). The Canal's surface salinity gradient in September 1954 was 18.9 ppt. After the cessation of the annual Nile flood following the completion of the Aswan High Dam, measurements taken on October 1966 showed that the Canal's surface salinity gradient was reduced to 11 ppt, though the surface salinity of the GBL was greater than 47 ppt, (Miller & Munns 1974) (Fig. 11).

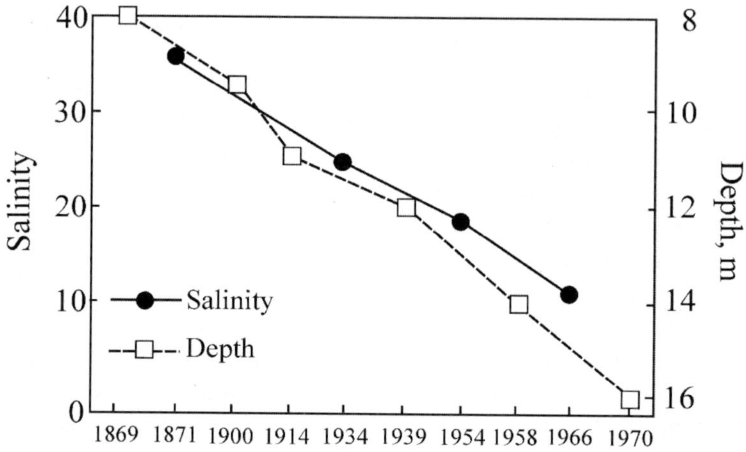

Fig. 11. Depth and surface salinity gradient in September – October in the Suez Canal.

The currents in the Canal are regulated by differences in the mean sea level (MSL) between the two ends of the Canal, tides, winds, and horizontal salinity gradients. The MSL at Port Said is nearly stable, with a range of less than 6 cm, whereas MSL at Port Taufiq may vary over 70 cm in the course of a year (*fide* Abril & Abdel-Aal 2000). During much of the year (October through June) MSL is higher in the Gulf of Suez than in the southeastern Levant, but it is reversed in the summer months, when the sea level in Port Said is slightly higher. The maximal tidal range is 60 cm at Port Said, 30 cm in Ismailia and 150 cm in the Gulf of Suez. High winds (over 10 m s^{-1}) at Port Said blow as a rule between 240 and 360° from the north (~ 65% of the time), and between 330 and 360° at Port Taufiq (~50% of the time). Insofar as much of the Canal runs along a north-south course, wind forced water circulation may be significant when MSLs at the entrances to the Canal are similar (Abril & Abdel-Aal 2000). Mean surface salinity in the Gulf of Suez is 42 ppt, as compared to 39.0 ppt in the southeastern Levant, whereas within the Canal five water masses were identified (Gerges & Stanley 1985), qualified by their salinity. Currents may change direction rapidly and irregularly, the net outflow determined by a complex hierarchy of forcing factors. However that may be, only fragmentary measurements of the direction, duration and velocity of the currents in the Canal are available, and most of the information is inferred from indirect indicators such as the salinity distribution and changes in the sedimentological assemblages. The assumption that the spatial change in salinity measurements reflects water masses circulation brought about a heated debate concerning the purported impact of the Aswan Dam on the current regime in the Suez Canal (Morcos 1967, El-Sabh 1968, Morcos & Messieh 1973a,b, Hassan & El-Sabh 1974, Morcos & Gerges 1974, Sharaf El Din 1974, Hassan & El-Sabh 1975,

Morcos 1975). A survey of the changes in mineral assemblages along the Canal revealed that the sediment provinces correlate well with the seasonal salinity distributions: the northward limit of Suez Bay-southern Canal sediment is in the northern GBL, driven by the net northward water transport from October through June; whereas the Mediterranean and Nile-derived sediment can still be found in the vicinity of Ballah bypass, indicative of the net southerly flow to the Bitter Lakes in summer. The characteristics of the mineral assemblages confirm that the southern Canal, south of Kabret, is undergoing erosion by strong tides that tend to flush unconsolidated material northward to the Bitter Lakes and southwards to Suez Bay. Still, a certain disparity between sediment provinces and hydrographic conditions was noted, suggesting that some net dispersal of sediment occurs in the direction of the ship traffic, particularly in areas of weak currents, contributing to the southward displacement of resuspended sediments (Stanley et al. 1982).

Few measurements were taken of current directions and velocities along the Canal, and those were made at a limited number of sites. Measurements were taken by two continuous recording current meters at two sites, north and south of the GBL, from 1933 to 1937 by the "Compagnie du canal maritime" (Baussan 1938). The data points to a regime where from winter to early summer the current flowed from the Gulf of Suez to the Mediterranean; from June to August, the current was reversed in the northern part of the Canal, with waters flowing into the GBL from both ends of the Canal; a complete reversal occurred in September when the current flowed through the Canal from the Mediterranean into the Gulf of Suez; in October the current was reversed in the southern part of the Canal, waters again flowing into the GBL from both ends of the Canal; the current turning northward in the northern Canal in November. Measurements from tide gauges installed in 1935-1937 in the northern and southern ends of the Bitter Lakes, and at Kabret, between the Great and little Bitter Lakes, indicated that from June to October the MSL at Kabret was higher than at both ends of the Bitter Lakes (Morcos 1975). These results are supported by Miller & Munns (1974: 302) who calculated an evaporation potential of 3.9-6.5 mm^3/day from the estimated 232,438 km^2 surface of the Bitter Lakes, and wrote (there, 306) "... net inflows into the Canal from both northern and southern ends of the Canal in the same period, together with extra volumes of Nile water entering Lake Timsah, are reconciled with the excessive evaporation potential of the Bitter Lakes". The net outflow into the Mediterranean was estimated at 5,250 million m^3, and the late summer southward net outflow at 150 million m^3 (*fide* Morcos 1975). A series of current measurements made in November 1973 midways between Lake Timsah and GBL recorded wind-dependent southward currents in the upper 2 m layer (up to maximum of 25 cm s^{-1}); the midlayer (4-8 m) current flowed northward at speeds varying between

10-25 cm s^{-1}; and the bottom layer (8-10 m) was either still, or flowed southward at 5 cm s^{-1} (A. Hecht, IOLR, pers. comm.). A survey carried out by the Japan International Cooperation Agency during August – October 1983, when the MSL was higher in Port Said than in Port Taufiq – recorded mostly tidally regulated low velocities at Port Said (>0.2 m s^{-1} less than 2%); a southerly current of higher velocity south of Lake Timsah (>0.4 m s^{-1} 17%), and south of the GBL (>0.4 m s^{-1} 35%); and tidally directed high velocities at Port Taufiq (>0.6 m s^{-1} 28%). The blockage of the Canal between October 1973 and June 1974 by an earth dam built north of the GBL greatly reduced circulation of the Lake's water. The increased residence time of the water precipitated the formation of a 2 m thick bottom brine layer with salinity exceeding 300 ppt (Meshal 1975). Conversely, it was postulated that enlargement of the Canal would effect higher velocities, and lower salinity extremes by reducing water residency in the GBL (Miller & Munns 1974, Sharaf El Din 1974). Following the great increase in the Canal's depth and width in the late 1970s (see above), a model of tidal currents using a recent (1994) bathymetric map supplemented with tidal currents and MSL measurements has shown that the "... deepening of the canal to depths more than 19.0 m has increased the M2 tidal current in the southern part of the canal from 37.0 cm/sec to more than 45.0 cm/sec. Hence the tidal current alone .... may reach values more than 100.0 cm/sec" (Soliman 1995b).

In 1876 Lesseps wrote "L'évaporation étant plus active dans le centre de l'isthme qu'aux deux entrées de Suez et de Port Said, le courant vient presque toujours du sud au nord à partir de Suez jusqu'aux Lacs Amers et du nord au sud à partir de Port Said". Long dismissed, Lesseps' observations were validated a century later. The data supports an evaporation and wind-driven current reversal during the summer months when water flows from both ends of the Canal into the Bitter Lakes.

8 Studies of the biota of the Suez Canal and the adjacent marine biotas 1869-2005

8.1 Studies of the canal biota prior to 1920

Even before the Suez Canal was fully excavated Vaillant (1865: 97) argued that "Le percement de l'isthme de Suez.... offrira... une occasion précieuse de constater les phénomènes que doivent amener l'émigration des espèces et le mélange des faunes". Vaillant advocated what today would be considered a 'baseline study', and raised provocative and prescient questions: "La mer Rouge et la mer Méditerranée montrent, quant à leurs mollusues, ... des différences considérables, mais il serait nécessaire de chercher à bien fixer d'avance quelles elles sont maintenant pour pouvoir mieux juger plus tard des changements qui pourront survenir. Sans aucun doute il va y avoir transport des espèces, celles-ci, en changeant de milieu, vont-elles conserver tous leurs caractères ou subir quelques modifications? Celles qui ont entre elles certaines affinities génériques vont-elles se croiser et donner lieu à de nouvelles variétés? Ces dernières seront-elles transitoires ou stables? Ce sont autant de questions qu'il importerait beaucoup de pouvoir résoudre, et de longtemps, sans doute, une occasion aussi favorable d'aborder quelques uns de ces points ne pourra se rencontrer". To this end Vaillant studied the malacofauna in the vicinity of Port Suez (24.i.-7.v.1864), and intended to examine the environs of Port Said (not accomplished for want of time), for he believed those localities "à recevoir les premiers colons, à fournir les premiers emigrants" (Vaillant 1865:98).

Most of the earlier studies of the canal and canal-proximate biotas concerned mollusks, and no wonder: the handsome shells of the Red Sea were depicted in the early natural history books that were at once scientific studies and decorative works of art, and were sought after by collectors for their "cabinets de curiosité". The first scientific collection from Suez was assembled by the ill-fated Danish Expedition to Arabia [1761-1767] in October 1762. The bulk of the material collected by the expedition's naturalist, P. Forsskål, was shipped, after his death in Yemen, by a roundabout way to Copenhagen, where the molluscan collections were dispersed among the royal "Kunstkammer" and private collections, and the original labels lost. Forsskål's posthumous "*Descriptiones animalium*..." (1775) was culled from 1800 sheets of his field notes that were subsequently lost (Yaron & al. 1986). The Expedition d'Egypte [1798-1801] fared no better: Savigny, the 'savant' in charge of the invertebrate studies, collected in Damietta and Lake Menzaleh (November 1798), in Suez (December 1799-January, 1800), and followed Napoleon to Syria [present day Israel] (February-June, 1799). On his return to Paris in 1802, Savigny began working on the collections, producing several manuscripts, and supervising the illustration and engraving of the plates (Savigny 1817), but in 1817 his deteriorating health and failing eyesight put an end to his studies. Audouin, charged with completing Savigny's study of Red Sea shells, but having no access to the original notes, now jealously guarded by the offended Savigny, introduced many errors, and was unable to provide locality information for the specimens (Audouin 1826, Bouchet & Danrigal 1982). His work was severely criticized: "un tissue d'erreurs et de fautes grossières" (Fischer 1865: 242); "questo lavoro può dirsi destituito di qualsiasi valore scientifico e come tale ne terremo poco conto nei cataloghi che formano parte della nostra memoria" (Issel 1869: 7). A similar misfortune befell the collections made by Hemprich and Ehrenberg who took part in the calamitous German scientific expedition to the Levant in the 1820s: "the shells from both seas [Mediterranean and Red Sea] were mixed when brought home; ... consequently the list of species common to the Red Sea and the Mediterranean, given by R. A. Philippi ... based upon these materials, does not deserve credit" (Martens 1865: 237). At the time, Philippi's (1836) argument that the mollusk fauna of both seas has much in common, was uncontested, and widely cited (Woodward 1866). It was thus of great importance to establish an accurate record of the fauna on both ends of the Canal prior to its opening. Fischer (1865, 1870, 1871), who studied material from Port Said, Suez and the Bitter Lakes, some of it forwarded by the "Compagnie du canal maritime", determined that "La comparaison des faunes ne laisse pas l'ombre d'un doute dans notre esprit, et nous établirons notre conclusion en quelques mots: *il n'existe aucune coquille commune à la mer Rouge et à la Méditerranée*" (1865: 245). Fischer was certain "que l'achèvement du canal maritime aura peut-être pour résultat de modifier les faunes, actuelement si distinctes, des deux rivages méditerranéen et érythréen de l'isthme" (1865: 241). In '*Malacologia del Mar Rosso*' Issel (1869) reviewed previous records and analyzed the material he himself had collected in the spring of 1865 in the northern Gulf of Suez, and Italian collections of fossil molluscs. An invaluable contribution to the knowledge of the Suez malacofauna prior to the opening of the canal was made by MacAndrew, a wealthy Liverpool merchant, who, supplied with "all the requirements for encampment in the desert, including a good cook", went on a "dredging excursion" in February and March of 1869 (MacAndrew 1870: 429). Though the dredging was cut short as "our stores were becoming exhausted, and our water putrid" (there, 430), MacAndrew assembled an important collection, including many small-sized species that had been poorly represented in previous collections. He too commented on "The extraordinary dissimilarity between the fauna of the Red Sea and the Mediterranean... which appears to be confirmed by further researches on both sides of the isthmus" (MacAndrew 1870: 431).

The opening of the Suez Canal engendered debates on its impact on the Erythrean and Mediterranean biotas, yet for the next 50 years the documentation of the biota in the Canal itself and the changes in the biota in the bordering marine environments were left to "Forschungsreisender" – a now extinct breed of learned amateurs and natural scientists who styled themselves 'scientific travelers'.

Keller, a professor of Zoology in Zurich, considered the canal "... auch als Karawanenstrasse für die thierischen Bewohner beider Meere benutzt" (1883:3). Keller traveled to Egypt in 1882 and 1886 and collected between Port Said and Suez, seeking evidence for the presence of Red Sea and Mediterranean species in the Canal.

Fig. 12. "Migrations-Schema" documenting the advance of the Red Sea and Mediterranean species along the Canal (Source: Keller 1883).

Keller believed that species proceeded from both ends of the canal, and produced a "Migrations-Schema" that charted the advance of the Red Sea and Mediterranean species along the Canal (1883, Fig. 12). He identified 10 Mediterranean species and 17 Red Sea species from the canal itself, 5 Mediterranean species in the Red Sea, and 4 Red Sea species in the Mediterranean. Doubts have been raised concerning some of his identifications, as well as his sources: his record for the presence of Common sole, *Solea vulgaris*, in the Red Sea was purchased in the Suez fish market! However, the importance of his work lies in confirming the role of the canal as a 'corridor' for the movement of species and in the transformation of the adjacent marine biotas: "Es unterliegt heute keinem Zweifel mehr, dass in den ersten zwölf Jahren seit der Eröffnung des Lesseps'schen Kanales bereits fühlbare Veränderungen in den beiden Faunen aufgetreten sind, und dass an den beiden Kanal-Enden für die Zukunft noch weitere Veränderungen zu erwarten stehen" (Keller 1883: 36). Keller was well aware that his work was the first scientific account on the subject and boasted: "Wenn spätere Beobachter meine Ergebnisse zu vervollstandigen im Falle sein werden, so glaube ich immerhin einen Einblick in die wichtigsten Migrationserscheinunen im Lesseps'schen Kanal erlangt zu haben" (Keller 1883: 4). When a younger colleague, Professor

Krukenberg of Jena, traveled to Egypt in 1886/7, and subsequently published his own analysis of the problems of bioinvasions across the Isthmus, criticizing Keller's premises concerning the possibility of an earlier contact between the two seas (Krukenberg 1888), Keller was greatly incensed. Keller published an extended rebuttal, and moreover, accused Krukenberg of appropriating his Canal record of the Indo-Pacific 'upside-down' jellyfish, *Cassiopea andromeda*: "Krukenberg meine Belehrung ohne Bedenken angenommen" (Keller 1888: 390). Very shortly after that scathing exchange Krukenberg died at the age of 37 of an overdose of cocaine (Anonymous 1889).

Valuable studies were conducted by the employees of the "Compagnie du canal maritime". Fuchs, who conducted the geological survey of the Isthmus in 1864/5, published on the subfossil molluscs of the region, and collected, on a visit in 1876, a few living molluscs from the Menzaleh and Timsah lakes. Two avid amateurs, A. Bavay, 'Pharmacien en chef de la Marine', and L. Tillier, 'directeur du transit du Canal', were convinced "C'est certainment une étude de zoologie géographique des plus intéressantes" (Bavay 1897: 199). They are to be thanked for assiduously assembling data on the molluscs and fish of the canal. Their samples were collected by divers and by dredgings, and the catches of the native fishing boats operating along the canal were inspected, as well as the merchendise offered at the Ismailia fish market (Bavay 1897, 1898, Tillier 1902, Tillier & Bavay 1905, 1906).

Efforts by travelers to sample the biota provided scientists with sporadic glimpses into the movement of species along the Canal: Martens (1887) received molluscs collected by Drs. Gottsche (xii.1881) and Pastor (vi.1882); Giesbrecht's records (1897) stem from a single plankton haul taken in the Bitter Lakes by a naval physician, Dr. Krämer; Jordan & Hubbs's record of *Atherinomorus lacunosus* (as *Atherina forskali*) (1917) is based on a day's collection at Port Said. The first planktonic Copepoda described from the Canal were collected in the bathroom of a passing steamer – Professor Herdman's ingenuous collecting method consisted of "a tow-net being fixed to the tap so that sea-water was running through it day and night" (Thompson & Scott 1903).

8.2 Studies of the canal biota 1920-present

Fifty-five years after the opening of the Canal, 'The Cambridge Expedition to the Suez Canal', consisting of three research students and an amateur field naturalist "of the good old school" (Calman 1950:587), embarked on an investigation of the "intermingling of the Mediterranean and Red Sea organisms in the Suez Canal" (Gardiner 1924:520). The expedition was charged with ascertaining "what forms have passed through the Canal zone from the Mediterranean to the Red Sea and vice versa, when they passed through, whether in the prehistoric period, in the times of the earlier or of the present canal, how they passed through, whether by swimming, by drifting, by attachment to ships, or by other means. All these facts lead up to the question why some forms of life can get through the Canal and others cannot, and it is hoped that the expedition will throw light on marine migrations in general, the area being one which can be periodically investigated at small cost" (there, 521). Never was a graduate student charged with such an ambitious agenda, but Fox, the research student who headed the expedition, was uniquely suited to the task – during WWI he was stationed in the Levant, and afterwards he spent 4 years in Cairo as a lecturer in biology. Fox not only made excellent use of the time (less than three months) the frugally-funded expedition spent in the field, but accomplished an even more demanding task – a timely publication of the scientific results. Twenty-nine authors contributed 36 chapters, besides Fox's introduction and summary, for a hefty volume (874 pp) of the "Transactions of the Zoological Society of London", published in six sections between December 1926 and January 1929. This 'tour-de-force' is the only thorough study on the biota of the Suez Canal to date. Subsequent studies have not dealt with the range of biota considered by 'The Cambridge Expedition', or were of limited scope.

Gruvel, a fisheries expert with the Laboratoire des Pêches et Productions colonials d'origine animale, Muséum National d'histoire Naturelle, Paris, who was well familiar with the Levantine fisheries, recognized the economic importance of the entry of the Erythrean species into the Mediterranean: "Mais si l'ensemble de ces mouvements d'espèces animals et végétales constitue, pour la Science, une fait fort intéressant, les passages définitifs de ces espèces à traverse la totalité du Canal présentent un résultat économique également très important" (Gruvel 1936:228). The collections of fish, molluscs, and algae he assembled on his visits to the Bitter Lakes in the spring of 1932, and Lake Timsah the following year, and his observations on the fishery methods and catch composition along the Canal and its lakes (Chabanaud 1932, 1933a,b, 1934, Gruvel & Chabanaud 1937, Gruvel & Moazzo 1933, Lami 1932, Moazzo 1939), served a basis for his study on the passage of species through the Canal (Gruvel 1936). The importance of the plankton as a food source for commercially important fishes, and its significance as a water mass characteristic, prompted the surveys of the Suez Canal plankton executed by the Egyptian Fisheries Administration (Macdonald 1933, Ghazzawi 1939). Tortonese, who spent some time on the shores of the Timsah and Bitter lakes in 1944-45 "regretting very much that the war conditions of those years prevented me from a more intensive biological study" (1954:1), assembled valuable information on fish and echinoderms (Tortonese 1947, 1948, 1952), and a "three weeks' holiday dredging in the Great Bitter Lake" (Beets 1953:97), yielded data on its crustaceans and molluscs (Holthuis 1956, Hoenselaar & Dekker 1998).

Between 1967 and 1973 collections were made by Israeli scientists, mostly along the east shore of the Bitter Lakes. Though of limited duration, the expeditions provided interesting data on the progress of Erythrean biota through the Canal. The contributions of 22 scientists who identified much of the material, from macrophytes to fishes, were assembled in a special volume of the "Israel Journal of Zoology", published in 1972. A short visit in January 1975, by Norwegian scientists, to sample fouling organisms on the ships trapped in the Bitter Lake since 1967, yielded new data on the phytoplankton (Heimdal et al. 1977).

The 1980s-1990s saw a rise in studies conducted by Egyptian scientists and students: the species composition and population dynamics of the plankton have been studied (Dowidar 1974, El Serehy 1992, El-Serehy & Shalaby 1994, El-Serehy et al. 2000a,b), polychaetes (Ghobashy & El–Komi 1980a,b, Selim 1997, Shalla & Holt 1999), molluscs (Fouda & Abou Zeid 1990, Abdel-Fattah et al. 1992, Mohamed et al. 1992, Mohammed 1997), and the biology of fishes (El-Etreby 1986, Fouda 1995, Fouda et al. 1993).

8.3 Studies of the Erythrean biota in the Levantine Sea

In his concluding remarks Fox (1929: 844) noted " ..there is a need of an investigation of the coastal faunas of Northern Egypt and Palestine". But even before the "Cambridge Expedition" departed Britain, an internist from Breslau, W. Steinitz, recognized the scientific significance of the movement of biota through the Canal and published a pamphlet at his own expense (Steinitz 1919), where he pointed out that no scientific institute had taken on a comprehensive study of biotic transfer: "The Suez Canal … connects two oceans having each a very different fauna from the other, which throws up a number of new problems: Does an exchange of fauna take place between the two oceans by means of this canal? Do the immigrants from the other side flourish on this or die? Has such an immigration caused important changes in the fauna of the eastern basin of the Mediterranean?". Steinitz studied the marine biota of Palestine in 1924 and 1925, and noting 15 species of Indo-Pacific origin, he urged the creation of a Marine Station in the Levant dedicated to the study of the transformation of the Levantine biota (Steinitz 1929, 1936). The advent of

Erythrean biota in the Mediterranean was noted initially in the lists of the Levantine malacofauna (Monterosato 1899, Pallary 1912a,b, 1919, 1938, Gruvel & Moazzo 1931, Haas 1937, 1948, Steuer 1939, Carmin 1946); decapod crustaceans (Gruvel 1928, 1929, 1930a,b, 1931, Monod. 1930, 1931, 1932) and fish (Gruvel 1929, 1931, Liebman 1934).

In 1933 the Marine Laboratory of Alexandria initiated a survey of 'The Fishery Grounds near Alexandria' by inviting Steuer, co-director of the Institute of Marine Biology of Rovinj. Steuer distributed the material he collected during his three months' stay in Alexandria among his colleagues. Their reports, published between 1935 and 1939, in the 'Notes and Memoirs, Fisheries Research Directorate' are "a welcome contribution to our knowledge of the fauna of that basin in which we find particularly interesting biological conditions on account of …the communication with the Red Sea by the Suez Canal" (Balss 1936).

The intensive investigations of the biota of the continental shelf of Israel by the Sea Fisheries Research Station in 1946-1956 resulted in a series of publications that highlighted the extent the Erythrean taxa have been established along the coast (Ben-Tuvia 1953a,b, Gottlieb 1953, 1960, Holthuis & Gottlieb 1958, Pérés 1958, Tebble 1959, Oren & Steinitz 1959). In 1967, a joint program by the Smithsonian Institution, the Hebrew University of Jerusalem, and the Sea Fisheries Research Station, Haifa, was established to investigate the spread of the Erythrean biota in the Levant (Israel, Cyprus, Rhodes) and its impact on the native biota. The scope of the program was expanded following the 'Six Days War' to include the Sinai coasts and the Suez Canal itself (see above). H. Steinitz, fulfilling his father's dream, headed that program with W. Aron. Steinitz, listed 140 Erythrean and Indo-Pacific species known to have crossed the Suez Canal into the Mediterranean (H. Steinitz 1970). Steinitz´s work was continued by Por (Por 1978).

The advent and spread of Erythrean species in the Levantine Sea were repeatedly recorded over the past three decades, and it was widely perceived that the littoral and infralittoral biota of the sea is undergoing a rapid and profound change (Por 1978). Yet, no concerted effort had been undertaken since the early 1970s, and most of the records stem from fortuitous finds. Recently, a series of Atlases that summarizes the extant knowledge of the scale and impact of 'Exotic species in the Mediterranean' documents the prevalence of Erythrean taxa (www.ciesm.org/atlas/). A basin-wide targeted effort to survey the presence and abundance of the Erythrean species, and study their biology and ecology is wanting.

9 The drivers of the Erythrean invasion

The Suez Canal was opened in 1869, yet Erythrean species newly appear in the Mediterranean decades and centuries later. What drove the invasion in the first place, and what allows it to continue long after the trans-isthmian corridor had been well established?

Following the 'The Cambridge Expedition to the Suez Canal', sent out fifty-five years after the opening of the Canal, to study its biota and to assess why some taxa passed through the Canal and others cannot, Fox (1926: 16) summed up the "Factors favouring and hindering migration through the Canal", and identified impediments such as the canal's salinity and temperatures extremes, turbidity, and width; the last, he declared, to be "the first factor which would act as an hindrance to migration" (1926: 17). He recognized the decreasing salinity of the GBL: if " .. a high salinity barrier to migration is in reality formed by the Bitter Lakes, this barrier was more pronounced in the past than it is to-day", and wondered whether the "great preponderance of Red Sea over Mediterranean species" in the Canal may be correlated with "a general set of currents in the Canal from south to north which would account for this excess of erythraean species among the forms distributed by currents"(Fox 1926: 17). Bodenheimer (1935: 457), who studied the ecology and zoogeography of the SE levant, maintained that "Most of the animals which manage to penetrate the Canal successfully are, as a rule, killed by these changes of salinity. … For ten months the current in the Suez Canal is directed towards the Mediterranean Sea and only for two months (August, September) towards the Red Sea. This fact explains the prevalence of the migration of Indo-Pacific forms into the Mediterranean as against the migration in the opposite direction". Gruvel (1936) listed the causes that favor, impede or arrest the movement of species through the Canal. Among the "causes favorisant le penetration des espèces" he cited shipping and currents, whereas oil pollution, turbidity, temperature and salinity gradients, and unfavorable currents impede transport. Ben Tuvia too (1966) referred to the extreme conditions in terms of temperature and salinity in the Canal itself, and to the seasonal salinity and temperature gradients in the Levantine basin. But he (Ben Tuvia 1966: 257) was the first to suggest that the nearly unidirectional invasion may relate to the impoverishment of the eastern Mediterranean biota: "The Red Sea fauna is much richer than that of the Mediterranean both in density of population and in the number of species". He also realized that the "peculiar hydrographic conditions of the Levant Basin" and in particular, the hydrographic changes, like the rise in surface temperature in the Levantine Basin, and the salinity drop in the GBL, "may have facilitated the recent wave of migration through the Suez Canal" (1966: 258). Baskets of living pearl oysters offered for sale in the fish market of Beirut brought the magnitude of the phenomenon to the attention of Gunnar Thorson, the noted Danish marine biologist (Christensen 1972). Thorson proposed that the chances of Erythrean taxa to pass the Canal and settle in the Mediterranean rose because the "Nasser plan" to deepen and widen the Canal (see above) increased the speed of tidal currents through the Canal, the bottom layer of salt in the GBL had been dissolved, and the cessation of the Nile outflow with the completion of the Aswan High Dam caused the stabilization of salinity in the SE Levantine Basin (Thorson 1968, 1971).

The Erythrean invasion is predicated on the region's geological history, environmental factors and anthropogenic activities. The latter include physical changes in the Canal that impacted its hydrography and hydrology, and may have increased its potential as a "corridor"; and changes to the Levantine marine environment that have made it more susceptible to invasion by modifying its hydrological properties and species diversity, and destabilizing the shelf's community structure.

9.1 Faunal impoverishment in the Levantine Sea

About five million years ago Atlantic waters, bearing tropical and subtropical Atlantic species, refilled the Mediterranean Sea. During the Pliocene (5.3-1.6 mya), the cooling of the sea prompted the substitution of temperate for tropical biota (the 'atlanto-mediterranean' biota). During the Quaternary glacial periods – characterized by reduced salinity and temperature, and changes in the stratification patterns – a north-eastern Atlantic boreal biota entered the Mediterranean, whereas in the interglacial periods warmer 'Senegalian' elements were added. The Mediterranean Sea has only a fraction of the biota found in the temperate eastern Atlantic Ocean, and the Levantine Sea has less than half the number of the benthic species found in the whole of the Mediterranean (Fredj 1974). This profound faunal impoverishment has been attributed to the limited access through the narrow and shallow straits of Gibraltar and the Siculu-Tunisian sill, the relatively recent recolonisation after the Messinian crisis, to Quaternary climatic fluctuations and to the sea's extreme oligotrophy (Fredj 1974, Yacobi et al. 1995). Sarà (1985) proposed that the high temperature and salinity prevailing in the Levantine Basin made it unsuitable for many atlanto-mediterranean taxa, and those atlanto-mediterranean organisms present, are presumably at the limit of their ecological tolerance. Williamson & Fitter (1996) postulated that a successful invader occupies a vacant niche, be it a trophical niche or a functional niche. Thus, the niches unoccupied by the missing atlanto-mediterranean taxa and only partially occupied with tropical Atlantic species, are vulnerable to invasion. Consequently, when the tropical Erythrean species arrived, there were few ecological obstacles to prevent their successful implantation.

9.2 Rising seawater temperature

The Erythrean aliens are thermophilic, originating in tropical waters, and to establish populations they are thought to require "temperatures high enough for the reproductive processes and development of eggs, and minimum winter temperatures above their lethal limits" (Ben Tuvia 1966: 254). Thus, it stands to reason that rising Mediterranean sea-water temperature enhances the reproduction, growth, and survival of the Erythrean aliens, and provides them with a distinct advantage in interspecific interactions with the native Mediterranean biota. In a few cases there is clear concurrence between rising water temperatures and changes in the composition of resident biota that lead us to believe they facilitate invasions by favoring alien over native biota.

For some of the most successful Erythrean invasive species, the initiation of the explosive population growth coincided with a rise in winter water temperatures. The "abrupt rise in catch of the lizard fish *Saurida undosquamis*, taken by otter trawlers with the usual gear on the regular fishing grounds" (Oren 1957b: 1) was attributed to a rise of 1-1.5°C in sea temperature during the winter months of 1955 (Ben Yami 1955, Chervinsky 1959). Few individuals were caught before, and the fish had been only "of taxonomic and zoogeographic interest" (there, 1), yet, "In the summer of 1955, unusal numbers of fingerlings were found in the cod ends of trawl nets. The bulk of them consisted of two Red Sea species, the yellow-striped goatfish and the lizardfish" (Ben Yami & Glaser 1974: 364). In 1955-56 the lizardfish became commercially important, constituting for a few years up to one fifth of the total annual trawl catch along the Mediterranean coast of Israel, and over half of the total catch on the shallow shelf opposite El-Arish.

In the late 1940s the Erythrean goldband goatfish, *Upeneus moluccensis*, made up 10-15% of the total mullid catches off the Israeli coast (Wirszubski 1953). Following the exceptionally warm winter of 1954-55, its percentages increased to 83% of the catch, replacing the native red mullet, *Mullus barbatus*. Both native and Erythrean mullids have a similar diet, and occupy muddy bottoms shallower than 75 m, but whereas the red mullet spawns from April to June with a peak in May, the goldband goatfish spawns from June to September (Wirszubski 1953). The considerably higher water temperatures at depth of 75 m in May of 1955 and 1956 may have resulted in poor survival of the red mullet spawn; that year the goldband goatfish has had the same temperatures during its spawning period as in previous years, yet an unusually large year class survived. In previous years, the young red mullets would settle to the bottom during July through September, where they have had a distict size advantage over the later-spawned goldband goatfish. The failure of the 1955 red mullet year class may have left their niche only partly occupied, to the advantage of the Erythrean species, and the unusually warm waters enhanced the latter species' survival rate (Oren 1957b). The minimum seawater winter temperatures over the SE Levantine shelf never returned to their early 1950s value of 15°C (Ben Yami & Glaser 1974), suggesting that increasing the minimum temperature favors thermophilic aliens by increasing their recruitment relative to native biota.

During the last 20 years considerable changes in the Adriatic ichthyofauna were observed, and a rise in the numbers of thermophilic species was noted, including six Erythrean taxa – *Hemiramphus far, Parexocetus mento, Sphyraena chrysotaenia, Leiognathus klunzingeri, Saurida undosquamis* and *Epinephelus coiodes* (Dulčić & Grbec 2000, Dulčić & Lipej 2002). Long-term sea surface temperature measurements in the Adriatic point to two periods of elevated temperatures, 1985-1987 and 1990-1995 (+0.15°C and 0.30°C respectively), with extremely high surface temperature along the eastern Adriatic in the summer of 1992 (Zore-Armanda et al. 1999), together with a salinity increase. The presence of the Erythrean fish have been documented only since the mid 1980s, though the Adriatic ichthyofauna has been well studied. It seems that the changes in the pattern of water exchange between the Adriatic Sea and the Mediterranean and a rise in the eastern Adriatic surface temperatures in 1985-1987 and 1990-1995 is correlated with the appearance of Erythrean species and the increase in populations of previously rare thermophilic Mediterranean biota.

9.3 The decreasing salinity in the Suez Canal

The Suez Canal traversed five man-made lakes on its course from Port Said to Suez, which salinity in the first decades of the Canal's existence ranged from brackish to hypersaline: the salinity in the Bitter Lakes, overlaying an immense salt bed, was as high as 161 ppt right after opening the canal in 1869, whereas the shallow marshy Lake Menzaleh was inundated by the Nile flood waters and its salinity ranged from 4 to 13 ppt.

As sea water flowed through the Canal the salinity gradient diminished precipitously: the salinity in the Bitter Lakes was reduced to 68 ppt in 1872-1874, less than 53.5 ppt in 1924-1925, and 44 ppt in the late 1960s (Thorson

1971). The dissolution of the Bitter Lakes' salt bed was complete by the 1960s, and its present salinity is due to evaporation. Already 75 years ago Fox (1929: 843) remarked that once the salt bed is dissolved the "barrier to the passage of certain organisms through the Canal will have disappeared".

As protection against the erosion generated by the transit of ships, the banks of the Canal were surfaced by revetments of stone and steel piles, that also serve to isolate its waters from the seasonally fluctuating salinity of the brackish Lake Menzaleh, leaving a single opening through the El-Raswah Canal.

The amelioration of the salinity extremes in the Canal "… means there will be no hindrance for the passage of Red Sea animals that are already adapted to such high salinities" (Thorson 1971: 842).

9.4 Damming the Nile

The annual Nile floods had for millennia disgorged an enormous volume of fresh water and sediments into the Levantine Basin. The sluices of the 'Delta Barrage' built in 1861 at the head of the Nile delta to improve irrigation, were regularly opened in early August and closed in December. Maximum flow occurred during September when the rain waters from the East African highlands reached the Nile Delta, and the salinity off the delta dropped below 1 ppt (Halim 1960). The average volume (for the years 1959-1963) was 43 km^3, but in 1964, just before the diversion of the river, an exceptionally generous flood was recorded, reaching nearly 53 km^3 (Oren 1970). The flood waters formed a distinct "Nile Stream", a turbid coastal current of diminished salinity along the southeastern Levantine coast, its dimensions subject to interannual fluctuations in rainfall upstream. The effect of the Nile flood regularly extended beyond Haifa Bay, where late summer sea surface salinity was reduced to 34 ppt (Oren 1969). The Nile floods coincided with the summer current inversion and the inflow of Mediterranean water into the canal, resulting in markedly decreased salinity, as low as 26 ppt in the northern portion of the canal at the peak of the Nile flood. Since the completion of the Aswan High Dam, the Nile flow has been drastically reduced, and there is no appreciable seasonal decline in salinity.

The Nile waters and sediments (140 million tons annually) were the major source of nutrients in the severely oligotrophic Levantine Basin (Halim 1960). The 1964 flood carried dissolved and silt-adsorbed nutrients estimated at 8200 tons of dissolved phosphate, five times as much silt-adsorbed, and 410,000 tons of silicate (Halim et al. 1995). The nutrient load spurred a prodigious phytoplankton bloom, increasing the standing crop 200 fold, that attracted grazers and planktivors.

Since the completion of the Aswan Dam in 1966, the outflow of Nile waters into the Mediterranean has nearly ceased. The mean sea surface salinity along the Egyptian coast increased by an average of 2-3% (Al Kholy & El-Wakeel 1975), and along the Israeli coast the mean salinity promptly increased from 38.76 ppt to 39.07 ppt, and surface seawater salinity later rose to 39.5 ppt. Whereas due to the nilotic dilution surface waters were in the past the least saline layer, from 1965 the least saline layer is that influenced by the Atlantic current at depth of 75 m (Oren 1970). That "salting of the coastal waters" was blamed for the "penetration of Red Sea types" into the Mediterranean (Al Kholy & El-Wakeel 1975: 168).

With the sediments entrapped in lake Nasser, the nutrients were sharply reduced, the phosphates to 4%, silicates to 18% of their pre-damming levels, inevitably the entire primary production off the Delta was decimated, phytoplankton levels were reduced by 100 fold (Al Kholy & El-Wakeel 1975, Halim et al. 1995). Al Kholy & El-Wakeel (there, 244) acknowledge "There is a direct ecological relationship between the sardine stocks and the volume of river discharge. The discharge reduction is the main reason for the drop in the sardine population", and then (there, 246) "The changes …. have adversely affected the population, distribution and biology of the majority of fishery items". Indeed, the sardine fisheries declined from 18,200 tons in 1962 to 1,200 tons in 1966 (only 554 tons according to Aleem 1969), whereas the penaeid catch was down from 7200 tons in 1962 to 2700 tons in 1966 (Al Kholy & El-Wakeel 1975). But "Not only the pelagic fish stock has been affected but the implications of the phenomenon [reduction of the Nile flow] on the benthos and the demersal fish communities cannot be overlooked" (Aleem 1969).

Bebars & Lasserre (1983) analysed the Egyptian Mediterranean fisheries statistics from 1962 to 1976 and found that the damming of the Nile and the decimation of the fauna have simplified community structure, modified species diversity and resource utlization strategies and lessened interspecific competition. For certain, before the damming of the Nile the clupeid catch was composed of the native *Sardina pilchardus* and *Sardinella aurita*, but in September 1971 the Erythrean *Dussumieria elopsoides* made up more than half of the clupeid yield on the Egyptian shelf. Bebars et al. (1997), showed that in 1963 the sardines accounted for 40% of the total catch, and the penaeids for 26%, but only 12% and 7% respectively in 1987; whereas the 'diverse' category, consisting of mostly less commercially favored taxa, including Erythrean aliens, increased at that period from 3% to 47% of the landings.

9.5 Increased fishing intensity along the Levantine Coast

It had been commonly believed that the Egyptian sea fisheries crashed due the environmental changes which followed the damming of the Nile (see above). But a careful examination of the fisheries statistics belie this: the records show a rapid increase in the catch between 1958 and 1960, followed by a decline, that turned catastrophic indeed by 1966. The causes are to be found in the effects of the Egyptian governmental policy initiated in the mid 1950s aimed at creating and supporting fishermen's cooperatives, promoting the mechanization of the fishing fleet and subsidizing the purchase of modern fishing gear (Halim et al. 1995). As a consequence, the number of licensed motorized fishing boats working off the Mediterranean coast of Egypt increased from 30 in the late 1930s (Wimpenny 1934) to 428 in 1958, and 622 in 1961 (El-Zarka & Koura 1965). The increased fishing intensity and the introduction of mechanization, as a matter of course increased the yield from the Mediterranean coast of Egypt from 5,600 tons in 1928 to 35,147 tons in 1958, and to 51,484 tons two year later (El-Zarka & Koura 1965: 228). The pressure on the fish stocks intensified considerably, and landings, size of principal fish and shrimps, and yield per effort slipped steeply after the peak of 1960. The Egyptian fisheries experts recognized that the enlargement and modernization of the fishing fleet, its dependence on sardines and shrimps which seasonal fluctuations were tied on the Nile flow, and the concentration of the trawl fisheries in the shallow fishing grounds, would "...create a difficult situation in our sea fisheries in the near future" (El-Zarka & Koura 1965: 228). It was foreseen that the decline of the Nile-dependent seasonal fisheries upon the completion of the Aswan High Dam, would place greater pressure on the benthic and shore fisheries, as indeed happened. The fishing pressure crested in the years following the "Six Days War" (1967-1978) when Egyptian fishing vessels were not permitted east of Port Said, aggravating the fishing pressure elsewhere. The drastic reduction in the fishing grounds and the decreased fertility of the coastal zone due to the damming of the Nile, caused the dwindling of the yield to between 13 – 26% of its 1962 value (Halim et al. 1995). A joint Soviet-Egyptian expedition called on to evaluate the state of marine fisheries resources in 1970-71 recorded a more varied shelf population, the "enrichment" stemming in part from the addition of Erythrean "trans-isthmic" aliens.

Similar results, though far less catastrophic, were recorded alont the Levantine coast. The fisheries off the Israeli coast up to the 1930s were "purely of an inshore character and their activities do not extend beyond the 10 fathom line" (Liebman 1934). Their modernization, first under the British Mandate, and then by the Fisheries Department of Israel, meant the introduction of ever larger, more powerful vessels, more efficient gear and techniques, and consequently, the obliteration of much of the sessile shelf biota. In the Iskenderun and Mersin bays, Turkey, where the "destructive fishing intensity on the region is well known" (Gücü and Bingel 1994: 92), where trawling "severely exploited for decades" local resources, and have denuded the shallow shelf of the native biota (Bingel et al. 1993, Gücü & Bingel 1994: 92), it was estimated that Erythrean invaders constitute 62% of the demersal fish biomass, whereas on the less intensively exploited narrow shelf just west of Mersin, between Incekum and Anamur, they made up only 27% of the biomass.

In his seminal work Elton (1958) suggested that anthopogenic interference, particularly when resulting in depletion of native biota, exacerbates the vulnerability to invasion. It is proposed that fishing pressure may be an important factor in the colonization success of Erythrean aliens by repeatedly removing the native biota, thus benefiting highly reproductive "opportunistic" taxa, particularly those which spawn in the summer months, since along the Levant the traditional trawling season is from September to May.

9.6 The increasing depth of the Suez Canal

The depth of the Suez Canal in 1869 was 8 m deep, 22 m wide at the bottom, and 60-90 m wide at the surface. Repeatedly deepened and widened, the canal is at present between 300 and 365 m wide and its maximum permissible draft is 17.5 m. The ongoing expansion of the channel to permit passage of super tankers will increase the canal width to 400 m and the permissible draft of ships to 22 m by 2010.

If earlier progress through the canal might have been restricted to euryhaline and generally hardy littoral species, it is now mainly depth-restricted. Whereas in the past most Erythrean invaders occupied the Mediterranean littoral and infralittoral to depth of 60 m, and with few exceptions were hardly found in deeper waters (Galil 1989, Golani 1996, Bilecenoglu & Taskavak 1999). The recent finds of the typically deepwater Erythrean molluscs, *Ergalatex contracta* Huart 1996, and *Mactrinula tryphera* Melvill 1899, off the Levantine coast, conceivably augur the entry of deepwater invaders (Mienis 2004). An effective increase in canal depth of 5 m will allow invasion of species whose upper depth range (as adults or larvae) did not permit passage until now, and cohorts of new invaders will gain admittance to the Levantine Sea.

The extensive dredging of the Canal in the early 1960s increased water flow by half. Following the great increase in the Canal's depth and width in the late 1970s (see above), a model of tidal currents has shown that the "... deepening of the canal to depths more than 19.0 m has increased the M2 tidal current in the southern part of the canal from 37.0 cm/sec to more than 45.0 cm/sec. Hence the tidal current alone …. may reach values more than 100.0 cm/sec" (Soliman 1995a,b).

The implications of higher current velocities on transport of Erythrean biota through the Canal was clear: "… today, when the Canal has been deepened to comprise 1½ times the waterflow of the year 1961 …. the speed of the current has increased (relatively less friction against the banks of the Canal), the chance for larvae to pass seems better than any time before. If this is so, this will greatly increase the chances for many new species to pass. …I regard studies of the larval migration as especially important, since this will probably allow many more species to settle in the Mediterranean than a transport of adult animals." (Gunnar Thorson‘s letter to Heinz Steinitz, 18.7.1967, cited in Israel Journal of Zoology 1972: 145). Many benthic invertebrates and fish have a pelagic larval stage of duration sufficient for passive passage through the Canal (Thorson 1971). With the amelioration of barriers to transport of pelagic larvae, the chances for entirely new taxa to invade the Mediterranean have greatly increased: "… huge larval swarms may successfully pass the Canal and invade the Mediterranean. Even if such an invasion succeeds only once every 5 or 10 years, it could be responsible for the establishment of several new species. With gradually improving chances for planktonic larvae to pass the Canal a steeply increasing invasion of Red Sea animals into the Mediterranean can be expected – an immigration which in a not too far future might radically change the whole faunal composition of its eastern basin" (Thorson 1971: 846).

10 Records of invasion

The concept of native species and their natural ranges and, *inter alia*, of alien species, is largely dependent on the scientific knowledge of the biota. In the Mediterranean Sea extensive biological surveys were conducted since the beginning of the 20th century, allowing a reasonable measure of confidence in separating the alien from the native biota in the better-known taxa. An 'alien' is defined as an organism occurring outside of its known or consensual range (as documented in scientific publications), and an 'invasive alien' as an alien whose population has been rapidly extending its range (Occhipinti-Ambrogi & Galil 2004).

An Erythrean alien species is defined as a species that had originated in the Red Sea, entered the Mediterranean through the Suez Canal and established a population, at least temporarily, in the Mediterranean Sea. Since the likelihood of encountering a stray incursion from the Red Sea in the Mediterranean is diminishingly small, most recorded alien species are considered as 'established' species that have self-maintaining populations of some duration in the Mediterranean Sea. It is recognized that some alien species may fail to maintain populations over time and thus a single record dating back several decades may be considered an ephemeral entry. The distinction between the 'established' and 'ephemeral' aliens can vary spatially and temporally, and is sometimes difficult to discern and circumscribed in large part by our ignorance.

A distinct size bias is apparent in the lists of Erythrean aliens, and data is entirely absent for many of the small-sized invertebrate taxa. For lack of study or expertise, some alien species were not recognized for decades on end: the tiny (4 -5 mm) pyramidellid gastropods *Chrysallida maiae* and *Syrnola fasciata* were collected off the Israeli coast in 1935 and 1949 and identified only in 1992 (van der Linden & Eikenboom 1992), and 1995 (Mienis 1995) respectively; the tube worm Hydroides brachyacantha was collected in Jaffa, Israel, in 1933, but identified only nearly 60 years later (Ben-Eliahu 1991); the 20 mm long ogyridid shrimp *Ogyrides mjobergi* was found along the Israeli coast already in 1947, but identified and recorded only a decade later (Holthuis & Gottlieb 1958), and even the rabbitfish *Siganus luridus* was described in 1964 from material collected a decade earlier (Ben Tuvia 1964). The data is presumably most accurate for large and conspicuous species, which are easily distinguished from the native biota, but even data concerning these species may be occasionally off by a decade (e.g. *S. luridus*). Still, a large, conspicuous species occurring along a frequently sampled or fished coast where taxonomic expertise is readily available will be noted earlier than a small, obscure species present in a rarely examined habitat.

The date of the establishment of the first population is of significance when we come to study the patterns and processes of invasion (Ruiz et al. 2000). However, in many, if not most cases, comprehending the patterns of spread of Erythrean aliens is hampered by our ignorance of the species' arrival date. Often reports of new records depend upon intensity of research effort, whether by international expeditions, governmental agencies seeking to exploit marine resources, or by enthusiastic individuals (see Studies Of The Biota Of The Suez Canal And The Adjacent Marine Biotas 1869-2004).

Since research efforts vary greatly along the coasts of the eastern Mediterranean, and even the better studied locales suffer temporal and taxonomical lacunae, there are occasionally justified doubts concerning the actual entry date. Consequently, it is commonly agreed that the date of collection (or, when missing, the date of publication) of the earliest specimens provides an upper bound date of entry.

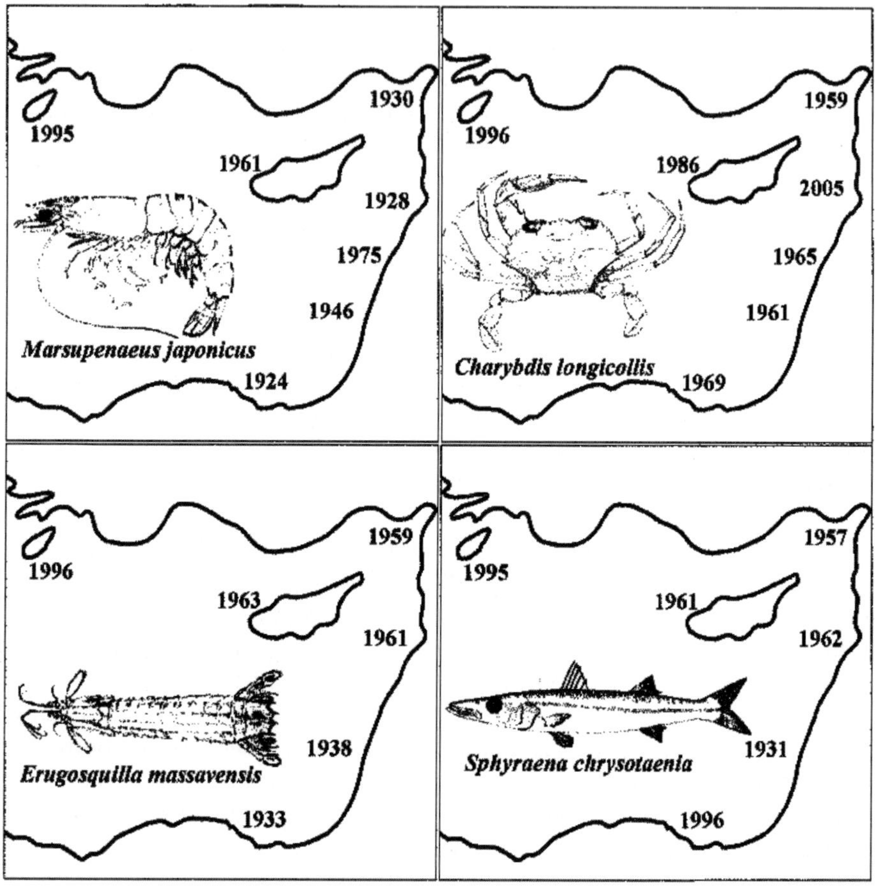

Fig. 13. Spatial and temporal spread of four Erythrean invasive species in the Levantine Sea, *Marsupenaeus japonicus, Charybdis longicollis, Erugosquilla massavensis, Sphyraena chrysotaenia.* The year of the first record in each country is indicated.

It is has been assumed that Erythrean aliens progress through the Suez Canal and along the coasts of the Levant as a result of "natural" dispersal, by autochthonous active or passive larval or adult movements, unaided further either directly or indirectly by human activity. Indeed, a temporal succession of directional ("stepping stones") records from the Red Sea, the Suez Canal, and along the coasts of the Levant confirms a species status as a naturally dispersing Erythrean alien. However, dispersal could also result from anthropogenic translocation, already Fox (1926:20) wrote "It is, of course, well known that ships have in more than one instance dispersed marine organisms from one part of the world to another. This must apply equally to transport through the Suez Canal. .... Possibly tugs and barges permanently employed in the Canal may take a larger share than other vessels in this transport from one end of the Canal to the other. There are coal barges, for instance, which remain for some months at one end of the Canal and then are towed through to stay for some months more at the other end. The time spent at either end would permit on the one hand of the settling of larvae on the bottom of the barges, and on the other hand of the liberation of eggs or larvae from mature individuals". Shipping is considered the largest single vector for the movement of alien marine species across the globe (Ruiz et al. 1997). So it is entirely conceivable that shipping may serve to transport Erythrean aliens further on. Even where records are consistent with long-shore autochthonous dispersal, there might be a degree of uncertainty where fouling organisms (such as serpulid polychaetes or mussels) are concerned, as they are more susceptible to shipping-mediated transfer. In some cases we suspect simultaneous mechanisms of transport.

The spread pattern of four relatively large, conspicuous and successful Erythrean aliens is illustrated (Fig. 13) these presumably have a better chance of attracting attention and being recorded. The dates of their collection, rather than the publication date of the new record, are used here, as the latter are largely dependent on local scientific interest and on the availability of taxonomic expertise.

Indeed, the hiatus between the dates of collection and publication can last more than a decade – *Charybdis longicollis* had been collected off Lebanon already in 1965, but reported only in 1981. *Marsupenaeus japonicus* had been collected off Israel as early as 1946, but the record was published in 1953 for lack of local expertise, whereas later records from Lebanon and Rhodes were published within a year of collecting the specimens. The chronological sequence too is largely dependant on local scientific interest. Though *M. japonicus* was collected in the Mediterranean coast of Egypt already in 1924, and off Syria and Turkey a few years later, it was collected off Israel in 1946, and Lebanon in 1975. Clearly, the last two records reflect the state of local research rather than the sequence of the prawn's spread. But even when the chronological sequence does not follow the increasing distance from the Suez Canal, it is of great value in tracing the time-span of the invasion: the rapid expansion of *Ch. longicollis* along the Levantine coast is obvious, with all first records collected in a single decade.

The list of Erythrean species (mainly fish, molluscs, decapod and stomatopod crustaceans, polychaete worms and macrophytes) includes, for each species, the year of publication of the first record in the Suez Canal, and in each country along the Mediterranean, and where known, the year of first collection (Table 1). The list was culled from myriad sources. The primary sources include research papers, biota surveys, ecological studies, and fisheries management reports. Earlier records were reassessed and some were deleted, such as the seven sponge species and three species of hydroids listed by Por (1978), that, as tactfully suggested by Zibrowius (1992), were in need of "critical reevaluation". Por (1978) also included in his list seven species of ascidiaceans, that are either widely distributed, or of "problematic identification. The list thus provides an inexact impression of lessepsian migration among ascidians" (Zibrowius, 1992: 96). For the Israeli coast additional data was gleaned from M.Sc. and Ph.D. theses, the local Fishermen's Bulletin, and unpublished monitoring studies conducted over the past 20 years by the author. These records were supplemented and amended by personal communications with scientists of particular taxonomic expertise.

The opening of the Suez Canal precedes studies of all taxa, save mollusks and fishes, by at least half a century, thus prejudicing invasion patterns based on recorded data. 'The Cambridge Expedition' in 1924 was the only thorough study of the biota of the Suez Canal, but though the extensive collections were expertly identified, the excursion was limited to three winter months, inevitably missing the seasonal variation of the biota. Subsequent studies have not dealt with as wide a range of taxa or were of limited spatial and temporal focus.

A basin-wide targeted effort to survey the presence and abundance of the Erythrean aliens in the eastern Mediterranean has never materialized. The closest to this was the joint program by the Smithsonian Institution, the Hebrew University of Jerusalem, and the Sea Fisheries Research Station, Haifa, that investigated the spread of the Erythrean biota along the coasts of Israel, Cyprus, Rhodes in the late 1960s and early 1970s. Most of the records used in this analysis stem from fortuitous finds, and therefore are uneven in their methodology, timing, extent, taxonomic expertise, and choice of habitats. Recently, a series of Atlases that summarizes the extant knowledge of the scale and impact of 'Exotic species in the Mediterranean' documents the prevalence of Erythrean taxa (www.ciesm.org/atlas/). The choice of taxa treated in this series – fishes, decapod and stomatopod crustaceans, and mollusks – emphasizes the precept that taxonomic and biogeographic data are biased in favor of larger taxa of economic importance.

Table 1. Temporal and spatial spread of Erythrean alien species in the Mediterranean Sea. The first published record in each country for a given species, and , if known, the year the specimen was collected [in square brackets].

TAXON	SPECIES	SUEZ CANAL	EGYPT	ISRAEL	LEBANON	SYRIA	TURKEY	CYPRUS	RHODES	OTHER
Macrophytes	*Acetabularia calyculus*	[24]1926	1973	[43]1955						Majorca, Spain [57]1968; Morocco [89]1994
	Botrycladia madagascariensis						2000			Lampedusa, Italy [91] 1992; Malta [94]1997
	Caulerpa mexicana			[39]1941	1941					
	Caulerpa racemosa var. *lamourouxii* f. *requienii*	[24]1926	[44]1948	[51]1960	[99]2001		[92]2000	[80]1991		
	Caulerpa scalpelliformis	1930		[29]1934	1930			[95]1998		
	Chondria collinsiana	[69]1973					1993			Greece [82]1987
	Chondria pygmaea									Sicily, Italy [91]1992; Albania [74]1995; Malta [94]1997
	Cladophoropsis javanica		[48]1948	1955						
	Derbesia boergesenii					1976				
	Hypnea cornuta		1948	1964						Italy [00]2004
	Hypnea valentia	[24]1926	1993			1976			[94]1898	France [96]2001
	Lophocladia lallemandii		1993	1976	[96-97]2000	1976	1993			Greece [08]1918; Libya [10]1918; Algeria, Tunisia 1938; Italy [69]1971; Spain [84]1989
	Padina boergesenii			1965	2000					Italy [63-66]1967; Libya [74]1981
	Padina boryana		1993							
	Rhodymenia erythraea		1948							
	Sarconema filiforme		1948	[45]1963		1976				
	Sarconema scinaioides									Greece [80]1985
	Solieria dura	[44]1950	[44]1950	[57]1963		1976				
	Stypopodium schimperi		[82]1991	[73]1980	[96-97]2000	[79]1991	[90]1991	[89]1991		Libya [86]1989; Greece [96]1999
Phanerogam	*Halophila stipulacea*	[24]1926	1962		[66]1975		[67]1975		[94]1895	Crete, Greece, 1958; Malta [70]1970; Albania [74]1995; Italy [88]1988; Tunisia [03]2003
Scyphozoa	*Rhopilema nomadica*		1988	[76]1990	1989			[95]1995		
	Cassiopeia andromeda	[86]1888		[88]1990	1988			1903		Thira I., Greece [55]1955
	Phyllorhiza punctata			[65]1990						

Table 1. continued.

TAXON	SPECIES	SUEZ CANAL	EGYPT	ISRAEL	LEBANON	SYRIA	TURKEY	CYPRUS	RHODES	OTHER
Octocorallia	*Acabaria erythrea*	[24]1929		[99]2005						
Polychaeta	*Branchiomma boholense*	[24]1928	[33]1937	[27]1991	[65]1966			[98]2005		Malta [29]1991
	Branchiomma luctuosum							[98]2005		Naples, Italy [78-79]1991
	Ceratonereis mirabillis	[24]1927	[69]1976	[74-75]1976				[97]2005		
	Eunice cf. *indica*	[24]1927	1973	1973						
	Hydroides cf. *brachyacanthus*			[33]1991						
	Hydroides heteroceros	[24]1928		[90]1992	[65]1966			[98]1999		
	Hydroides homoceros	1991		[55]1991						
	Hydroides minax			[90]1991	[78]1981					
	Hydroides operculatus			[59]1991	[78]1981					
	Leonnates decipiens	[24]1927	[67]1970	1989						
	Leonnates indicus	1991		[91]1994						
	Leonnates persicus	[24]1927	[68]1939	1991				[01]2002		
	Lepidonotus tenuisetosus		[67]2003	[69]2003						
	Linopherus acarunculata	1973		1976	[65]1966		[93]1997	[97]2005		
	Lysidice collaris			1992			[93]1997	[68]1970		
	Metasychis gotoi		1973	1973				[97]2005		Croatia 1934
	Nereis gilchristi		[68]1973							
	Nereis persica	[24]1927	[24]1927	[37]1989						
	Neanthes willeyi	[24]1927	[24]1927							
	Notomastus mossambicus							[97]2005		
	Oenone cf. *fulgida*	[67]1973						[97]2005		
	Paradyte cf. *crinoidicola*									Straits of Kassos, Greece [64]2003
	Perinereis nuntia	[24]1927	[24]1927							
	Pista unibranchiata			[74-75]1976				[97]2005		
	Pomatoleios kraussii	1990?		[58]1991						
	Prionospio saccifera						[95]1999			
	Pseudonereis anomala	[24]1927	[33]1937	[37]1989	[65]1966		[75]1989	[69]1972		
	Sphaerosyllis longipapillata							[97]2005		

Table 1. continued.

TAXON	SPECIES	SUEZ CANAL	EGYPT	ISRAEL	LEBANON	SYRIA	TURKEY	CYPRUS	RHODES	OTHER
	Spirobranchus tetraceros	[24]1928			[65]1966			[98]1999	[70]1989	
	Terebella ehrenbergi	[24]1928		[76]1977				[69]1972		Corsica, France 1952
Crustacea										
Decapoda										
Penaeidae	*Maesupenaeus japonicus*	[24]1927	[24]1927	[46-51]1953	[75]1976	1928	1930	[61]1969	[95]1996	Italy 1986; France 1972
	Melicertus hathor	[24]1927		[97]1999			[02]2002			
	Metapenaeopsis aegyptia			[87]1990					[96]1998	
	Metapenaeopsis consobrina	[1895?]1915		[96]1997					[95]1998	
	Metapenaeus monoceros	[24]1927	[24]1927	[46-51]1953	[75]1976	1980	[59]1961	[61]1969		Tunisia [94/5]1995
	Metapenaeus stebbingi	1888	[24]1927	[58]1964	[75]1976		1981			
	Penaeus semisulcatus	1936	1930	[46-51]1953	[75]1976	1928	1930			
	Trachysalambria palaestinensis	[24]1927	[69-71]1975	[24-25]1929			[68]1968		[95]1998	Tunisia [93]1993
Solenoceridae	*Solenocera crassicornis*		[71]1975							
Luciferidae	*Lucifer hanseni*	[24]1927	[24]1927							
Pasiphaeidae	*Leptochela aculeocaudata*		[33]1936							
	Leptochela pugnax			[47]1958			1981			
Palaemonidae	*Palaemonella rotumana*			[48]1958						
	Periclimenes calmani	[24]1926	[24]1927							
Alpheidae	*Alpheus audouini*	[24]1926	[24]1926	[51]1956						
	Alpheus inopinatus	1936	[33]1936	[51]1958			1969			Tunisia [56]1956
	Alpheus migrans			[75]1978						
	Alpheus rapacida			[60]1964			1981			
Ogyrididae	*Ogyrides mjobergi*			[47]1958						
Palinuridae	*Palinurus ornatus*			[88]1989						
Raninidae	*Notopus dorsipes*			[62]1964						
Dorippidae	*Dorippe quadridens*	[29]1937		[04]2004						
Matutidae	*Ashtoret lunaris*			[87]1990						
Calappidae	*Calappa hepatica*						[92]2003			
Leucosiidae	*Ixa monodi*			[79-82]1983		2005	[55]1956		[99]2002	
	Leucosia signata	[24]1926	[69]1976	[53]1956	[77]1981	2002	1981			

Table 1. continued.

TAXON	SPECIES	SUEZ CANAL	EGYPT	ISRAEL	LEBANON	SYRIA	TURKEY	CYPRUS	RHODES	OTHER
	Myra subgranulata	[24]1927	[33]1936	[29]1930	[76]1981	2002	1930			
Majidae	*Hyastenus hilgendorfi*	[24]1927	[69-71]1976	[60]1964						
	Micippa thalia				[99]2002		[94]1995			
Portunidae	*Carupa tenuipes*			[02]2004			[02]2004			
	Charybdis helleri		[33]1936	[24-25]1929	[64]1981	[92]1993	1981	1999		
	Charybdis longicollis	1937	[69-71]1976	[61]1964	[65]1981	2005	[59]1961	1986	[96]2002	
	Portunus pelagicus	[86]1888	[1898]1924	1924	1929	1930	1928	[61]1969	[91]2003	Sicily 1966
	Thalamita poissonii	[24]1927		[52]1958	[75]1981	2005	[59]1961	1978		Aegean Sea [83]1984; Crete [87]1994; Peloponnesos [84]1994
Goneplacidae	*Eucrate crenata*	[24]1927	[24]1927	[94]1997			[87]1992			Tunisia [85]1993
Xanthidae	*Atergatis roseus*			[61]1964	[75]1981	2002	[87]1992			
	Halimede tyche			[98]2000						
Dairidae	*Daira perlata*						[88]1995			
Pilumnidae	*Heteropanope laevis*	[24]1927	[24]1927							Tyrrhenian Sea 1957
	Pilumnopeus vauquelini	[83]1888	[24]1927	[51]1956			1981	1963		Tunisia [02]2003
	Pilumnus hirsutus		1936							
Eriphiidae	*Sphaerozius nitidus*	[69]1976	[69]1976							
Ocypodidae	*Macrophthalmus graeffei*			[02]2004			[94]1995			
Stomatopoda	*Erugosquilla massavensis*	1936	[33]1936	1938	1961		[59]1961	1963	[96]2002	Crete [91]1994
Cirripedia	*Heterosaccus dollfusi*			[92]1995			[94]1997			
Amphipoda	*Bemlos leptocheirus*	[24]1928	1998							
	Elasmopus pectenicrus	[24]1928	[24]1928	[46-50]1959						Venice, Italy 1983
	Gammaropsis togoensis			[46-50]1979						
	Photis lamellifera	[24]1928		[46-50]1959						
Mollusca										
Gastropoda Prosobranchia										
Nacellidae	*Cellana rota*	1905	1994	[61]1967	[54]1973					Aegean 1999
Haliotidae	*Haliotis pustulata*			[61]1971						Libya 1994
Fissurellidae	*Diodora funiculata*			[98]2002						
	Diodora ruppellii	1905		1948			[85-89] 1990			Tunisia 2003
Neritidae	*Smaragadia souverbiana*						[92]1994	[95]1997		Dodecanese [93]1994

Table 1. continued.

TAXON	SPECIES	SUEZ CANAL	EGYPT	ISRAEL	LEBANON	SYRIA	TURKEY	CYPRUS	RHODES	OTHER
	Nerita sanguinolenta	1905								Dodecanese[69] 1973; Libya 1994
Trochidae	*Trochus erithraeus*	1905	1974	[68]1973	[86]1987		[92]1995	1987		Crete[94]1997
	Pseudominolia nedyma	[50]1984		[66]1973	[86]1987		[92]1995			
	Stomatella impertusa						[99]2000			
Cerithiidae	*Cerithium nesioticum*			[71]1977				[85]1989		
	Cerithium scabridum	[76]1878	1883?	1899	1938	1938	[85-89]1990	[83-91]1993		Italy [72]1978; Tunisia [98]2001
	Cerithium columna			[66]2003						
	Cerithium egenum			[71]2001						
	Rhinoklavis kochi		[75]1977	[63]1973	[86]1987		[85-89]1990	1976		
	Rhinoklavis sinensis			[03]2004						
	Clypeomorus bifasciatus	1905	1994	[83]1985						Libya 1994
Planaxidae	*Angiola punctostriata*			[50]1980						
	Planaxis griseus	1905	1905							
Litiopidae	*Gibborissoa virgata*	2002		[70]1989			[97]2002			
Obtortionidae	*Finella pupoides*	[50]1998		[58]1977	[86]1987		1990	1996		
	Ceritidium diplax	1939		[70]1977			1995	1994		
Dialidae	*Diala varia*	1905	[70]1977	[35]1984	[86]1987					
Cerithiopsidae	*Cerithopsis pulvis*			[78]1992	[86]1987		1990	[85]1989		
	Cerithopsis tenthrenois			[82]1989			1990	[95]1997		
Triphoridae	*Metaxia bacillium*			[78]1985			[92]1995	[92]1995		
Rissoidae	*Rissoina ambigua*						[03]2004			
	Rissoina bertholleti	[50]1998		[58]1963			[85-89]1990	[85]1989		
	Rissoina spirata									Tyrrhenian [74]1984
	Voorwindia tiberiana	1939	1997	[97]1999						
Costellariidae	*Vexillum depexum*			[64]2004						
Turridae	*Lienardia mighelsi*						2003			
Stromibidae	*Strombus persicus*			[83]1986	[86]1987	[85]1986	[78]1986	1985		Dodecanese [83]1984
	Strombus mutabilis			[91]2001						
Hipponicidae	*Sabia conica*			[80]1986						
Cypraeida	*Erosaria turdus*	1939		[80]1986						Tunisia
	Palmadusta lentiginosa			[90]1990						

Table 1. continued.

TAXON	SPECIES	SUEZ CANAL	EGYPT	ISRAEL	LEBANON	SYRIA	TURKEY	CYPRUS	RHODES	OTHER
	Purpuradusta gracilis notata			[81]1983	1999		[82]1983	[00]2003		
Naticidae	*Natica gualteriana*		1997	[66]2000						
Epitoniidae	*Cycloscala hyalina*						[86]1999	[92]1994		
Eulimidae	*Sticteulima* cf. *lentiginosa*						[89]1994	[95]1997		
Muricidae	*Ergalatax contracta*			[01]2004	2004					
	Ergalatax obscura						[92]1995	[95]1997		
	Thais lacera	[24]1927	1939	[28]1977			[92]1995	2001		Italy 1983
	Thais sacellum				[00]2003			[00]2003		
	Murex forskoehli	[96-97]1897	1905	[32]1967	[14]1938					
Buccinidae	*Cantharus tranquebaricus*			[03]2004						
	Engina mendicaria			[01-03]2004						
Nassariidae	*Nassarius obvelatus*		[74]1977	[68]1977						
Fasciolariidae	*Fusinus verrucosus*	[82]1883	1905	1929						
Columbellidae	*Zafra savignyi*	[50]1998		[54]1976	[86]1987		[86]1997	[95]1997		
	Zafra selasphora			[80]1997			[86]1997	[95]1997		
Conidae	*Conus fumigatus*									Libya
Gastropoda Heterobranchia										
Anisocyclidae	*Murchisonella columna*			[97]1999			[93]1995			
Pyramidellidae	*Chrysallida fischeri*			[74]1979			[88/9]1992			Adriatic [93]2002
	Chrysallida maiae		1992	[35]1992	[86]1987	1992	[63]1992	[95]1997		
	Chrysallida pirintella			[84]1989			[88/9]1992			
	Monotygma lauta		[79]1984	[67]1981			[88/9]1992			
	Leucotina natalensis			[78]1985			[88/9]1992	1996		
	Styloptygma beatrix			[94]1997			[88/9]1992			
	Cingulina isseli			[80]1983	[86]1987		[88]1992	[98]2003		
	Tubonilla edgarii			[84]1989			[88/9]1992	1996		
	Syrnola fasciata			[49]1995	[86]1987		[63]1989	1995		
	Syrnola cinctella						[94]1998			
	Odostomia lorioli			[74]1987						
	Oscilla jocosa			[84]1989						
	Iolaea neofelixoides						[92-95]2001			
	Hinemoa cylindrica						[92-95]2001			

Table 1. continued.

TAXON	SPECIES	SUEZ CANAL	EGYPT	ISRAEL	LEBANON	SYRIA	TURKEY	CYPRUS	RHODES	OTHER
	Leucontina cfr. *eva*						[95]2001			
Gastropoda Opisthobranchia										
Cylichnidae	*Acteocina crithodes*						[03]2004			
	Acteocina mucronata			[86]1999	[86]1990		1990	[92]1994		Greece [91]1997
Retusidae	*Cylichnina girardi*	1939	[74]1976		[86]1987		1990	1996		Crete [94]1997
	Pyrunculus fourierii	[50]1998		1989	[86]1987		1989	[95]1997		
	Retusa desgenettii	1939		[97]2002						
Bullidae	*Bulla ampula*	[88-89]1992		[78]1982			[01]2004	[00]2003		Greece [98]1999
Haminoeidae	*Atys cylindricus*	[50]1998		[02]2004						
	Haminoea cyanomarginata						[02]2004			Greece [01]03
Aglajidae	*Chelidonura fulvipunctata*			[86]1987			[59]1961			Malta
Oxynoidae	*Oxynoe viridis*						[02]2004			
Elysiidae	*Elysia grandifolia*				2004		[01]2004			
	Elysia tomentosa						[01]2004			
Aplysiidae	*Syphonota geographica*						[02]2004			Italy
Pleurobran-chidae	*Pleurobranchus forskalii*			[75]1977						
Triophidae	*Plocamopherus ocellatus*	[24]1929		[77]1982	2002		[98]2004			
Discondrididae	*Discordis liliacina*			[74]1977	2002					Tunisia [2003]
Chromodorididae	*Hypselodoris infucata*			[65]1977	2002		[99]2001			
	Chromodoris quadricolor									Italy,Liguria [82]1986; Tunisia [03]2004
Dendrodorididae	*Dendrodoris fumata*	[24]1929		[80]1986						
Tethyidae	*Melibe viridis*						[97]2003			Greece,Ionian [70]1986; Tunisia 1990; Italy,Messina 1998; Croatia [01]2002
Flabellinidae	*Flabellina rubrolineata*	[24]1929		[88]1993			[99]2003			
Facelinidae	*Caloria indica*			[86]1993						

Table 1. continued.

TAXON	SPECIES	SUEZ CANAL	EGYPT	ISRAEL	LEBANON	SYRIA	TURKEY	CYPRUS	RHODES	OTHER
Gastropoda Pulmonata										
Siphonaridae	*Siphonaria crenata*	1905		[65]1973			[99]2001			
Bivalvia, Pteromorphia										
Arcidae	*Acar plicata*	1939		[78]1986						
	Anadara natalensis	1939	1939	[35]1937			[85-89]1990			
Glycymeridae	*Glycymeris arabicus*			[66]1977						
Limopsidae	*Limopsis multistriata*	[24]1927		[65]1977						
Mytilidae	*Musculista perfragilis*	1939	1939	[60]1973						
	Musculista senhousia	[50]1998	[69]1971	[60]1971						
	Modiolus auriculatus	1905		[35]1937						
	Brachiodontes pharaonis	[76]1878	[76]1878	1937	1929	[78]1985	[75]1980	[69]1977		Sicily [69]1971; Croatia 1977; Greece 1979; Rhodes 1989; Corsica 1999
	Septifer forskali	1939					[01]2001			
Ostreidae	*Saccostera cucullata*	[82]1883	2003				[98-99]2003			
	Dendrostrea frons						[98]2001			
Pteriidae	*Pinctada radiata*	[82]1883	[74]1878	1899	[65]1973	[75]1985	[82]1985	1899		Tunisia [1890] 1929; Malta 1912; Karpathos [63] 1985; France 1989; Adriatic 1996; Corsica 1999
Malleidae	*Malvufundus regulus*	1905	[75]1977	1937	1929	[81]1985	[73]1974	[70]1971		Libya 2001; Greece 2001
Spondylidae	*Spondylus spinosus*	1905		[88]1993	2003		[91]2001			
	Spondylus groschi			1998						
	Spondylus nicobaricus			[02]2004						
Bivalvia, Heterodonta										
Lucinidae	*Divalinga arabica*		[56]1986	[76]1979						
Trapeziidae	*Trapezium oblongum*		[80]1980							
Ungulinidae	*Diplodonta bogii*	1939		[82]2001						
Chamidae	*Chama pacifica*	1905	1905	[88]1993	2003		[00]2001	2003		
	Pseudochama crobieri	[82]1883		[68]1973				1971		Greece 1946

Table 1. continued.

TAXON	SPECIES	SUEZ CANAL	EGYPT	ISRAEL	LEBANON	SYRIA	TURKEY	CYPRUS	RHODES	OTHER
Cardiidae	*Fulvia australis*			1948						
	Fulvia fragilis	[82] 1883	1939	[55]1973			[85-89]1990	1988		Tunisia [90]1993; Spain [91]2003; Greece 1999
	Afrocardium richardi	[24]1927		[97]1999			2000			
Tellinidae	*Angulus flacca*			[97]2004						
	Tellina valtonis	1905	[70]1977				2001			
	Pasammotreta praerupta	[50]1998					[92]1999			
Psammobiidae	*Hiatula ruppelliana*	[86]1888	1905							
Mactridae	*Mactra olorina*	[76]1878	1889	1973	1931					
	Mactra lilacea		[65]2002	[01]2002						
Meso-desmatidae	*Atactodea glabrata*	1905		[73]1977						
Veneridae	*Gafrarium pectinatum*	[82]1883	1905	[84]1999	1999		1987			
	Circenita callipyga	1905		[72]2000						
	Clementia papyracea	1939		[37]1948			[85-89]1990			
	Paphia textile	1939		[46]1948		[92-93]1999	[85-89]1990			
	Antigona lamellaris	1992					[88]1999			
	Dosinia erythraea	[96-97]1898	1905							
	Redicrice sulcata			[70]2004						
	Timoclea marica	[96-97]1898		[97]1999						
	Callista florida	1939	[39]2005							
Myidae	*Sphenia rueppelli*	1939		[78]1986						
Gastrochaenidae	*Gastrochaena cymbium*	[33]1933		[54]1973			[90]1991			Greece 1989
Bivalvia, Anomalodesmata										
Laternulidae	*Laternula anatina*	[82]1883	1905	[58]1973	[86]1987		[92]1995			
Polyplacophora										
	Chiton hululensis			[34]1974						
Cephalopoda										
Sepiidae	*Sepia pharaonis*			[03]2003						
Loliginidae	*Sepioteuthis lessoniana*			[04]2004			[02]2002			
Octopodidae	*Octopus aegina*			[34]2004			[92]1999			
	Octopus cyaneus			[97-98]2003						

Table 1. continued.

TAXON	SPECIES	SUEZ CANAL	EGYPT	ISRAEL	LEBANON	SYRIA	TURKEY	CYPRUS	RHODES	OTHER
Holothuria	*Synaptula reciprocans*	[24]1926		1986			1986	2004		
Asteroidea	*Asterina burtoni*	[24]1926		[53-5]1966	[65]1966					
	Ophiactis parva	[24]1926		[68]1991						
Ophiuroidea	*Ophiactis savignyi*	[24]1926	[24]1926	[53-5]1954	[65]1966					
Chondrichthyes										
Dasyatidae	*Himantura uarnak*	1902	1994	[54]1955	[65]1966	1999	1966			
	Torpedo sinuspersici					[00-03]2004				
Osteichthyes										
Clupeidae	*Dussumieria elopsoides*	1902	[70]1975	1949	[62]1963		[52]1953			
	Etrumeus teres		1994	[61]1963			[94-96]1997	[99]2000	2004	
	Herklotsichthys puncatatus	1902	[74]1975	1943	[74]1977					
	Spratelloides delicatulus			[73]1978						
Congridae	*Rhynchoconger trewavasae*			1993						
Muraenesocidae	*Muraenesox cinereus*			[79]1982						
Plotosidae	*Plotosus lineatus*			[01]2002						
Synodontidae	*Saurida undosquamis*	1902	1966	[52]1978	[62]1963	1999	[52]1966	[60]1962		Aegean, Naxos 1971; Libya 1982; Albania 1995; Crete 1991; Tunisia [04]2005
Exocoetidae	*Parexocoetus mento*			1935	[62]1963				[37]1937/ 38?	Libya 1966; Albania [85]1986; Ionian 1987; Tunisia [99]2004
Belonidae	*Tylosurus choram*	[1900]1902		[01]2005	[62]1963					
Hemirapmhidae	*Hemiramphus far*	1902	1978	[24-25]1927	[62]1963	1929	[42]1950	[61]1969	[43]1946	Albania [85]1986; Tunisia 2003
	Hyporhamphus affinis	[24]1927			[64]1964					
Fistulariidae	*Fistularia commersonii*			[00]2000			[01]2002	[99]2005	[01]2002	Tunisia [02]2004; Italy [02]2004; Libya [04]2005
Syngnathidae	*Hippocampus fuscus*	[67]2002		[01]2002			[03]2004			
Atherinidae	*Atherinomorus lacunosus*	1902	1902	[24]1927	[62]1963	1999	[49]1950		1986	Libya 1982

Table 1. continued.

TAXON	SPECIES	SUEZ CANAL	EGYPT	ISRAEL	LEBANON	SYRIA	TURKEY	CYPRUS	RHODES	OTHER
Holocentridae	*Sargocentron rubrum*	[32-34]1937		[47]1947	[62]1963	[92-94]1996	[49]1950	[61]1969	[47]1948	Libya [68]1973
Scorpaenidae	*Pterois miles*			[91]1992						
Platycephalidae	*Papilloculiceps longiceps*	[32-34]1936		[86]1990						
	Sorsogona prionota			[46]1947						
	Platycephalus indicus	1882	1963	1953	1977					
Serranidae	*Epinephelus cioides*			[66]1969						Italy [98]2001
	Epinephelus malabaricus			[66]1969						
Teraponidae	*Pelates quadrilineatus*		1978	[69]1970	1977		[84-85]1987			
	Terapon puta		[73]1977		1977					
Apogonidae	*Apogon pharaonis*	[24]1927	[24]1924	1947	[62]1963	[92]1995	[84-85]1987	[61]1969	[02]2004	
Sillaginidae	*Sillago sihama*	1902	1994	[77]1985	[76]1977		[83]1994			
Rachycentridae	*Rachycentron canadum*			[78]1985						
Carangidae	*Alepes djedaba*	[1900]1902		[24-25]1927	[62]1963		[55]1957	[61]1969		Libya [90]2005
Leiognathidae	*Leiognathus klunzingeri*	1902	[29]1929	1934	[62]1963	1931	[42]1943	[61]1969	[37]1937	Lampedusa I. [56] 1966;Croatia [00] 2002; Aegean, Southern Adriatic 1989
Lutjanidae	*Lutjanus aregentimaculatus*				[77]1979					
Mullidae	*Upeneus moluccensis*	[32-34]1936	1978	1947	[62]1963	1931	[42]1950	[61]1969	[47]1947	Libya [68]1973
Haemulidae	*Pomadasys stridens*	[1900]1902	[73]1977	[71]1977	[74]1977					Italy [68]1969; Libya [81]1982
Sparidae	*Crenidens crenidens*	[1900]1902	1902	1992						Libya 1999
	Rhabdosargus haffara	1902		[91]1992						
	Upeneus pori	[24]1927	1989	1953	[63]1966	1999	[42]1950		2004	Tunisia [03]2005; Libya [94]2005
Pempheridae	*Pemperis vanicolensis*			[79]1985	[78]1979	[91]1992	[83]1994		[86]1986	Tunisia [01]2004; Libya [04]2005
Chaetodontidae	*Heniochus intermedius*						[02]2003			
Pomacentridae	*Abudefduf vaigiensis*	[32-34]1937		[97]1998			[05]2002			Italy 1959
Scaridae	*Scarus ghobhan*			[01]2002	[99]2005					
Mugilidae	*Liza carinata*	[1871]1929	1929				[55]1957			
Labridae	*Pteragogus pelycus*			[91]1992	1999		[98]2000	[97]2000	[92]1999	

Table 1. continued.

TAXON	SPECIES	SUEZ CANAL	EGYPT	ISRAEL	LEBANON	SYRIA	TURKEY	CYPRUS	RHODES	OTHER
Sphyraenidae	*Sphyraena chrysotaenia*	1933	1996	1931	[62]1963		1957	[61]1969	[95]1999	Libya,Tobruk [68]1973; Malta 1961; Croatia [00]2001
	S. flavicauda			[91]1992		1999	[01]2002		2004	Libya [98]2005
Blenniidae	*Omobranchus punctatus*			2003						
	Petroscirtes ancylodon	1915		[88]1989			[97]2000		2004	
Gobiidae	*Coryogalops ochetica*		[24]1927							
	Oxyurichthys petersi			[82]1983		[91]1992	[91]1992			
	Silhouetta aegyptia	[24]1986	[72]1986	1998		[93]1995				
Callionymidae	*Callionymus filamentosus*	[29]1929	1996	1953	[62]1963	[93]1995	[83]1994		2004	
Siganidae	*Siganus rivulatus*	1902	1978	[24]1927	[62]1963	1929	[42]1950	[28]1929	[32]1934	Malta 1961; Libya [68] 1970; Aegean 1971; Tunisia [74]1974; Croatia [02]2004
	Siganus luridus	1902	1978	[55]1964	[62]1963	1999	1973	[61]1969	[63]1963	Libya,Tobruk Greece, Cyclades 1968; Tunisia, Salammbo [69]1971; Italy, Linosa [03]2004
Scombridae	*Scomberomorus commerson*		1994	1935	[64]1965	1999	[81]1994			Tunisia [03-04]2005; Libya [74]2005
	Rastrelliger kanagurta			[67]1970						
Cynoglossidae	*Cynoglossus sinussarabici*	[32]1932	1978	1953	1977	[93]1995	[55]1957			
Monacantidae	*Stephanolepis diaspros*	1902	1996	[24-25]1927	[62]1963	1929	[49]1950	1935	[43]1946	Greece,Crete [74]1976; Tunisia [65]1966; Italy1967; Libya [65] 1966; Malta [80]1993
Ostracionidae	*Tetrosomus gibbosus*	[32-34]1937		[87]1988						
Tetraodonti-dae	*Lagocephalus sceleratus*			[04]2005			[03]2005			
	Lagocephalus spadiceus	1902	1978	[51-52]1953	[63]1963	2000	[49]1950			Dodecanese 1930
	Lagocephalus suezensis			[87]1996	1977		[98]1999		2004	
	Torquigener flavimaculosus	[32-34]1937		[87]1987			[02]2004			
	Tylerius spinosissimus								2004	
Diodontidae	*Chilomycterus spilostylus*			[92]1993						

11 The spatial and temporal patterns of the Erythrean invasion

Our ignorance of the marine biota leads to "massive underreporting and thus understatement of….. the altered distributions of nonindigenous species" (Carlton 2000). Despite the biases and limitations inherent in the records, and the constraints they impose on interpretion, they constitute an important tool for examining the temporal and spatial patterns of the Erythrean invasion.

A considerable increase in the number of Erythrean fish, decapods and molluscs along the Turkish and Cypriot coasts, and in the southern Aegean Sea in the 1990s can not be assigned solely to the increased research effort. A persistent drought in the period 1988-92 and changes in the water mass pathways initiated a 1-4 times increase in salt transport from the Levantine into the Aegean in the upper 200 m layer between 1987 and 1994 (Theocharis et al. 1999). In 1991, the source of the Eastern Mediterranean Deep Water shifted from the Adriatic to the southern Aegean Sea (Theocharis et al. 1992, Nittis & Lascaratos 1999), though the process might have started as early as 1987. The increased outflow of the newly formed, denser water through the Cretan Arc Straits into the eastern Mediterranean has been compensated for by inflowing Levantine surface and intermediate water (Wu et al. 2000). The significant changes in the South Aegean water mass characteristics, which have considerably influenced the thermohaline circulation of the eastern Mediterranean, have been termed the Eastern Mediterranean Transient (Lascaratos et al. 1999, Theocharis & Lascaratos 2000). The sudden influx of Erythrean aliens westwards along the Turkish Mediterranean coast and into the southeastern Aegean in the past decade is attributed to the augmented salinity and to the more extensive inflow of the Asia Minor Current, that runs along the Anatolian coastline carrying westwards warm, salty water and their biota from the Levantine Sea and passing northward through the eastern Cretan Arc Straits, mainly the Rhodes and Karapathos Straits (Galil & Kevrekidis 2002, Yokes & Galil 2004, Pancucci-Papadopoulou et al. 2005).

The bulk of the Erythrean biota is confined to the easternmost Mediterranean, from Port Said to the Bay of Mersin in southeastern Turkey, where the infralittoral is overrun by Erythrean aliens. As late as the 1990s it was believed the Erythrean invasion will be contained in the eastern Mediterranean, east of Sicily, and south of the Aegean and Adriatic Seas (Por 1990). However, six Erythrean fish species in the Adriatic Sea (*Epinephelus coioides, Hemiramphus far, Leiognathus klunzingeri, Parexocoetus mento, Sphyraena chrysotaenia, Siganus rivulatus*) were recorded between 1986 and 2004; ten of the 16 Erythrean mollusc species reported from the Aegean Sea were collected in the past 15 years; and of late, some mollusks were recorded from the western

Mediterranean – *Brachidontes pharaonis* and *Pinctada radiata* in Corsica (Boudouresque 1999), and *Fulvia fragilis* in Valencia, Spain (Gofas & Zenetos 2003). The recent slew of records from both the Aegean and the Adriatic Seas calls for a significant expansion of the geographic limits of the Erythrean invasion.

It is assumed that contiguous distributions stem from natural dispersal, whereas remote records mark anthropogenic dispersal, most probably by shipping. The small Erythrean mytilid gastropod, *B. pharaonis*, common in the Levantine basin, where it settles in dense clusters on midlittoral and infralittoral rocks, piers and debris (Barash & Danin 1992), has spread as far west as Sicily (Di Geronimo 1971) and Corsica (see above), probably in ship fouling. Similarly, the gastropod *Cerithium scabridum*, reported from Naples, Sicily and Tunisia, was possibly transported there by shipping. The pearl oyster, *Pinctada radiata,* was one of the first Erythrean molluscs recorded in the Mediterranean (Monterosato 1878, as *Meleagrina* sp.) and has spread as far west as Tunisia, Malta, Sicily, and France (Vassel 1896, Pallary 1912, Di Natale 1982, Zibrowius 1979). Its rapid dispersal is attributed to ship-borne individuals (Zibrowius 1992), or marine turtles – it was recorded as an epibiont on a loggerhead turtle off Lampedusa Island (Oliverio et al. 1992). Other species may have spread with ballast water, or entangled in fishing gear. Anthropogenic dispersal may take place either from the source populations, or from established Erythrean alien populations within the Mediterranean (Zibrowius 1979, De Min & Vio 1997), and may cast the Erythrean alien upon a distant shore. The Indo-West Pacific portunid crab *Charybdis hellerii* was first sighted in the Mediterranean in 1920s (Steinitz 1929), and has since been reported along the Levantine coast from Egypt to Cyprus, following a spread chronology denoting autochthonous dispersal. In 1987 *C. hellerii* was collected in Cuba (Gómez & Martinez-Iglesias 1990), and in rapid succession in Venezuela, Colombia, Florida, and Brazil (Campos & Türkay 1989, Lemaitre 1995, Carqueija & Gouvea 1996, Mantelatto & Dias 1999). Transport in ballast tanks is the most probable mode of dispersal because the crab's arrival corresponds to increased shipping contacts with the eastern Mediterranean (Campos & Türkay 1989).

The majority of the farthest spread species are early settlers. The six species of decapod crustaceans recorded in Tunis or Sicily (*Metapenaeus monoceros, Trachysalambria palaestinensis, Alpheus inopinatus, Portunus pelagicus, Eucrate crenata, Pilumnopeus vauquelini*) have first established populations in the Mediterranean an average of 84 years ago, and the nine fish species recorded in Tunis, Malta or Sicily (*Parexocoetus mento, Hemiramphus far, Fistularia commersonii, Leiognathus klunzingeri, Pempheris vanicolensis,*

Sphyraena chrysotaenia Siganus rivulatus, S. luridus, Stephanolepis diaspros) 61 years ago, as compared with an average Mediterranean residence of 43 and 33 years respectively for decapods and fish that are recorded only within the Levantine Sea. Similar figures are given by Gofas & Zenetos (2003) for mollusks.

Once established in the Mediterranean the temporal dynamics of the Erythrean species are markedly varied. In some cases the interval between the initial establishment and rapid population growth and expansion had been exceedingly short: the spiny oyster, *Spondylus spinosus*, was first collected in Haifa Bay, Israel, in 1988, and in the Gulf of Iskenderun, Turkey, in 1991; the conch, *Strombus persicus*, spread throughout the Levant between 1978 and 1985; the sweeper, *Pempheris vanicolensis*, was first collected off Israel and Lebanon in 1978-79, and in 1983 off Mersin, Turkey; and more recently, the cornetfish, *Fistularia commersonii*, spread, within two years, from Israel to the island of Lampedusa and Tunisia. However, a time lag, sometimes extending over half a century, is a more common pattern: the jack, *Alepes djedaba*, and the swimming crab *Charybdis hellerii*, were both collected in Haifa Bay in 1924, but respectively, nearly three decades and six decades later in Iskenderun. The cerithiid and muricid gastropods *Cerithium scabridum* and *Thais lacera* were collected off Jaffa in the 1899 and 1928 respectively, though only 90 and 64 years later along the the southeastern coast of Turkey; the hammer oyster, *Malvufundus regulus*, and the mussel, *B. pharaonis*, both sighted off Israel in 1937, were collected off Turkey in 1973 and 1975 respectively. Those are all conspicuous species, easily distinguished from the native Mediterranean taxa, and collected along intensively fished coasts, and so unlikely to be overlooked, yet the actual time lag is probably longer than the records imply, as chances of detecting the earliest members of the colonizing population are slim.

The last mentioned species, a mytilid widely spread throughout the Red Sea, had been an early entrant into the Mediterranean – already in 1876 it was recorded from Port Said, where it soon became abundant (Fuchs 1878, Pallary 1912). However, for decades after its arrival to the Israeli coast it remained quite scarce "… peu d'exemplaires seulement – , surtout en compagnie de *M.*[*Mytilaster*] *minimus* Phil., qui couvre les faces plates des rochers de grès…" (Haas 1948). It remained quite rare through the 1970s (Safriel et al. 1980 a,b, Safriel & Sasson-Frostig 1988), but during that decade it was first reported from Turkey (1978), Greece (1979), and Croatia (1977). When the wave-washed intertidal rocks off the Israeli coast were re-examined in 1994, they were covered with extensive swaths of *B. pharaonis* (Rilov et al. 2004). As the presence of algae is negatively correlated with the presence of the mytilid, and is considered to impede the settlement of its postlarvae (Safriel & Sasson-Frostig 1988), a shift in habitat conditions that reduces algal cover might have benefited *Brachidontes*. As it happens, in the rocky shores of the southeastern Levant few herbivors existed prior to the arrival of the Erythrean siganid fish. During high tide schools of siganids invade the intertidal platforms to feed. Indeed, algae identified from the gut contents of the (mostly young) platform-dwelling siganids indicate they graze on the upper surface of the platforms (Lundberg et al. 2004). It is suggested that the multitudinous siganids may have triggered the population increase, and consequently the propagule supply of *Brachidontes*, by clearing the intertidal platform of algae.

A fetching hypothesis, but many other decades-long lags remain without explanation. The lantern clam, *Laternula anatina*, was described from Port Said a century ago (Tillier & Bavay 1905). It is a large (75 mm), readily distinguished clam, and it is highly unlikely one would fail to notice it, yet it was collected in Haifa Bay only in 1958 (Barash & Danin 1973), and from the Gulf of Iskenderun in 1992 (Engl 1995). The jewel box oyster, *Chama pacifica*, another massive (70 mm) and distinctive species, was collected in Alexandria in 1905 (Tillier & Bavay 1905), but only in 1988 in Haifa Bay (Mienis et al. 1993a), and then within a decade to Lebanon, Turkey and Cyprus. The venerid clam *Gafrarium pectinatum* too was described from Port Said in 1905 (Tillier & Bavay 1905), but only in 1984 from the Israeli coast (Mienis 1999), and in 1987 from Turkey (Lindner 1987). These hiatuses can not be considered artifacts stemming from poor collection and identification, as the mollusks off the Israeli coast had been assiduously studied throughout that period.

Perusal of the dates when Erythrean aliens had been collected, or lacking that, first reported, in the eastern Mediterranean discloses a distinct temporal lag pattern. The Israeli and Turkish records for fish, mollusks, and decapod crustaceans were examined, because those are the better known taxa in the better investigated Levantine coasts. It appears that whereas off Israel the influx of Erythrean species surged in the 1950s, in was delayed untill the 1980s in Turkey. Even if we leave out the mollusk records, since few studies of the malacofauna were conducted prior to the 1980s (Ozturk et al. 2002), the data indicate a significant time lag. Of the decapod crustaceans reported between 1940 and 2000, ten species were recorded first off Israel, with time lags extending from 5 years for the penaeid prawn *Melicertus hathor*, to 34 years for *Leptochela pugnax*, and an average lag of 20 years. In the same period, four species were first encountered off the Turkish coast: *Ixa monodi*, *Charybdis longicollis*, *Eucrate crenata*, and *Macrophthalmus graffei*, with an average time lag of 10 years. Of the fish reported between 1940 and 2000, 20 species were recorded first off Israel with time lags extending from one year for the cornetfish (see above) to 37 years for the terapon *Pelates quadrilineatus*, and the cardinalfish *Apogon nigripinnis*, and an average lag of 13 years. Three species – *Upeneus moluccensis*, *U. pori*, *Lagocephalus spadiceus* – were reported earlier in Turkey, with an average lag of 6 years.

The time lag may reflect a requisite interval required for population growth needed to increase propagule abundance, possibly affected by biotic or abiotic modifications in the receiving environment; for adaptive evolution to overcome genetic constraints, such as reduced genetic variation; or to accrue repeated introductions, promoting both propagule pressure and genetic variation. Already in the 1970s attempts were made to determine whether the Suez Canal constitutes a genetic sieve, permitting only a fraction of the genetic variability extant in the Red Sea through. Pashtan & Ritte (1977) and Lavee (1981) used electrophoretic analysis of enzymes to compare the genetic variability in Red Sea and Mediterranean populations of the molluscs *Cerithium scabridum* and *B. pharaonis*. The authors found that more than 90% of the genetic variability of the source populations had been sampled in the Mediterranean, but that some alleles identified from the Mediterranean populations, had not been sampled in the Red Sea. Golani & Ritte (1999), found "no discernible genetic distance" in allozymes between the Red Sea and Mediterranean populations of the mullid fish *Upeneus moluccensis* and *U. pori*. Recent studies, utilizing DNA sequencing to study mitochondrial differentiation between Red Sea and Mediterranean populations of the silverside *Atherinomorus lacunosus* and the rabbitfish species *Siganus rivulatus* and *S. luridus*, concluded that the establishment of the Erythrean aliens in the Mediterranean involved either large founding populations or recurrent/continuous introductions, or possibly both (Bucciarelli et al. 2002, Bonhomme et al. 2003, Hassan et al. 2003). Though the genetics of Erythrean alien populations have drawn far too small attention, the studies cited indicate that lag times are not caused by reduced genetic variation.

The dynamics of the invasion since 1950 (earlier records are too local and fragmentary) indicate a steady rise in numbers of Erythrean aliens at each locale with time, and a decline with increasing distance from the southeastern Levant (Fig. 14).

The paucity of surveys focusing on mollusks along the Turkish coast till the 1980s is manifest in the scanty records of that phylum at the time. However, the ratio of Erytrean alien mollusks:fish:decapods in Turkey in the period 1981-2004 is similar to the records obtained for the periods 1951-1980 and 1981-2004 in the better surveyed Israeli coast, possibly signifying a natural trend.

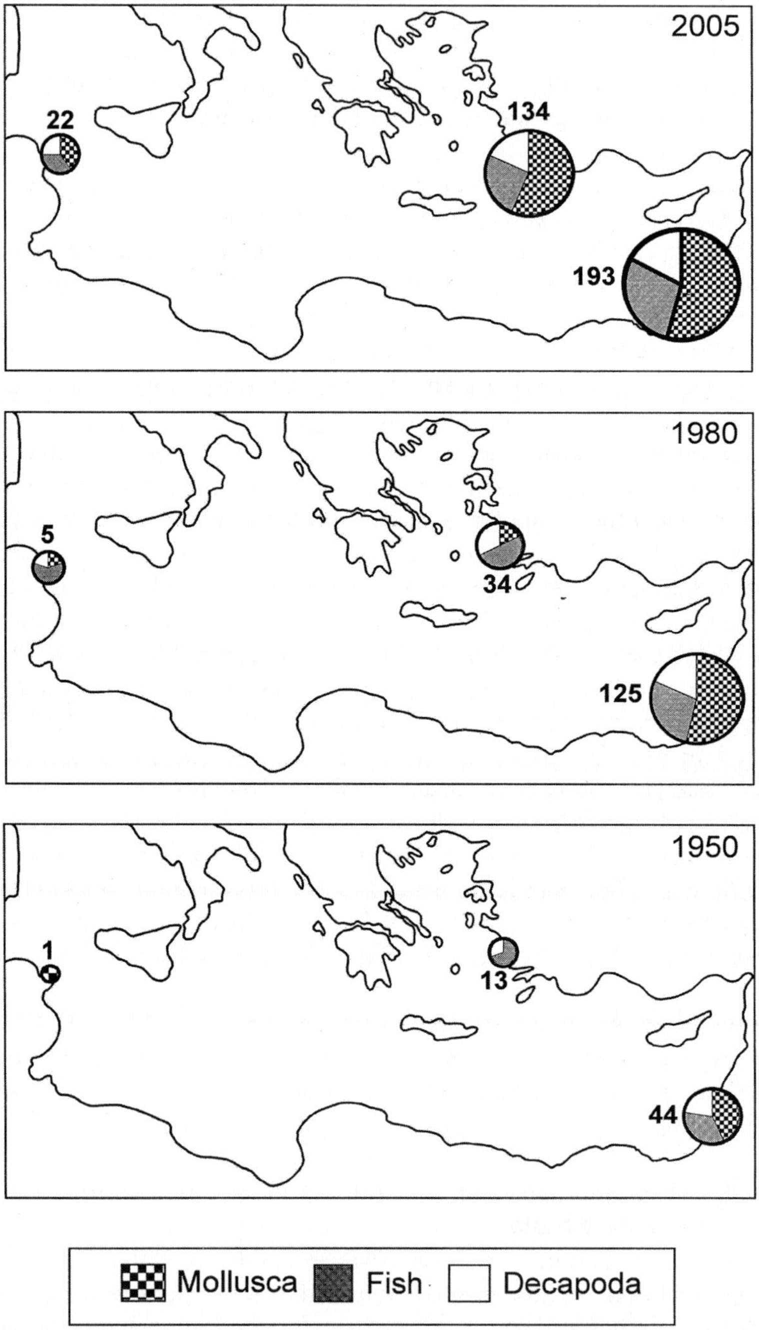

Fig. 14. The dynamics of the Erythrean invasion between 1950-2005 in the eastern Mediterranean: the number of alien molluscs, fish and decapods off Israel, Turkey and Tunisia.

The fraction of alien species that spread following establishment is considered one of the measures of invasion success. The 'tens rule' postulates that 1 in 10 of species "accidentally brought into the country" appear in the wild; 1 in 10 of those become established; "and that 1 in 10 of those established become a pest", 'pest' defined as an alien with "high population density" (Williamson & Fitter 1996). Recently the term was reinterpreted to mean an invasive alien i.e. a species that spreads beyond its point of entry, though "not necessarily harmful" (Jeschke & Strayer 2005). Our ignorance of the biota of the Red Sea, and more significantly, of the 'selector' dynamics of the Suez Canal, precludes analysis of the first step in the 'tens rule'. The scant and patchy surveys of the Levantine marine biota mean that chances of recording an ephemeral alien are rather low, barring analysis of the second step as well. However, the records are robust enough to examine what proportion of the established Erythrean aliens are abundant and/or expand beyond their point of entry. To test this, we arranged our data in invasion sequence, excluding all species known from a single record. Recognizing that some of the latter are recently recorded species, we eliminated them according the average time lag for the taxon (see above), 20 years for decapods, 13 years for fish.

The numbers of established Erythrean fish, decapods and molluscs in the Mediterranean are 47, 31, 116 respectively; of these 32, 24, 50 respectively, spread beyond their original region of establishment. The numbers of established Erythrean fish and decapods in the Mediterranean corrected for time-lags (restricting it to records before 1992 for fish, and 1985 for decapods; mollusks are excluded because few studies were conducted till the 1980s, see above) are respectively 42 and 25; of these, 30 and 18 spread beyond their original region of establishment. These figures manifest an invasion success of nearly 70% for fish and decapod crustaceans and over 40% for mollusks. Mollusk records differ from those of fish and decapods in being based on live material as well as the recent thanatocoenosis, and so may reveal a higher number of established species, thus accounting for the lower fraction of invasive mollusks. Even so the 'spread success' of Erytrean aliens far exceeds the 'tens rule', signifying that once established in the Mediterranean, they highly likely to turn invasive.

12 "Good neighbours"? – The impact of Erythrean aliens on the native biota

The Indo West Pacific starfish *Asterina burtoni* Gray, 1840, was first collected in the Mediterranean in 1955 (Achituv 1973). Within a decade its numbers increased greatly and it was found in abundance under rocks in shallow waters off Egypt, Israel, and Lebanon (Achituv 1969, 1973, Tortonese 1966). It was suggested that its successful invasion was mainly due to rapid proliferation by fissiparity (Achituv & Sher 1991, Karako et al. 2002). The decimation of the native sea star *Asterina gibbosa* (Pennant 1777) populations along the Israeli coast coincided with the rapid increase of its Erythrean congener and gave rise to consideration of some relation between the two. Por (1978: 149) cited this case when dismissing the risks of the Erythrean invasion: "Other than ….. *Asterina gibbosa* there is no known case in which a Lessepsian immigrant species has completely replaced a local one". He was wrong.

The Erythrean killifish, *Aphanius dispar* (Rüppell), is markedly euryhaline, occurring in freshwater, and in a range of salinities up to 4 times as high as seawater (Lotan 1982), whereas its peri-Mediterranean endemic congener, *A. fasciatus* (Nardo 1827), occurs mostly in brackish lagoons (Maltagliati 1999). Both species were described from the Suez Canal and its lakes (Tillier 1902, Norman 1927). In fact, specimens of *A. fasciatus* were collected in the brackish lagoons of Lake Timsah as early as 1871. Norman (1927: 386) suspected the species interbred: "Both species occur together in Ismailia Lagoon, and I have found it difficult to identify certain specimens from this locality with one species or the other; these examples may perhaps represent hybrids"; and so did Tortonese, who spent some time on the shores of the Timsah and Bitter lakes in 1944-45 (1954: 2): " Their [*A. fasciatus* and *A. dispar*] hybridization does not appear unlike". Indeed, of the 4,600 killifish collected, between 1973 and 1975, in the hypersaline Bardawil Lagoon on the Mediterranean coast of Sinai, Egypt, 45% were *A. dispar*, 17% *A. fasciatus*, and 38% were hybrids (Lotan & Ben Tuvia 1996). Naturally occurring hybrids of the two killifish species were described as " ... common, and in some localities (Ashdod Harbour [southern coast of Israel], parts of Bardawil Lagune etc.), these populations comprise mostly hybrids" (Goren & Rychwalski 1978: 261). Villwock (1985, 1987) too described hybrids from Bardawil Lagoon and al-Qanatir, Egypt. The first Erythrean killifish was collected off Tel Aviv, Israel, in the winter of 1943/44 (Mendelssohn 1947), and the last specimens of the native killifish were collected at Dor, on August 13, 1976 (M. Goren, pers. comm.; preserved in the National Collections, Tel Aviv University, access number 6319) – within a generation an endemic genotype was locally lost through hybridization, and *A. dispar* replaced *A. fasciatus* along the Mediterranean coast of Israel (Goren & Galil, 2005).

The Levantine Basin continues to admit new Erythrean biota at relatively high rates. Very little is known on the kaleidoscopic inter-relationships of native and invasive species necessary to demonstrate direct competition leading to niche limitation, displacement or local extirpation. The many documented instances of sudden changes in abundance, where a native Mediterranean species have been outcompeted wholly or partially displaced from their habitat space by an Erythrean alien could be attributed to competition for different resources or direct interference between the newcomers and the native species, as part of a profound anthropogenic alteration of the marine ecosystem through habitat destruction and pollution (Bariche et al. 2004, Gurevitch & Padilla 2004).

In the early 1970s the Erythrean mytilid *Brachidontes pharaonis* was "*c*. 250 times rarer" than the native mytilid *Mytilaster minimus* (Poli 1795), that formed dense '*Mytilaster* beds' on intertidal rocky ledges, with up to 26 specimens per cm^2 (at Palmahim, Israel, Safriel et al. 1980a, table 4). Safriel et al. added that "*B. variabilis* [*pharaonis*] never forms 'beds' in the eastern Mediterranean, and is singly or in small groups dispersed within *M. minimus*", and "... it did not displace *M. minimus*" (Safriel et al. 1980a: 39, 59). By the end of the 1980s they were less sanguine. Following a series of experiments it was determined that *Brachidontes* interferes with recruitment of *Mytilaster*, yet the authors maintained the Erytrean mytilid was not likely to outcompete the native mytilid: "The two species can coexist, both locally and regionally" (Safriel & Sasson-Frostig 1988: 225). Nowadays "the same rocks are .... completely covered with the Eritrean *B. pharaonis*, while *M. minimus* is only rarely encountered." (Mienis 2003: 15). Indeed, by the late 1990s it formed dense beds (>300 individuals/100cm^2) on rocky platforms "where mussel beds were absent in the past" (Rilov et al. 2004: 347). This competitive displacement is known from the Syrian coast as well (Saad 2000).

An Erythrean limpet, *Cellana rota* (Gmelin 1791), commonly found in the Red Sea and Suez Canal (Tillier & Bavay 1905, Sharabati 1984), was first collected in the Mediterranean in 1961 (Christiaens 1967). By 2000, *C. rota* spread along the southeastern Levantine coast, occupying the more protected sites in the intertidal zone, whereas the native Mediterranean limpet, *Patella caerulea* Linnaeus 1758, inhabited rocks exposed to the surf. A recent survey along the Mediterranean coast of Israel found that *C. rota* has come to dominate the upper rocky littoral and has been replacing the native limpet when "a few years ago *Patella caerulea* was the only limpet inhabiting the same rocks at that locality [Ashdod]" (Mienis 2002: 275). Along the southern coast of Israel it has already completely replaced the native limpet, along the central coast it "has taken possession of 40-50% of the available space, while even further north, around Haifa-Acco, the newcomer has displaced 10% of the local *Patella* species." (Mienis 2003: 15).

The Erythrean Spiny oyster, *Spondylus spinosus* Schreibers, 1793, was first noticed in 1988 in Haifa Bay, Israel, and within a five years has established thriving populations that have overlaid the submerged sandstone ridges at the western end of the bay (Mienis et al. 1993b). The Erythrean Spiny oyster has completely out-competed the smaller native oyster *Spondylus gaederopus* Linnaeus, 1758. The jewel box oyster, *Chama pacifica* Broderip, 1834, had been recorded off Alexandria, Egypt, in the early 20th century (Tillier & Bavay 1905), but only in 1988 was it noticed off the Israeli coast (Mienis et al. 1993a). It is found in abundance (30 specimens/m^2), together with the Erythrean Spiny oyster, on the submerged ridges off the northern Israeli coast, but also settles in great numbers on artificial structures such as piers and an offshore gas production platform (Mienis 2004). The Erythrean jewel box oyster outcompeted it much smaller native congener, *Chama gryphoides* Linnaeus, 1758: "The local *S. gaederopus* and *C. gryphoides* are hardly even encountered as epibionts on the new immigrant species." (Mienis 2003: 15). The native Mediterranean cerithiid gastropods, *Cerithium vulgatum* Bruguière, 1792 and *Cerithium lividulum* Risso, 1826 (*C. rupestre* in Barash & Danin, 1992), were respectively common and abundant in shallow water along the coast of Israel until the 1970s. Since then they have beeen supplanted by the Erythrean cerithiids *Cerithium scabridum* Philippi, 1848 and *Rhinoclavis kochi* (Philippi, 1848) (Mienis 2003: 15).

A native penaeid prawn, *Melicertus kerathurus*, was commonly caught by trawlers along the Israeli coastal shelf on sandy or sandy mud bottoms, and supported a commercial fishery throughout the 1950s (Holthuis & Gottlieb 1958). This native prawn has since nearly disappeared, and its habitat overrun by the Erythrean penaeid prawns. Geldiay & Kocatas (1972) reported that off the southern coast of Turkey, too, *Marsupenaeus japonicus* has been replacing *M. kerathurus* in fisheries catches. d'Udekem d'Acoz (1999) reported that *M. japonicus* "has almost evicted the native *P. kerathurus* from the easternmost part of the Mediterranean". The rapid advent of *Metapenaeus monoceros* into the Gulf of Gabes, Tunisia, has raised concerns over the fate of *M. kerathurus* fisheries there (Chaouachi et al. 1998). Similarly, the Erythrean snapping shrimps *Alpheus inopinatus* Holthuis & Gottlieb, 1958, and *A. audouini* Coutière, 1905, are more common in the south-eastern Levantine rocky littoral than the native *A. dentipes* Guérin-Méneville, 1832, and *A. rapacida* de Man, 1908, much more common than the native *A. glaber* (Olivi, 1792) on the muddy bottoms of the shallow shelf (Lewinsohn & Galil 1982, Galil 1986). The Erythrean dragonet, *Callionymus filamentosus*, has replaced the native callionymids *C. pusillus* Delaroche, 1809, and *C. risso* LeSueur, 1814, along the Levantine upper shelf. The populations of the Erythrean narrow-barred mackerel, *Scomberomorus commerson*, have greatly increased in the 1980s, coincidently with the decline in the populations of the one-time common native meager, *Argyrosomus regius*, to the point where the latter is rarely encountered along the Israeli coast.

Competitive displacement may also modify bathymetric ranges in populations of Erythrean and indigenous species. The native red mullet, *Mullus barbatus* and the native hake, *Merluccius merluccius* were both displaced into deeper, cooler waters by their respective Erythrean competitors, *Upeneus moluccensis* and *Saurida undosquamis* (Oren 1957a,b). The Erythrean mantis shrimp, *Erugosquilla massavensis*, was recorded from Alexandria, Egypt in 1933 (as *Squilla africana*) (Steuer 1936), and is commonly found off the Levantine coastline, Cyprus (Ingle 1963), Crete (Dounas & Steudel, 1994), Rhodes (Galil & Kevrekidis 2002), and the Sea of Marmara (Katağan et al. 2004). Though on occasion it is taken together with the native Spottail mantis shrimp, *Squilla mantis*, generally the latter occurs in deeper waters than *E. massavensis*. The Spottail mantis shrimp is generally fished commercially in 10-25 m in the northwestern Mediterranean and in the Adriatic Sea (Lewinsohn & Manning 1980), but off the Israeli coast it is found at greater depths, mostly between 70 and 90 m. Lewinsohn & Manning (1980) questioned "whether temperature, bottom type, or pressure from *O. massavensis*, or a combination of these is responsible for its depth distribution".

The changes caused by Erythrean alien species have gone beyond local decimation or replacement of native species, or modification of their ranges, some Erythrean invasives have altered the native food web. The two Erythrean siganid fish that settled in the Mediterranean: *Siganus rivulatus* (Forsskål, 1775) and *S. luridus* (Rüppell, 1828), were first recorded off the coast of Israel in 1924 (Steinitz 1927) and 1955 (Ben Tuvia 1964), respectively. Both species formed thriving populations along the Levant coast and are found as far west as Tunisia (Ktari-Chakroun & Bahloul 1971, Ktari & Ktari 1974). An analysis of gut contents of siganids, in conjunction with the spatial and seasonal composition of the local algal community at one rocky site off the Israeli coast, showed that their diet has also a significant impact on the structure of the local algal community: it seems that by feeding selectively they have nearly eradicated some of their favorite algae locally (Lundberg et al. 2004). The multitudinous siganids have replaced native herbivorous fish: "... along the Lybian coast, *S. rivulatus* seems to outcompete *Boops boops*, reducing the abundance of the latter, since both species feed upon algae.... Quite likely, the same is true for the S.E. Aegean Sea" (Papaconstantinou 1987). Likewise, along the Lebanese coast, it has replaced the native *Sarpa salpa* (Linnaeus, 1758), that had been abundant in trawl catches early in the 20th century (Gruvel 1931). The siganids comprise a third of the fish biomass in rocky habitats along the Israeli coast (Goren and Galil 2001), and 80% of the abundance of the herbivorous fish in shallow coastal sites in Lebanon (Bariche et al. 2004). Prior to the arrival of the siganids in the Mediterranean, there were few herbivorous fish and invertebrates and their role in the food web off the Levantine rocky habitats had been negligible. The algal contribution to the web was mainly through the decomposers. The algivourous siganids increased the rate large amounts of algal material were recycled (in hours through the fish gastrointestinal system rather than weeks or months of decomposition).

13 The parasitofauna of the Erythrean aliens in the Mediterranean

Absence of natural enemies, be it competitors, predators, pathogens, or parasites, is one of the oft-repeated explanations given for the success of alien biota (Wolfe 2002, Torchin et al. 2003). Diminished parasitization of alien species may be attributed to reduced probability of infestation due to the small numbers of the founding population, its being composed of uninfected larval stages, or, in the case of heteroxenous parasites, the absence of intermediate hosts in the new locale, and on the other hand, the host-specificity of some native parasites forestalling infection of alien hosts.

Fischthal (1980) examined parasite loads of 500 fish of 59 species collected by gill-netting in 1977 off the Mediterranean coast of Israel. He found that 29 of the 43 (67%) native Mediterranean species were infected with digenetic trematodes, but only 5 of the 12 (42%) Erythrean species. Interestingly, many of the trematodes infecting the Erythrean hosts were of Atlanto-Mediterranean origin and were probably acquired in the Mediterranean from native fish species (Table 1). These results support the 'enemy release hypothesis' – alien species generally import only a subset of their parasitofauna, thus releasing themselves from the parasites of their native habitat during introduction, and though subsequently accumulating parasites native to their new environment, they harbor fewer parasites in their new locale than in their native range (Torchin et al. 2001).

Meticulous parasitological studies of siganid fish in the Red Sea (Diamant 1985, Diamant & Paperna 1985, 1986, Diamant & Wilbert 1985) have shown that the fish harbor rich and diverse parasitofauna including members of Protozoa, Monogenea, Acanthocephala, cestoda, Digenea, Nematoda, Copepoda, Isopoda and Hirudinea. In 1974 a single specimen each of *S. luridus* and *S. rivulatus* were bought in the fish market of Sfax, Tunisia (Ktari & Ktari 1974), each harboring an Erythrean species-specific monogenean parasite, *Pseudohaliotrematoides polymorphus* and *P. suezicus* respectively. In 1971 and again in 1981-1983 the parasitofauna of the two Red Sea siganids that had entered the Mediterranean, *Siganus luridus* and *S. rivulatus*, was studied (Paperna 1972, Diamant 1989). The specimens examined were collected off the Mediterranean coast of Israel and in the Bitter Lake in the Suez Canal (Table 2).

Table 2. List of parasites found on Erythrean fish in the Mediterranean. AM – Atlanto-Mediterranean; IP – Indo-Pacific.

Fish species	Parasite	Origin of parasite
Alepes djedaba	*Ancylocoelium israelense*	AM
	Aponurus lagunculus	AM
	Chrisomon israelensis	
	Ectenurus lepidus	AM
	Lecithochirium haifense	
	Lecithocladium excisum	AM
	Opechona polonii	AM
Atherinomorus lacunosus	*Ancyrocephalus atherinae*	IP
Callionymus filamentosus	*Genitocotyle atlantica*	AM
Crenidens crenidens	*Lamellodiscus elegans*	AM
Pelates quadrilineatus	Didymozoid larva	
	Podocotyle (Pedunculotrema) israelensis	
Saurida undosquamis	Didymozoid larva	
	Diphtherostomum israelense	AM
	Lecithochirium magnicaudatum	IP
	Monilicaecum ventricosum	IP
	Plerocercoid (larval cestodes)	AM
Sargocentron ruber	Gnathiid	AM
Siganus luridus	*Octomitus* sp.	IP
	Entamoeba sp.	IP
	Ceratomyxa sp.	IP
	Pseudohaliotrematoides polymorphus	IP
	Tetrancistrum plectocirra	IP
	Polylabris cf. *mamaevi*	IP
	Gnathia sp. larva	
Siganus rivulatus	*Nosema ceratomyxae*	IP
	Octomitus sp.	IP
	Entamoeba sp.	IP
	Balantidium sigani	IP
	Ceratomyxa sp.	IP
	Zschokkella icterica	AM
	Ortholinea sp.	IP

Table 2 continued.

Fish species	Parasite	Origin of parasite
	Cryptocaryon irritans	AM
	Pseudohaliotrematoides polymorphus suezicus	IP
	Pseudohaliotrematoides polymorphus "nagatyi"	IP
	Tetrancistrum plectocirra	IP
	Polylabris cf. mamaevi	IP
	Hemiurus appendiculatus	AM
	Hysterolecitha sigani	IP
	Gnathia sp. larva	
Sillago sihama	–	
Sphyraena chrysotaenia	*Bucephalus labracis*	AM
Upeneus moluccensis	Aegiid	AM

Eleven parasite species were identified: five ectoparasitic monogeneans (including *Pseudohaliotrematoides polymorphus* and *P. suezicus*), five endoparasitic protozoans and the larval stages of a gnathiid isopod – all monoxenous parasites with direct life cycles requiring no intermediate host. However, Fischthal (1980) identified two species of heteroxenous digenetic trematodes with complex life cycles requiring more than one host in *S. rivulatus* specimens from the Mediterranean: *Hemiurus appendiculatus* which is distributed in the Atlantic, the Mediterranean and the Red Sea, and infects native Mediterranean fish, and *Hysterolecitha sigani*, known previously from Australia and New Caledonia, that probably arrived with its host from the Red Sea. Paperna (1972: 3) suggested that endoparasitic helminthes spread into a new area only if they are not highly host-specific or if the suitable intermediate host has arrived as well, though "in most cases the migration of the fish host does not coincide with similar migration by the intermediate invertebrate hosts". Indeed, the goatfishes *Upeneus moluccensis* and *U. pori* harbor many species of trematodes in their native habitat in the Indian and Pacific Oceans, but an examination of those species off Lebanon carried out in 1985 failed to find any, nor were they infected by the native Mediterranean trematodes (Maillard & Raibaut 1990). However, it is quite certain that the presence of *H. sigani* in a siganid collected off Haifa, a distance of some 470 kms from the Suez canal, implies the fish had been infected inside the Mediterranean. The intermediate hosts for many digeneans are gastropods or bivalves, one of the ninety Red Sea molluscs already recorded from the Levantine Basin may be its native intermediate host, or a yet unidentified Erythrean mollusc. In a later study (1995-1996), nine species of monogenean and gut parasites were identified from 152 specimens of *S. rivulatus* that were obtained from the port of Ashdod, next to fish farming cages (Diamant et al. 1999). The parasitofauna of the Mediterranean siganids was composed of nearly all the monoxenous parasites present in the Red Sea siganids, with the addition of a single cosmopolitan ciliate, but none of the Red Sea heteroxenous species. This result may reflect the polluted, degraded port environment, where the Mediterranean lot of siganids was collected.

The Levantine populations of the Erythrean swimming crab *Charybdis longicollis* have been parasitized recently by the sacculinid rhizocephalan, *Heterosaccus dollfusi* Boschma (Galil & Lütsen 1995). The swimming crab, found in the Red Sea, the Persian Gulf and Madagascar, was first recorded in the Mediterranean in 1954 (Lewinsohn & Holthuis, 1964). Since then, it has been recorded all along the Levant coast, from Egypt to Rhodes, where it is common on sandy-mud bottoms at 25-60 m and occasionally deeper. It can form as much as 70% of the benthic biomass on sandy-silt bottoms off the Israeli coast (Galil 1986). Release from parasites may have contributed to the success of invasive species – of the thousands of *C. longicollis* collected off the Israeli coast in over three decades, none were parasitized until 1992, when a few specimens were discovered carrying the externae (reproductive part of the parasite) of the castrating sacculinid, itself an Erythrean alien. Subsequent collections confirmed its presence on the easternmost part of the Anatolian coast (Oksnebjerg et al. 1997). *Charybdis longicollis* had entered the Mediterranean over half a century ago, but it seems that individuals harboring the sacculinid arrived after the host crab Mediterranean populations increased in density, affording the parasite increased transmission efficiency.

Heterosaccus dollfusi penetrates through the gills of young *C. longicollis* and forms a vegetative interna consisting of numerous microscopic roots. The late stage of the interna forms the reproductive part of the parasite (externa), that breaks through the host's skin assuming a sacciform shape and occupying the position normally taken by the eggs in berried females. Infection causes degeneration of the gonads in both sexes, in effect sterilizing the hosts. It also modifies morphological traits: female hosts lose the swimmerets; male hosts lose the copulatory appendages, and their abdomen broadens to afford the parasite's externa optimal protection. The presence of the externa induces the host to simulate egg-grooming behaviour, even in male crabs, in which grooming and fanning the abdomen do not occur normally (Innocenti et al. 1998). The presence of the parasite also modifies digging behaviour and inhibits burying in the sand, a common activity in non-parasitized crabs, and reduces belligerence in male crabs (Innocenti et al. 2003). Eighty-six percent of mature externae were found to be ovigerous. Since one or two days pass between emission of nauplii and the next oviposition, this means that practically all externae were reproducing from spring to fall. Multiple infections occur most frequently among the younger crabs, presumably because by placing a high nutritional demand on the host, they weaken it, and in many cases, cause its untimely death. Along the Israeli coast infection rates of up to 77% were recorded, with up to 57.6% of the infected hosts bearing more than one externa (Galil & Innocenti 1999). Multiple infections rise with increased incidence of infection, and may be ascribed to the aggregated pattern of distribution of the host. The rapid spread and the high prevalence of *H. dollfusi* infestation may be related to the dense population of the host and the year-round reproduction of the parasite, causing recurrent infection. The Erythrean sacculinid had not been detected in any of the other portunid crabs, alien or native, inhabiting the Levantine sublittoral, including the cogeneric *C. helerii.*

It has been suggested that biocontrol, involving the introduction of a predator, parasite or pathogen, may present an option for marine pests management, in particular invasive species (Thresher et al. 2000). Kuris (1974, 1997) suggested parasitic castrators, specifically rhizocephalan barnacles, may be important regulators of host population density, since they inflict reproductive death on their hosts, and proposed their use as control agents for invasive marine decapod crustaceans, provided they are specific to the target host. The Erythrean invasion presented in the last decade an interesting "field experiment": *H. dollfusi* had not been detected so far in any other portunid crab species inhabiting the Levantine sublittoral, alien or native, including the cogeneric *C. helerii*, but is it an efficient control agent? The prevalence of infection and occurrence of multiple externae were studied between 2003 and 2005 at a previously studied site, and compared with earlier studies in order to estimate the dynamics of host-parasite populations in the second decade of their presence in the Levantine Sea (Innocenti & Galil, 2005).

Though *H. dollfusi* has limited dispersal capabilities, it rapidly infected the Levantine populations, infecting within two years of its initial detection 62.6% of the Palmahim (south of Tel Aviv, Israel) host population with 25-47 CW (Galil & Lützen, 1995, table 1), and 67.4% in June 1996 (Galil & Innocenti, 1999). These values are quite similar to the incidence of infestation in recent years at the same site. It is clearly established that heavily parasitized host populations bring about higher rates of multiple infestation: 48% of the externa-bearing hosts in the May 1994 sample bore more than one externa (Galil & Lützen, 1995, Table 1), and 57.6% in June 1996 (Galil & Innocenti, 1999), in line with the rate of multiple externae in May 2002 and 2005, whereas the lower number of multiple externae bearing crabs (37.1%) in May 2004 is related to the dip in infestation (44.4 %) that year.

Whereas males in the pre-infected population of *C. longicollis* were significantly larger than females (Galil & Innocenti, 1999), the size gap diminished significantly in infected crabs. In post-infection populations the average and maximal size of uninfected males is increasingly reduced, though they are still larger than infected males. In female crabs, the average and maximal size of infected specimens, especially externa-bearing specimens, is higher than uninfected ones. Phillips and Canon (1978) attributed the stunted host size to molt inhibition by the parasite, O'Brien and Van Wyk (1985) to fewer molts, Hawkes et al. (1987) to smaller molt increments. However, parasite-induced molt interruption fails to explain the size increase in parasitized female specimens. Since the fecundity of the parasite is related to externa size, and the latter is dependent on the size of the host (Galil & Innocenti, 1999), it is proposed that *H. dollfusi*, capable of modifying its host morphology, physiology and behaviour, regulates the host size to best endure the significant metabolic costs of the reproductive externa, and provide it with optimal physical protection.

In its second decade in the Mediterranean, the population of *H. dollfusi* seems stable: despite the high prevalence of the parasite and its injurious impact on the host reproduction, there is no noticeable reduction in the host population. It is suggested that the high fecundity of the host females, the "size refuge" formed by parasite-free larger males, and the "open" recruitment dynamics of *C. longicollis*, keep its population density high enough for *H. dollfusi*, with its "closed" recruitment dynamics, to maintain its pandemic infection rates. The "field experiment" acted out along the Levantine coast since 1992 demonstrates clearly the ineffectuality of using a rhizocephalan barnacle to reduce populations of an invasive decapod with widely dispersed planktonic larvae (Innocenti & Galil, 2005).

14 The "silver lining"? – The economic impact of Erythrean aliens

Some Erythrean aliens have been exploited commercially almost as soon as they entered the Suez Canal. Already early on the Suez Canal Company sought to exploit the biota in the Canal, and hired Gruvel, a fisheries expert who was familiar with the Levantine fisheries, as 'chef de mission' to identify possible commercially advantageous products. The resulting report, 'Contribution à l'étude de la bionomie générale et de l'exploitation de la faune du Canal de Suez' (Gruvel 1936), identified fish, decapods and molluscs of economic interest. Among the Erythrean fish fished commercially along the Suez Canal Gruvel mentions the clupeids *Herklotsichthys punctatus* (as *Harengula punctata*) and *Dussumieria elopsoides* (as *D. productissima*) "qui se rencontre en assez grande abondance jusque dans le lac Timsah et qui est utilisée dans l'alimentation générale." (Gruvel 1936:153). Also abundant are the carangid *Alepes djedaba* (as *Caranx djedaba*), *Sphyraena obtusata*, *Monishia ochetica, Petroscirtes ancylodon* (the last two "sont vendus sur tous les marches locaux, de même qu'au Caire et Alexandrie, comme poisons de friture." (there, 173), while the Erythrean mullids – *Upeneus pori* (as *Upenoides vittatus*) and *Upeneus moluccensis* (as *Mulloides flavolineatus*) were known, but uncommon, in the local market. Gruvel, a Frenchman, could not resist adding some gustatory advice concerning the filefish *Stephanolepis diaspros* (as *Monacanthus setifer*): "La peau enlevée, après ebullition dans un court bouillon, il reste une chair blanc rosé, ferme et d'un gout excellent." (there, 168). Though Gruvel was disappointed by the lack of "les grands Crustacés comestibles, comme Langoustes, Homards et Scyllares" (there, 176), he conceded that the crustaceans were "de beaucoup le plus intéressant". Gruvel noticed the presence of the Erythrean mantis shrimp *Erugosquilla massavensis* (as *Squilla massavensis*) in the local markets: "On la trouve communément, sur les marchés de Suez et d'Ismaïlia". An early Erythrean invader, the swimming crab *Portunus pelagicus*, was recorded from Port Said in 1898, where it soon became abundant, and by the early 1900s it was offered in the markets of Port Said, Alexandria and Haifa (Fox 1924, Calman 1927). Gruvel reported that the crab was "véritablement importante au point de vue économique", and was found in Port Said and along the entire Egyptian Mediterranean coast, as well as along the Levantine coasts up to the Gulf of Iskenderun in southeastern Turkey. During the British Mandate the Arab fishermen of Haifa and Acre annually sold 20 tons of *P. pelagicus*, a crab "most common in the open sea, particularly in the Acre Bay region" (Perlmutter 1956:18). Off the Egyptian coast the crabs were fished by the Italian and Egyptian trawl fishermen, and by local fishermen using beach seines, and were sold 'en grand quantité' in the markets of Port Said, Ismaïlia, Suez, Cairo and Alexandria. Gruvel, doubtlessly a gourmand, praised the crabs: "On sert ces crabes, bouillis ou farcis, dans à peu près tous les restaurants. Farcis et cuits au four, ils constituent un mets excellent et, en général, très apprécié" (there, 178). The Erythrean penaeids, and especially the tiger prawn, *Marsupenaeus japonicus*, were greatly valued : "C'est le plus magnifique exemplaire de crevette que nous ayons jamais rencontré, au cours de nos différentes recherches dans les Colonies." (there, 181). Gruvel, however, was disappointed by the local molluscs: "Le nombre des espèces utilisées dans l'alimentation locale et que l'on trouve sur les différents marchés égyptiens est …extrêmement restreint. L'Indigene, en général, consommé peu ces animaux; ce sont surtout les Européens de diverses races, les Italiens en particulier, qui les utilisent en plus ou moins grande quantité." (there, 183).

Gruvel realized the economic importance of the Erythrean invasion: "..les passages définitifs de ces espèces à travers la totalité du Canal présentent un résultat *économique* [sic] également très important. En effet, nous avons vu plus haut, que l'industrie de la pêche s'est dèveloppèe, dans le Canal et sourtout dans le lac Timsah et dans le Grand lac Amer, d'une façon inattendue pour ceux qui ne sont pas au courant de cette question. Cette pêche nous paraît, même, beaucoup trop intensive par rapport à l'importance de la faune industrielle. Elle ne manqué, cepandant, pas d'intérêt puisqu'elle permet de contribuer, pour une assez large part, à l'alimentation des populations européennes et indigenes locales en fournissant les elements necessaries, non seulement aux marches de Port Said, Ismaïlia et Suez, mais aussi, en envoyant une partie des produits de cette pêche sue ceux d'Alexandrie et, sourtout, du Caire." (there, 228). The importance of the Erythrean-based fisheries was not limited to the Canal and the Egyptian coast, they "...constituent, aujourd'hui, pour les marches palestiniens et syriens, un appoint non négligeable et particulièrement intéressant, par consequent, pour l'ensemble des populations de ces deux Pays" (there, 229).

Bodenheimer (1935: 457) witnessed the early "penetration of Red Sea forms through the Suez Canal" and noticed that "Quite a number of fishes have not only reached our [SE Levant] shores, but some of them have even increased in such numbers, that they appear regularly in the fish market". By mid-century the Erythrean fishes were an important part of the Levantine fisheries. In fact, the proliferation of those invasive populations to the point that they are harvested commercially is an excellent index of how prevalent they have become (Ben Yami & Glaser 1974). The rabbitfish, *S. rivulatus*, that was first reported from the Levantine coast in 1929, was by 1937 an important and staple of the coastal fishery off Port Said and Alexandria (Faouzi 1951). Insofar as the Israeli and southeastern Turkish (Gulf of Iskenderun) fishing grounds were concerned, the bulk of the trawler catch from 1950 to 1955, was comprised of three species – the native red bream, *Pagellus erythrinus*, and hake, *Merluccius merluccius,* and the Erythrean yellow striped mullet, *Upeneus moluccensis*. The latter were fished commercially in the early 1940s only along the southern coast of Israel, but by 1946-1947 were found all along the coast (Gottlieb 1957), and by the late 1940's constituted an estimated 10-15% of the total mullid catch (Wirszubski 1953). In 1955 Israeli fishermen noticed greater numbers of the yellow striped mullet, and data assembled by the Sea Fisheries Research Station, Haifa, indicated that their percentage in the mullid catch rose to 20%, and to over 83% in early 1956 (Oren 1957a,b), and was considered "the most important commercial fish in the Israel trawl catches" (Gottlieb 1957:20). Since the total mullid catch had remained constant, the yellow striped mullet had in the early 1950s "almost completely replaced the Mediterranean species, the red mullet, *Mullus barbatus* in the trawl catch" (Perlmutter 1956:4). In 1955, another Erythrean alien, the lizardfish, *Saurida undosquamis* became an important part of the trawl catch (Oren 1957b). In 1953 it was first recorded from the Mediterranean coast of Israel (Ben Tuvia 1953) as much rarer than the native Mediterranean lizardfish, *Synodus saurus*. Within two years it was no longer true, commercial catches increased steadily, in November and December 1955, 22 and 27.5 tons respectively were taken, swelling to 40 and 46.8 tons in January and February 1956 respectively, to a total of 266.5 tons for 1956 - 20% of the total annual trawl catch (Oren 1957a,b). By the summer of 1956 it was common in the trawl catches in the Bay of Mersin, Turkey. The fisheries grounds opposite El-Arish were dominated by *S. undosquamis* (misidentified as *S. tumbil*) as well, this species making up 53% of the total catch in the spring of 1959 (Gorgy 1966). In 1962, 652 tons of *S. undosquamis* were landed in the area stretching from Damietta eastward to Port Said (El-Zarka & Koura 1965). The sudden increase in the populations of the lizardfish, the yellow striped mullet, the red soldierfish, and Erythrean penaeids was attributed to a rise of 1-1.5°C in sea temperature during the winter months of 1955 (Ben Yami 1955, Chervinsky 1959).

The Mediterranean fisheries, inclusive of the coastal lagoons, constituted more than half the fish production of Egypt before the completion of the Saad-el-Ali, the High Dam of Aswan (El-Zarka & Koura 1965). Up to 1928 the trawl fishery along the Mediterranean coast of Egypt was conducted from Italian-crewed sailing vessels, the *balancelles*, whose total annual yield was 5600 tons. When motorized fishing boats were introduced in the 1930s (25 trawlers operated from Alexandria, 5 from Port Said) the annual yield increased from 13,700 tons in 1930, to 51,484 tons in 1960 (559 trawlers), and the percentage contribution of the Mediterranean fisheries increased from 12.3% in 1928, and 31% in 1930, to nearly 60% in 1960 (El-Zarka & Koura 1965). The increase in commercial production was based on the then highly abundant sardine fishery and on penaeids, the latter contributing nearly a fifth of the landings – over 7200 tons in 1962. Between 1959 and 1961, the eastern section of the Mediterranean coast of Egypt (Port Said to El-Arish) was surveyed to identify "des espèces marines comestibles susceptibles d'être exploitées." (Gorgy 1966: 27). The most common penaeids in shallow waters were the Erythrean prawns *Marsupenaeus japonicus, Metapenaeus monoceros, M. stebbingi*, and *Penaeus semisulcatus*. El-Zarka & Koura (1965) warned that the Aswan Dam will impact the sardine catch (48 % of total landings in 1962), and the coastal fisheries "will depend entirely on the bottom fish" (there, 228).

With the completion of the Aswan Dam in 1966, the outflow of Nile waters into the Mediterranean nearly ceased, reducing the phytoplankton bloom on which the sardines fed, and the sardine fisheries decreased from 18,200 tons in 1962 to 1,200 tons in 1966, whereas the penaeid catch was down from 7200 tons in 1962 to 2700 tons in 1966 (Al Kholy & El-Wakeel 1975). "Such a decline in the marine fisheries called for enhancing the fish investigations. Hence, in 1965-1966 the first Soviet-Egyptian Expendition [sic] took place" (there, 3). Perhaps national pride did not allow attributing this disaster fully to the brand new dam, and the report stated that "In the opinion of Egyptian specialists the main reason of the sharp decrease in the number of commercial units was the intensification of fisheries and utilization of more up-to-date fishing gear" (there, 3). A further drop in the annual yield to barely 8500 tons in 1969, compelled the authorities to reconsider, and in carefully couched terms they called on another joint Soviet-Egyptian expedition: "The 6th Session of the Soviet-Egyptian Commission on fisheries .... took a decision to set an expedition with a view to evaluate the state of marine fisheries resources in connection with the considerable change of the hydrological regime." (there, 3). Before the damming of the Nile the clupeid catch was composed of the native *Sardina pilchardus* and *Sardinella aurita*, but Investigations carried out by the Soviet research vessel *Ichthyolog* on the Egyptian shelf in 1970-71 showed that in September 1971 the Erythrean *Dussumieria elopsoides* (misidentified as *D. acuta*) accounted for 54% of the clupeid yield on the Egyptian shelf (Al Kholy & El-Wakeel 1975). The Erythrean penaeid *M. monoceros* dominated in January 1971 at depths up to 20 m constituting 92.5% of the penaeid catch off Rosetta, whereas in April 1971 *M. stebbingi* was the more abundant species (up to 82%) off Abukir and Rosetta. Off the Sinai coast, the proceeds from the penaeids' landings accounted for over a third of the total trawl catch taken at depths of up to 25 m - where the dominant species were *P. semisulcatus* and *M. japonicus* (Tom 1979). Again, *S. undosquamis* had "an important place of the total catch" (there, 215), and was considered a valuable fish. The remarkable "penetration of Red Sea types" was considered to be the result of the "salting of the coastal waters" (Al Kholy & El-Wakeel 1975: 168). Following a half-hearted recommendation to utilize "species which where [sic] considered previously as non-commercial such as *Leiognathus klunsingeri* [sic]."(there, 240), the members of the joint expedition forecasted that "Shrimp and fish culture will compensate for the marine fisheries decrease along the Mediterranean Coast of Egypt." (there, 249). A study of the fishery statistics of the Mediterranean coast of Egypt for the period from 1962 to 1989 (Bebars et al. 1997), showed that in 1963 the sardines accounted for 40% of the total catch, and the penaeids for 26%, but only 12% and 7% respectively in 1987; whereas the 'diverse' category, consisting of mostly less commercially favored taxa, increased at that period from 3% to 47% of the landings. Indeed, lately gastropods, previously considered unsuitable for consuption (see above), have been added to the commercially important taxa: "of commercial species the most important are.... *Gafrarium pectinatum*, *Murex tribulus* [*forskalli*] and *Fusinus marmoratus* [*verrucosus*]." (Mohamed et al. 1992), "... *Thais carnifera* is one of the most common edible molluscs in the Suez Canal area." (Hanafy 1996).

A survey of the *jaroofi* seine fishery in St. George Bay, Lebanon, was undertaken to "permit evaluation of the changes that are and will be taking place in the fish populations" following the High Dam completion, and the salinity decrease in the Bitter Lakes (George & Athanassiou 1967: 239). Of the 101 species identified, 15 were Erythrean aliens, yet their import in the fishery was much greater – the Erythrean fish dominated the Levantine near-shore fishery. George & Athanassiou (1967:254) describe *Siganus rivulatus* as "one of the most firmly established erythraean immigrants .... one of the commonest fishes of St. George Bay entering the catches of trammel net, hook and line and seine fishermen in large numbers". The Erythrean goatfish *Upeneus pori* (described as *U. asymmetricus*) was thought to have "already displaced a significant part of the indigenous *M. barbatus* population" (there, 250), *U. moluccensis* "found favor in the markets of Beirut and further north along the coast" (there, 258), and the Erythrean obtuse barracuda *Sphyraena chrysotaenia* was successfully established as well. Baskets of the Erythrean pearl oyster, *Pinctada radiata*, are offered for sale in the fish market of Beirut (Christensen 1972), and Syria, where it is "currently consumed and available in the local market in limited quantities" (Saad 2000). The Erythrean spiny oyster, *Spondylus spinosus*, is "served in restaurants in Jbail, Lebanon" (www.ciesm.org/atlas).

Examination of the Israeli fisheries statistics since the mid 1980s underscores the growing prominence of the Erythrean aliens along the Levantine coast. The Erythrean conch, *Strombus persicus*, and on occasion the Erythrean spiny oyster, are served in seafood restaurants in Israel. Erythrean penaeid prawns make up most of the shrimp catches along the SE Levantine coasts. The Erythrean prawns, in particular *M. japonicus* and *P. semisulcatus*, are highly prized and beginning in the 1970s a shrimp fishery developed off the Sinai coast, and since the mid 1980s off the Israeli coast where a small fleet of coastal "mini" trawlers has specialized in shrimping, bringing in a quarter of the total trawl catch volume and a third of the trawl gross income (Pisanty and Grofit 1991, Snovsky and Shapiro 1999). Nearly half of the trawl catches along the Israeli coast consist of Erythrean fish (Golani & Ben Tuvia 1995). The dominant fishes in the inshore fisheries (trammel-netting and hook-and-lining) are the siganids *S. rivulatus* and *S. luridus*, the obtuse barracuda, and the Erythrean jack, *Alepes djedaba*. The above species, together with *Sillago sihama* and *S. commerson*, two species that underwent population explosion in the early 1980s, are common in purse-seine landings. The annual catch of the Erythrean lizardfish which reached 400 tons in 1960 soon after its arrival (see above), declined to 100 tons in the mid 1960s, but has since increased, and catch fluctuations are correlated with CPUE. Catch statistics for mullids do not distinguish between the natives, *M. barbatus* and *M. surmuletus*, and the Erythrean aliens *Upeneus moluccensis* and *U. pori*, but a study of the frequency of the latter in trawl catches conducted in the mid 1980s showed they formed 87% of the mullid catch off the coast of Israel at depths of 20 m, and 50% at 55 m, whereas the native mullids are more abundant in deeper waters (Golani & Ben Tuvia 1995). The percentage of the Erythrean mullids in the total mullid catch has been increasing steadily, from 30% in 1980, 42% in 1984, to 47% in 1989 (Golani & Ben Tuvia 1995). Similarly, catch statistics of sphyraenids do not separate the Red Sea obtuse barracuda from the native Mediterranean species *S. Sphyraena* and *S. viridensis*. However, examination of the landed catch showed that the Erythrean barracuda had outnumbered the native sphyraenids in inshore trawl and purse-seine catches (Grofit 1987). In addition, two of the four species of Erythrean clupeids that established populations in the Levant – *Dussummieria elopsoides* and *Herklotsichthys punctatus* – are of importance in the inshore-pelagic fishery. The increasing exploitation of Erythrean aliens meant the shifting of the trawling grounds nearshore since their densest populations occur at depths up to 50 m – between 1980 and 1986 the Israeli trawlers doubled their activity (measured as hours fishing) in shallow waters (Pisanty & Grofit 1991). The shoreward displacement of the fishing grounds coupled with the inexorable gain of Erythrean aliens raise the ratio of alien to native taxa in the Levantine trawl landings. The prominence of Erythrean aliens in trawl hauls is not limited to the south-eastern Levant. Already in the mid 1940s the yellow striped mullet was common off the southern Aegean Turkish coast (Laskarides 1948), and by 1952 it constituted a significant portion of the trawl catch in the Bay of Mersin, on the southeastern Turkish coast (Gottlieb 1957), concurrently with dimunition in the numbers of the previously common native red mullet (Oren 1957a). By the mid 1960s *S. undosquamis* formed the "main catch of trawlers off Mersin" (Ben Tuvia 1966: 257). A study conducted in 1980-1984 in the Gulf of Iskenderun, Turkey, showed that the Erythrean fishes constituted up to 74.5% of fish landings in the fall months (Gücü & Bingel 1994). At depths of 14-59 m Erythrean fish (mainly *L. klunzingeri* and *S. undosquamis*) accounted for 51.9% and 67.6% of the biomass in October of 1983 and 1984 respectively. The importance of alien species for the Anatolian fishery is increasing: "*S. undosquamis and U. moluccensis* are the most abundant and commercially utilized fish species found in nearly every haul." (Gücü et al. 1994), and *P. pelagicus* made up a quarter of the crabs hauled in trawl surveys conducted in 2002-2003 in the Gulf of Iskenderun (Ozcan et al. 2005), and is offered in many restaurants catering to the thriving tourist industry along the southern Turkish coast.

But every silver lining is fastened to a cloud. Each summer since the mid 1980s huge swarms of the invading jellyfish, *Rhopilema nomadica*, have appeared along the Levantine coast. In 1995 the jellyfish was recorded off the south-eastern coast of Turkey (Kideys & Gücü 1995), and in 1998 a specimen was collected near Izmir (A. Karatas pers. comm.). These swarms of voracious planktotrophs, some stretching 100 km long, must play havoc with the limited resources of this oligotrophic sea, and when the shoals draw nearer shore, they adversely affect tourism, fisheries and coastal installations. Local municipalities report a decrease in holiday makers frequenting the beaches because of the public's concern over the painful stings inflicted by the jellyfish. Coastal trawling and purse-seine fishing are disrupted for the duration of the swarming due to net clogging and inability to sort yield "It is not uncommon that fishermen, especially purse seines, discard entire hauls due to the overwhelming presence of poisonous medusae in their nets" (Golani & Ben Tuvia 1995: 287). Jellyfish-blocked water intake pipes pose a threat to cooling systems of port-bound vessels and coastal power plants: in the summer of 2001 Israel Electric removed tons of jellyfish from its seawater intake pipes at its two largest power plants, at estimated costs of 50,000 US$ (M. Cohen, pers. comm.). Yet, the jellyfish shelters among its nematocyst-laden tentacles the juveniles of the Red Sea carangid fish, *Alepes djedaba* (Galil et al. 1990), and may have precipitated the sudden population increase of this commercially important species (Grofit 1987).

15 Coda

The profound changes wrought on the eastern Mediterranean biota commenced with the opening of the Suez Canal. The unabated influx of the Erythrean biota is rooted in the unceasing enlargement of the Canal that has altered its hydrography and hydrology, and enhanced its potential as a "corridor" allowing greater numbers of organisms through. Complex changes to the Levantine marine environment have made it more susceptible to invasion by modifying its hydrological properties, species diversity, and community structure. Many aspects of the Erythrean invasion remain unforseeable. It is unknown which particular species will pass through the Canal and establish themselves, and when, and what will be their interactions with the Mediterranean biota. It is clear though that once an Erythrean alien is established in the Mediterranean, it is likely to spread.

Global warming would likely have a significant influence on the Erythrean invasion. Higher temperatures may change the pool of species that could establish themselves in the Mediterranean, allow the temperature-limited

Erythrean alien species to expand beyond their present distributions, and may impact a suite of population characteristics (reproduction, survival) that determines interspecific interactions, and, therefore, dominance and prevalence patterns of alien species.

The biota across a wide swath of the Mediterranean Sea has been willfully altered, with dire ecological effects. Unless a salinity barrier (such as a hypersaline lock) is installed in the Suez Canal, which had served as a conduit for over 80% of the known alien fish, molluscs and decapods in the Mediterranean, the eastern Mediterranean countries would find the biota in their part of the sea fundamentally changed within a few decades. In an era of heightened environmental concern, it is surprising that plans to further expand the Suez Canal, have raised little or no attention, controversy, or a discussion on "environmental accountability".

"It is almost certain that the Indo-Pacific influx is still under way and it will be most interesting to study this process." (Bodenheimer 1935: 460).

Acknowledgements

I am grateful to Elisabeth Apt, the librarian of the Israel Oceanographic & Limnological Research Institute, Haifa; and to Mary de Jong and Tom Barione, librerians in the American Museum of Natural History, New York; for bibliographical help. I thank Dr. Arik Diamant for kindly advising me on the Erythrean parasite fauna, Guy Paz and Sima Usvyatsov for their help with the illustrations. The review had been partly supported by the EU FP6 STREP DAISIE.

16 References

Abdel-Fattah, A., A.F.A. Ghobashy, S.Z. Mohammed, H.R. Gabr, & A.R. Brand 1992. Community structure and seasonal variation of Mollusca at Lake Timsah (Suez Canal). Journal of the Egyptian-German Society of Zoology, 7: 145-160.

Abril, J.M., & M.M. Abdel-Aal 2000. Marine radioactivity studies in the Suez Canal, Part I: Hydrodynamics and transit times. Estuarine, Coastal and Shelf Science, 50: 489-502.

Achituv, Y. 1969. Studies on the reproduction and distribution of *Asterina burtoni* Gray and *A. wega* Perrier (Asteroidea) in the Red Sea and the eastern Mediterranean. Israel Journal of Zoology 18: 329-342.

Achituv, Y. 1973. On the distribution and variability of the Indo-Pacific sea star *Asterina wega* (Echinodermata, Asteroidea) in the Mediterranean Sea. Marine Biology 18: 333-336.

Achituv, Y. 1999. Recent changes in the distribution of two species of *Asterina* (Echinodermata: Asteroidea) along the Mediterranean shore of Israel. Israel Journal of Zoology 45(1): 121-125.

Achituv, Y., & E. Sher 1991. Sexsual reproduction and fission in the sea star *Asterina burtoni* from the Mediterranean coast of Israel. Bulletin of Marine Science 48(3): 670-678.

Ahmed, M.T., L. Naglaa, & Y.Y. Mosleh. 2001. Residues of chlorinated hydrocarbons, polycyclic aromatic hydrocarbons and polychlorinated biphenyl in some aquatic organisms in Lake Temsah around Ismailia. Journal of Aquatic Ecosystem Health and Management 4: 165-173.

Aillaud, L. 1868. Les eaux des Lacs amers et du Lac Timsah. Journal l'Isthme de Suez.

Aleem, A.A. 1969. Marine resources of the United Arab Republic. Studies and Reviews. General Fisheries Council for the Mediterranean. 43: 1-22.

Aleem, A.A. 1984. The Suez Canal as a habitat and pathway for marine algae and seagrasses. Deep Sea research. A. 31(6-8): 901-918.

Al-Kholy, A.A., & S.K. El-Wakeel 1975. Fisheries of the south-eastern Mediterranean Sea along the Egyptian coast, Soviet-Egyptian Expedition 1970-1971. Bulletin of the Institute of Oceanography and Fisheries 5: 1-300.

Alzieu C., P. Michel, I. Tolosa, E. Bacci, L.D. Mee, & J.W. Readman 1991. Organotin compounds in the Mediterranean: a continuing cause for concern. Marine environmental Research, 32: 261-270.

Anonymous, 1889. Professor Friedrich Krukenberg. Leopoldina, 25: 44.

Audouin, V. 1826. Explication sommaire des planches de Mollusques de l'Egypte et de la Syrie publiées par J. C. Savigny. Description de l'Egypte ou recueil des observations et des recherches qui ont éte faites en Egypte pendant l'expédition de l'armée francaise, publiée par les orders de sa majesté l'empereur Napoléon le grand. Histoire Naturelle, Animaux invertébrés 1(4): 7-56. Imprimerie impériale, Paris.

Balss, H. 1936. The Fishery Grounds near Alexandria. VII. Decapoda. Notes and Memoirs, Fisheries Research Directorate 15: 1-67.

Barakat, M.A.K., Shimy, T.M., & Y.M. Mostafa 1996. Relationships between sediment morphology and oil pollution along the Suez Canal beaches, Egypt. Energy Sources 18: 755-766.

Barash, A., & Z. Danin 1973. The Indo-Pacific species of Mollusca in the Mediterranean and notes on a collection from the Suez Canal. Israel Journal of Zoology 21(3-4): 301-374.

Barash, A., & Z. Danin 1992. Fauna Palaestina. Mollusca 1: annotated list of Mediterranean molluscs of Israel and Sinai. Israel Academy of Science and Humanities, Jerusalem. i-xii, 1-405.

Bariche, M., Letourneur, Y., & M. Harmelin-Vivien 2004. Temporal fluctuation and settlement patterns of native and Lessepsian herbivorous fishes on the Lebanese coast (eastern Mediterranean). Environmental Biology of Fishes 70: 81-90.

Baussan, J. 1938. Variations annuelles de niveau le long du Canal de Suez. Revue de Géographie Physique et de Géologie Dynamique 11: 293-321.

Bavay, A. 1897. Au sujet du passage d'un mollusque de la Mer Rouge dans la Méditerranée. Bulletin de la Société Zoologique de France, 22: 199.

Bavay, A. 1898. Note sur les mollusques du Canal de Suez. Bulletin de la Société Zoologique de France, 23: 161-164.

Bebars, M.I., & G. Lasserre 1983. Analyse des captures des pêcheries marines et lagunaires d'Egypte de 1962 à 1976, en liaison avec la construction du haut barrage d'Assouan achevé en 1969. Oceanologica Acta 6(4): 417-426.

Bebars, M.I., G. Lasserre, & T. Lam Hoai 1997. Analyse des captures des pêcheries marines et lagunaires d'Egypte en liaison avec la construction du haut barrage d'Assouan. Oceanologica Acta 20(2): 421-436.

Beets, C. 1953. Notes on dredging in the Great Bitter Lake of the Suez Canal. Zoologische Mededelingen, 32: 97-106.

Ben Eliahu, N.M. 1991. Red Sea serpulids (Polychaeta) in the eastern Mediterranean. In: M.E. Petersen and J.B. Kirkegaard (eds.), Systematics, Biology and morphology of World Polychaeta. Proceedings of the 2nd International Polychaete Conference, Copenhagen, 1986. Ophelia Supplement 5: 515-528.

Ben Tuvia, A. 1953. Mediterranean fishes of Israel. Bulletin of the Sea Fisheries Research Station, Haifa 8: 1-40.

Ben Tuvia, A. 1964. Two siganid fishes of Red Sea origin in the Eastern Mediterranean. Bulletin of the Sea Fisheries Research Station, Haifa 37: 3-9.

Ben Tuvia, A. 1966. Red Sea fishes recently found in the Mediterranean. Copeia 2: 254-275.

Ben Yami, M. 1955. 1954 – 1955: over-fishing or bad season? Fishermen's Bulletin Haifa 6: 10-14.[Hebrew].

Ben Yami, M., & T. Glaser 1974. The invasion of Saurida undosquamis (Richardson) into the Levant Basin – an example of biological effect of interoceanic canals. Fishery Bulletin 72(2): 359-373.

Bilecenoglu, M., & E. Taskavak 1999. General characteristics of the Turkish marine ichthyofauna. Zoology in the Middle East, 18: 41-56.

Bingel, F., E. Ozsoy, & U. Unulata 1993. A review of the state of the fisheries and the environment of the northeastern Mediterranean (northern Levantine Basin). Studies and Reviews, General Fisheries Council for the Mediterranean. 65: 1-74.

Bodenheimer, F.S. 1935. Animal life in Palestine. An introduction to the problems of animal ecology and zoogeography. L. Mayer, Jerusalem. 1-506.

Bonhomme, F., A. Baranes, D. Golani, & M. Harmelin-Vivien 2003. Lack of mitochondrial differentiation between Red Sea and Mediterranean populations of the Lessepsian rabbitfish, *Siganus rivulatus* (Perciformes: Siganidae). Scientia Marina 67(2): 215-217.

Bouchet, P., & F. Danrigal 1982. Napoleon's Egyptian campaign (1798 – 1801) and the Savigny collection of shells. The Nautilus 96(1): 9-24.

Boudouresque, C. F. 1999. The Red Sea – Mediterranean link: unwated effects of canals. In: Sandlund, O.T., Schei, P. J., Viken, A. (eds.) Invasive Species and Biodiversity Management, Kluwer Academic Publishers, Dordrecht, the Netherlands, 213-228.

Bucciarelli G, D. Golani, & G. Bernardi 2002. Genetic cryptic species as biological invaders: the case of a Lessepsian fish migrant, the hardyhead silverside *Aterinomorus lacunosus*. Journal of Experimental Marine Biology and Ecology 273: 143-149.

Butler, A.J. 1978. The Arab conquest of Egypt and the last thirty years of Roman dominion. Oxford University Press, Oxford. 563 pp.
Calman, W.T. 1927. Report on the Crustacea Decapoda (Brachyura). Zoological results of the Cambridge expedition to the Suez Canal. 1924. XIII. Transactions of the Zoological Society of London, 22: 211-217.
Calman, W.T. 1950. Obituary, Dr. Robert Gurney. Nature, 165: 587-588.
Campos, N.H., & M. Türkay 1989. On a record of *Charybdis helleri* from the Caribbean coast of Colombia. Senckenbergiana Maritima 20(3/4): 119-123
Carlton, J.T. 2000. Global change and biological invasions in the oceans. In:Mooney H.A., Hobbs, R. J. (eds.) Invasive species in a changing world. Island Press, Washington D.C., Covelo, California, 31-53.
Carmin, J. 1946. Mollusca of Palestinian shores, first report. Bulletin of the independent biological laboratories Palestine 4(38/39): 1-9.
Carqueija, C.R.G., & E.P. Gouvea , 1996. The occurrence in coastal Brazil of a Portunidae (Crustacea, Decapoda), originating from the Indo-Pacific and the Mediterranean. Nauplius 4: 105-112.
Chabanaud, P. 1932. Poissons recueillis dans le Grand Lac Amer (Isthme de Suez) par M. le Professeur A. Gruvel en 1932. Bulletin du Museum National d'Histoire Naturelle, Paris, Ser. 2, 4: 822-835.
Chabanaud, P. 1933a. Sur divers poisons de la Mer Rouge et du canal de suez. Description de deux éspèces nouvelles. Bulletin de l'Institut océanographique, Monaco, 30(627) : 1-12.
Chabanaud, P. 1933b. Contribution a l'etude de la faune ichthyologique du Canal de Suez. Bulletin de la Société Zoologique de France,58: 287-293.
Chaouachi B., A. Dos Santos, & O.K. Ben Hassine 1998. Occurrence of the shrimp *Metapenaeus monoceros* (Fabricius, 1798) in the Gulf of Gabes: a first record for Tunisian coasts. IV International Crustacean Congress, Amsterdam [Abst].
Chervinsky, J. 1959. A systematic and biological comparison between the lizardfish (*Saurida grandisquamis*) from the Mediterranean and the Red Sea. Fishermen's Bulletin, Haifa 19: 10-14.[English Abstract].
Christensen, A.M. 1972. Gunnar Thorson and the Mediterranean. Israel Journal of Zoology 21: 135-141.
Christiaens, J. 1967. Présence de la Pattele *Cellana rota* Gmelin en Méditerranée orientale. Bulletin du Museum National d'Histoire Naturelle, Paris. (2)38(6): 903.
De Min, R., & E. Vio 1997. Molluschi conchiferi del litorale sloveno. Annals for Istran and Mediterranean Studies, Koper, Historia naturalis, 11 (4): 241-258.
Diamant, A. 1985. Biology of the parasites of rabbitfish *Siganus* spp. (Pisces: Siganidae). Thesis submitted for Ph.D. degree, The Hebrew University of Jerusalem, 160 pp.
Diamant, A. 1989. Lessepsian migrants as hosts: a parasitological assessment of rabbitfish *Siganus luridus* and *Siganus rivulatus* (Siganidae) in their original and new zoogeographical regions. In: Spanier E., Steinberger Y., Luria, M. (eds.) Environmental quality and ecosystem stability, IV-B. ISEEQS Publ, Jerusalem, Israel, pp. 187-194.
Diamant, A., A. Banet, I. Paperna, H. v.Westernhagen, K. Broeg, G. Kruener, W. Koerting, & S. Zander 1999. The use of fish metabolic, pathological and parasitological indices in pollution monitoring. II. The Red sea and Mediterranean. Helgoland Marine Research 53: 195-208.
Diamant, A., & I. Paperna 1985. The development and ultrastructure of *Nosema ceratomyxae* sp. nov., a microsporidian hyperparasite of the myxosporean *Ceratomyxa* sp from Red Sea rabbitfish (Siganindae). Protistologica 21: 249-258.
Diamant, A., & I. Paperna 1986. The parasites of wild Red Sea rabbitfish (*Siganus* spp.) (Teleosti: Perciformes) as potential pathogens in mariculture. In: Pathology in marine aquaculture (C.P. Vivares et al., eds.). European Aquaculture Society, Special Publication No. 9, Bredene, Belgium, pp. 71-83.
Diamant, A., & N. Wilbert 1985. *Balantidium sigani* sp. nov., a trichostome ciliate from Red Sea rabbitfish . (Pisces: Siganidae). Archive für Protistenkunde 129: 13-17.
Di Geronimo, I. 1971. Prima segnalazione sulle costa italiane di *Brachiodontes variabilis* (Krauss). Bollettino delle sedute dell' Accademia Gioenia di scienze naturali in Catania 10: 847-852.
Di Natale, A. 1982. Extra-Mediterranean species of Mollusca along the southern Italian coasts. Malacologia 22(1-2): 571-580.
Dowidar, N.M. 1974. Tintinnids from the Suez Canal. Rapports et procès-verbaux des réunions, Commission internationale pour l'Exploration scientifique de la mer Méditerranée, 22(9) : 123-124.
Dounas, C., & C. Steudel 1994. Stomatopod Crustacea from the island of Crete. Crustaceana, 66(2): 252-254.
Dulčić, J., & B. Grbec 2002. Climatic change and Adriatic ichthyofauna. Fisheries Oceanography 9 (2): 187-191.
Dulčić, J., & L. Lipej 2002. Rare and little-known fishes in the Eastern Adriatic during last two decades (1980-2001). Periodicum Biologorum, 104(2): 185-194.
El-Etreby, S.G. 1986. Fishes of the Suez canal II. Survey and ecological study of fishes of Timsah Lake. Proceedings of the Zoological Society, Egypt 12: 199-214.
El-Sabh, M.I. 1968. Effect of the Aswan High Dam on the distribution of salinity in the Suez Canal. Nature, 218: 758-760.
El-Samra, M.I., K.Z. El-Zeeb, A.I. Askar, S.D. Wahby, & M.S. H. El-Shazly 1983. Preliminary study of petroleum hydrocarbon pollution along the Suez Canal. Bulletin of the Institute of Oceanography and fisheries, 9: 97-101.
El-Serehy, H.A. 1992. Lake Timsah as a barrier for planktonic tintinnid protozoan migration along the Suez Canal. Journal of the Egyptian-German Society of Zoology, 9(D): 237-246.

El-Serehy, H.A., S.M. Aboul-Ezz, A.A. Samaan, & N.S. El-Rahman 2000a. Migration activity of zooplankton between the Red Sea and Mediterranean across the suez Canal. Journal of the Egyptian-German Society of Zoology, 32: 1-14.

El-Serehy, H.A., S.M. Aboul-Ezz, A.A. Samaan, & N.S. El-Rahman 2000b. Zooplankton communities: their distribution and relationship to environmental variables in the Suez Canal marine ecosystem. Journal of the Egyptian-German Society of Zoology, 32: 85-101.

El-Serehy, H.A. & I.M. Shalaby 1994. Seasonal distribution and numerical abundance of zooplankton along the Suez Canal. Journal of the Egyptian-German Society of Zoology, 14: 379-395.

Elton, C.S. 1958. *The Ecology of Invasions by Animals and Plants*. Methuen & Co., London. 181 pp.

El-Wakeel, S.K., & S.D. Wahby 1970. Hydrography and chemistry of Lake Manzala, Egypt. Archiv für Hydrobiologie, 67: 173-200.

El-Zarka, S. El-Din, & R. Koura 1965. Seasonal fluctuations in production of the principal edible fish in the Mediterranean Sea off the United Arab Republic. Procceings of the General Fisheries Council for the Mediterranean, 8: 227-259.

Engl, W. 1995. Specie prevalentemente Lessepsiane attestate lungo le coste Turche. Bollettino Malacologico 31(1-4): 43-50.

Faouzi, H. 1951. Le Canal de Suez, voie d'échanges biologiques entre la Méditerranée et la Mer Rouge. Ann. Cent. Université de la Méditerranée Nice 5: 23-30.

Farnie, D.A. 1969. East and West of Suez, The Suez Canal in history 1854-1956. Clarendon Press, Oxford. 813 pp.

Fischer, P. 1865. Note sur les faunes conchyliologiques des deux rivages de l'isthme de Suez. Journal de Conchyliologie, 13: 241-248.

Fischer, P. 1870. Sur la faune conchyliologique marine des baies de Suez et de l'Akabah. Journal de Conchyliologie, 18: 161-179.

Fischer, P. 1871. Sur la faune conchyliologiquede marine de la baie de Suez. Journal de Conchyliologie, 19: 209-226.

Fischthal, J.H. 1980. Some digenetic trematodes of marine fishes from Israel's Mediterranean coast and their zoogeography, especially those from Red Sea immigrant fishes. Zoologica Scripta 9: 11-23.

Fontane, M. 1869. Le Canal maritime de Suez illustré. Histoire du Canal et des Travaux par M. Marius Fontane. Itinéraire de l'Isthme par. M. Riou. Paris. 191 pp.

Forsskål, P. 1775. *Descriptiones animalium avium, amphibiorum, piscicum, insectorum, vermium; quae in itinere oriental observavit Petrus Forskål. Prof. Haun. Post mortem auctoris edidit Carsten Niebuhr*. Møller, Copenhagen, xxxiv + 140 pp.

Fouda, M.M. 1995. Life history strategies of four small-size fishes in the Suez Canal, Egypt. Journal of Fish Biology, 46: 687-702.

Fouda, M.M. & M.M. Abou Zeid 1990. Bivalves of the Suez Canal Lakes. Proceedings of the zoological Society of Egypt 21: 231-240.

Fouda, M.M., M.Y. Hanna, & F. M. Fouda 1993. Reproductive biology of a Red Sea goby, *Silhouetta aegyptia*, and a Mediterranean goby, *Pomatoschistus marmoratus*, in Lake Timsah, Suez Canal. Journal of fish Biology, 43: 139-151.

Fox, H. M. 1924. The migration of a Red Sea crab through the Suez Canal. Nature, London, 113: 714-715.

Fox, H. M. 1926. General Part. Zoological results of the Cambridge expedition to the Suez Canal. 1924. I. Transactions of the Zoological Society of London, 22: 1-64.

Fox, H. M. 1929. Summary of results. XXXIX.Zoological results of the Cambridge expedition to the Suez Canal. 1924. I. Transactions of the Zoological Society of London, 22: 843-863.

Fredj, G. 1974. Stockage et exploitation des données en écologie marine. C. Considérations biogéographiques sur le peuplement benthique de la Méditerrannée. Mémoires de l'Institut Océanographique (Monaco) 7: 1-88

Fuchs, Th. 1878. Die geologische Beschaffenheit der Landenge von Suez. Denkschriften der Kaiserkichen Akademie der Wissenschaften, Mathematisch-Naturwissenschaftliche Classe, 38: 25.

Galil, B.S. 1986. Red Sea decapods along the Mediterranean coast of Israel: ecology and distribution. In: Dubinsky Z & Steinberger Y (eds) Environmental quality and ecosystem stability, vol. IIIA/B. Ramat Gan, Bar-Ilan University Press 179-183 pp

Galil, B.S. 1989. Bathymetric distribution and habitat preferences of Lessepsian migrant Decapoda along the Mediterranean coast of Israel, or: do decapods have cold feet. In: Spanier, E., Steinberger Y and M. Luria (eds) Environmental quality and ecosystem stability, vol. IV-B: 147-153.

Galil, B.S., & G. Innocenti 1999. Notes on the population structure of the portunid crab *Charybdis longicollis* Leene parasitized by the rhizocephalan *Heterosaccus dollfusi* Boschma, off the Mediterranean coast of Israel. Bulletin of Marine Science, 64(3): 451-463.

Galil B.S., & K. Kevrekidis 2002. Exotic decapods and a stomatopod off Rhodes Island (Greece) and the Eastern Mediterranean Transient. Crustaceana 75(7): 925-930

Galil, B.S., & J. Lütsen 1995. Biological observations on *Heterosaccus dollfusi* Boschma (Cirripedia: Rhizocephala), a parasite of *Charybdis longicollis* Leene (Decapoda: Brachyura), a lessepsian migrant to the Mediterranean. Journal of Crustacean Biology, 15(4): 659-670.

Gardiner, A.H. 1920. The ancient military road between Egypt and Palestine. J. Egyptian Archeology 5: 127.

Gardiner, J.S. 1924. The biology of the Suez Canal. Nature, 114: 520-521.
Geldiay, A., & A. Kocatas 1972. A report on the occurrence of Penaeidae (Decapoda, Crustacea) along the coast of Turkey from eastern Mediterranean to the vicinity of Izmir, as a result of migration and its factors. 17e Congrés International de Zoologie (Monte Carlo, 1972). 7 pp.
George, C.J., & V. Athanassiou 1967. A two year study of the fishes appearing in the seine fishery of St. George Bay, Lebanon. Annali del Museo Civico di Storia Naturale di Genova 76: 237-294.
Gerges, M.A., & D.J. Stanley 1985. Assessing hydrography and man's influence on sediments in the northern Suez Canal. Marine Geology 65: 325-331.
Ghazzawi, F.M. 1939. A study of the Suez Canal plankton. A. the Phytoplankton. Perliminary Report (with 11 Figures and 2 Charts in the Report and 65 Figures in the Appendix). Notes and Memoirs, Hydrobiological and Fisheries Directorate Government Press, Cairo. 24: 1-83.
Ghobashy, A.F.A., & M.M. El–Komi 1980a. Fouling in Lake Timsah (Egypt). Hydrobiol. Bull. 14(3): 169-178.
Ghobashy, A.F.A. & M.M. El–Komi 1980b. Fouling in the southern region of the Suez Canal. Hydrobiological Bulletin 14(3): 179-185.
Gibbs, P.E., G.W. Bryan, P.L. Pascoe, G.R. Burt 1987. The use of the dog-whelk, *Nucella lapillus*, as an indicator of tributyltin (TBT) contamination. Journal of the Marine Biological association of the U.K., 67: 507-523.
Giesbrecht, W. 1897. Ueber pelagische Copepoden des Rothen Meeres. Zoologischer Jahrbücher Abtheilung für Systematik, geographie und biologie der thiere, 9: 315-328.
Godley, A.D. 1975. Herodotus. In: G.P. Goold (ed.) The Loeb Classical Library. Harvard University Press, William Heinemann. I: 1-504.
Gofas, S., & A. Zenetos 2003. Exotic mollusks in the Mediterranean: current status and perspectives. In: R.N. Gibson, R. J. A. Atkinson (eds.) Oceanography and Marine Biology: an Annual Review, 41: 237-277.
Golani, D. 1996. The marine ichthyofauna of the eastern Levant – history, inventory and characterization. Israel Journal of Zoology 42: 15-55.
Golani, D., & A. Ben Tuvia 1995. Lessepsian migration and the Mediterranean Fisheries of Israel. In: Armantrout NB (ed) Condition of the world's aquatic habitats. Proceedings of the World Fisheries Congress. Theme 1. Science Publishers, Lebanon, New Hampshire 279-289 pp
Golani, D., & U. Ritte 1999. Genetic relationship in goatfishes (Mullidae: Perciformes) of the Red Sea and the Mediterranean, with remarks on Suez Canal migrants. Scientia Marina 63(2): 129-135.
Gómez, O., & J.C. Martinez-Iglesias 1990. Reciente hallazgo de la especie indopacifica *Charybdis helleri* (A.Milne Edwards, 1867) (Crustacea: Decapoda: Portunidae) en aguas cubanas. Caribbean Journal of Science 26(1-2): 70-72.
Goren, M., & B.S. Galil 2001: Fish biodiversity and dynamics in the vermetid reef of Shiqmona (Israel). Marine Ecology 22(4): 369-378.
Goren, M., & B.S. Galil 2005. A review of changes in the fish assemblages of Levantine inland and marine ecosystems following the introduction of non-native fishes. Journal of Applied Ichthyology 21(4): 364-370.
Goren, M., & E.M. Rychwalsky 1978. Hybrids of *Aphanius dispar* and *Aphanius mento*, (Cyprinodontidae: Pisces). Zoological Journal of the linnean Society 63: 259-264.
Gorgy, S. 1966. Les pêcheries et le milieu marin dans le secteur Méditerranéen de la République Arabe Unie. Revue des Travaux de l'Institut des pêches maritimes 30(1): 25-92.
Gottlieb, E. 1953. Decapod crustaceans in the collection of the Sea Fisheries Research Station, Caesarea, Israel. Bulletin of the Research Council of Israel, 2: 440-441.
Gottlieb, E. 1957. Age and growth studies of the yellow-striped mullet (*Upeneus moluccensis*). Fishermen's Bulletin, Haifa 11: 13-20. [English Abstract].
Gottlieb, E. 1960. The benthonic amphipoda of the Mediterranean coast of Israel. I. Notes on the geographical distribution. Bulletin of the Research Council of Israel 9B: 157-166.
Grofit, E. 1987. The trawl fishery in 1986. Fisheries and Fishbreeding in Israel 20(2): 6-42. .[English Abstract].
Gruvel, A. 1928. Répartition géographique de quelques crustacés comestibles sur les côtes d'Egypte et de Syrie. Compte rendu de la Société de biogeographie, Paris 39: 45-46.
Gruvel, A. 1929. De l'influence du percement du Canal du Suez sur la faune marine des côtes de Syrie. Communications. Académie des sciences coloniales. Paris. 188: 1697-1699.
Gruvel, A. 1931. Les Etats de Syrie. Richesses marines et fluviales, exportation actuelle – Avenir. Société d'Edition Geographiques, Maritimes et Coloniales, Paris. 453 pp.
Gruvel, A., 1934. Sur quelquesunes des causes qui arrêtent la pénétration des éspèces animales dans le Canal de Suez. Communications. Académie des sciences, Paris.
Gruvel, A. 1936. Contribution à l'étude de la bionomie générale et de l'exploitation de la faune du Canal de Suez. Mémoires présentés a l''institut d'Égypte 29: 1-229.
Gruvel, A., & P. Chabanaud 1937. Missions A. Gruvel dans le Canal de Suez. II. Poissons. Mémoires présentés a l'Institut d'Égypte, 35: 1-31.
Gruvel, A., & G. Moazzo 1931. Contribution à la faune malacologique marine des côtes Libano-Syriennes. Pp. 437-453. In : Gruvel, A., (ed.), Les États de Syrie. Richesses marines et fluviales, exportation actuelle – Avenir. Société d'Edition Geographiques, Maritimes et Coloniales, Paris.

Gruvel, A., & G. Moazzo 1933. Mollusques testacés du Grand Lac Amer (Canal de Suez) (Premiére liste). Bulletin du Museum National d'Histoire Naturelle, Paris. 5: 143-147.
Gücü AC., & F. Bingel 1994. Trawlable species assemblages on the continental shelf on the Northeastern Levant Sea (Mediterranean) with an emphasis on Lesseptian migration. Acta Adriatica 35(1/2): 83-100.
Gücü, A.C., F. Bingel, D. Avsar, & N. Uysal 1994. Distribution and occurrence of Red Sea fish at the Turkish Mediterranean coast – northern Cilician basin. Acta Adriatica 34 (1/2): 103-113.
Gurevitch J., & D.K. Padilla 2004. Are invasive species a major cause of extinctions? Trends in Ecology and Evolution, 19(9): 470-474.
Haas, G. 1937. Mollusca marina. Pp. 275-280. In : Bodenheimer, F.S. (ed.), Prodromus faunae Palestinae, Mémoires de l'Institut d'Egypte.
Haas, G. 1948. Sue l'immigration de mollusques indo-pacifiques dans les eaux cotières de la Palestine. Journal de Conchyliologie, 88: 141-144.
Halim, Y. 1960. Observations on the Nile bloom of phytoplankton in the Mediterranean. Rapports et procès-verbaux des réunions, Commission internationale pour l'Exploration scientifique de la mer Méditerranée 26(1): 57-67.
Halim, Y. 1990. On the potential migration of Indo-Pacific plankton through the Suez Canal. Bulletin de l'Institut océanographique, Monaco, nº special 7: 11-27.
Halim, Y., S.A. Morcos, S. Rizkalla, & M.K. El-Sayed 1995. The impact of the Nile and the Suez Canalon the living marine resources of the Egyptian Mediterranean waters (1958-1986). *In*: FAO Marine Resources Service 1995. Effects of riverine inputs on coastal ecosystems and fisheries resources, 349: 19-50.
Hanafy, M.H. 1996. Induction of imposex in the gastropod *Thais carnifera* from the Suez Canal and the Mediterranean Sea. Journal of the Egyptian-German Society of Zoology, 21(D): 135-152.
Hassan, E.M., & M.I. El-Sabh 1974. Change in the current regime in the Suez Canal after construction of Aswan High Dam. Nature, 248: 217-218.
Hassan, E.M., & M.I. El-Sabh 1975. Circulation and salinity distribution in the southern part of the Suez Canal. Limnology and Oceanography, 20: 667-671.
Hassan, M., M. Harmelin-Vivien, & F. Bonhomme 2003. Lessepsian invasion without bottleneck: example of two rabbitfish species (*Siganus rivulatus* and *Siganus luridus*). Journal of Experimental Marine Biology and Ecology 291: 219-232.
Hawkes, C.R., T.R. Meyers, & T.C. Shirley 1987. Growth of Alaskan blue king crabs, *Paralithodes platypus* (Brandt), parasitized by the rhizocephalan *Briarosaccus callosus* Boschma. Crustaceana, 52: 78-84.
Heimdal, B.R., J.P. Taasen, & M. Elbrächter 1977. Net phytoplankton of the Great Bitter Lake in the Suez Canal. Sarsia, 63: 75-83.
Herold, J.C. 1962. Bonaparte in Egypt. Harper & Row, New York. 424 pp.
Hoenselaar, H.J., & H. Dekker 1998. Molluscs of the Great Bitter Lake, Suez Canal, Egypt, collected by C. Beets in 1950. Basteria, 62: 197-214.
Holthuis, L.B. 1956. Notes on a collection of Crustacea Decapoda from the Great Bitter Lake, Egypt, with a list of the species of Decapoda known from the Suez Canal. Zoologische Mededelingen, 34: 301-332.
Holthuis, L.B. & E. Gottlieb 1958. An annotated list of the Decapod Crustacea of the Mediterranean coast of Israel, with an appendix listing the Decapoda of the Eastern Mediterranean. Bulletin of the Research Council of Israel 7B: 1-126
Ingle, R.W. 1963. Crustacea Stomatopoda from the Red Sea and the Gulf of Aden. In: Contributions to the knowledge of the Red Sea, 26. Bulletin Sea Fisheries Research Station (Haifa) 33: 1-69.
Innocenti, G., & B.S. Galil 2005. *Modus vivendi* - invasive host/parasite relations: the case of *Charybdis longicollis* Leene and *Heterosaccus dollfusi* Boschma. Abstracts of the Sixth International Crustacean Congress, Glasgow, Scotland. p. 194.
Innocenti G., N. Pinter, & B.S. Galil 2003. Observations on the agonistic behavior of the swimming crab *Charybdis longicollis* Leene infected by the rhizocephalan barnacle *Heterosaccus dollfusi* Boschma. Canadian Journal of Zoology, 81: 173-176.
Innocenti, G., M. Vannini, & B.S. Galil 1998. Notes on the behaviour of the portunid crab *Charybdis longicollis* Leene parasitized by the rhizocephalan *Heterosaccus dollfusi* Boschma. Journal of Natural History, 32: 1577-1585.
Issel, A. 1869. Malacologia del Mar Rosso, ricerche zoologiche e paleontologiche. Editori della Biblioteca Malacologica, Pisa. 387 pp.
Jeschke, J.M., & D.L. Strayer 2005. Invasion success of vertebrates in Europe and North America. Proceedings of the National Academy of Sciences, 102(20): 7198-7202.
Jones, H.L. 1967. The geography of Strabo. In: E.H. Warmington (ed.) The Loeb Classical Library. Harvard University Press, William Heinemann. VIII: 1-510.
Jordan, D.S. & C.L. Hubbs 1917. XVIII. Notes on a collection of fishes from Port Said, Egypt. Annals of the Carnegie Museum, 11: 461-468.
Karako, S., Y. Achituv, R. Perl-Treves, & D. Katcoff 2002. *Asterina burtoni* (Asteroidea: Echinodermata) in the Mediterranean and the Red Sea: does asexual reproduction facilitate colonization? Marine Ecology progress Series 234: 139-145.

Katağan, T., A. Kocataş, M. Zengin, & A.S. Ateş 2004. An Indo-Pacific stomatopod from the Sea of Marmara: *Erugosquilla massavensis* (Kossmann, 1980). Crustaceana, 77(3): 381-383.

Keller, C. 1883. Die Fauna im Suez-Kanal und die Diffusion der Mediterranean und Erythräischen Tierwelt. Neue Denkschriften der allgemeinen Schweizerischen Gesellschaft für die gesamten naturwissenschaften, ser. 3, 28: 1-39.

Keller, C. 1888. Die Wanderung der marinen Thierwelt im Suezcanal. Zoologischer Anzeiger, 11: 359-364; 389-395.

Kideys, A.E., & A.C. Gücü 1995. *Rhopilema nomadica*: a Lessepsian scyphomedusan new to the Mediterranean coast of Turkey. Israel Journal of Zoology 41: 615-617

Krauss, W. 1958. Eine Bemerkung zur Abnahme des Salzgehaltes im Suezkanal. Deutsche Hydrographische Zeitschrift 11(3): 109-112.

Krukenberg, C.Fr.W. 1888. Die Durchfluthung des Isthmus von Suez in chorologischer, hydrographischer und historischer Beziehung. Wissenschaftliche Ergebnisse meiner Reise vom Etang de Berre über Marseille und Triest nach Suakim und Massaua. II. Theil. *Vergleichend- physiologische Studien. Experimentele Untersuchungen,* ser. 2, 5 (1): 1-156.

Ktari, F., & M.H. Ktari 1974. Presence dans le Golfe de Gabes de *Siganus luridus* (Rüppel, 1829) et de *Siganus rivulatus* (Forsskal, 1775) (Poissons, siganids) parasites par *Pseudohaliotrematodides polymorphus*. Bulletin de l'Institut National Scientifique et Technique d'Océanographie et de Pêche de Salammbô, 3(14): 95-98.

Ktari-Chakroun, F., & M. Bahloul 1971. Capture de *Siganus luridus* (Rüppel) dans le golfe de Tunis. Bulletin de l'Institut National Scientifique et Technique d'Océanographie et de Pêche de Salammbô, 2(1): 49-52.

Kuris, A.M. 1974. Trophic interactions: similarity of parasitic castrators to parasitoids. Quarterly Review of Biology, 49: 129-148.

Kuris, A.M. 1997. Conceptual framework for biocontrol of introduced marine pests. In: Thresher, R.E. (ed) Proceedings of the first International Workshop on the Demography, Impacts and Management of Introduced Populations of the European crab, *Carcinus maenas*. Centre for Research on Introduced Marine Pests Technical Report (11): 66-68.

Lami, R. 1932. Quelques algues du Grand Lac Amer (Basse Egypte) récoltées par M. le Professeur Gruvel en Avril 1932. Revue Algologique 6: 355-356.

Lascaratos, A., W. Roether, K. Nittis & B. Klein 1999. Recent changes in deep water formation and spreading in the eastern Mediterranean Sea: a review. Progress in Oceanography, 44: 1-36.

Laskarides, C. 1948. Contribution to the biology of the fish *Mulloides auriflamma* (Forsk.). Praktika Hellenikon Hydrobiologikon Institutouton, 2(1): 127-129.

Lavee, D. 1981. Genetic variability in a colonisatory system: comparison between the mussels Brachidontes variabilis and Modiolus auriculatus (Bivalvia: Mytilidae). Ph.D. Thesis, Hebrew University, Jerusalem. [Hebrew].

Lemaitre, R. 1995. *Charybdis helleri* (Milne Edwards, 1867), a nonindigenous portunid crab (Crustacea: Decapoda: Brachyura) discovered in the Indian River lagoon system of Florida. Proceedings of the Biological Society of Washington 108(4): 643-648.

Le Père, J.M. 1822. Mémoire sur la communication de la mer des Indes a la Méditerranée par la mer Rouge et l'Isthme de Soueys. In: Description de L'Égypte ou Recueil des observations et des recherches qui ont été faites en Égypte pendant l'Expédition de l'Armée Francaise. 2nd Edition. C.L.F. Panckoucke, Paris. XI: 37-280.

Lesseps, F. de 1874. Communication sur les Lacs amers de l'Isthme de Suez. Comptes Rendus Hebd. des Séances de l'Académie des Sciences Paris78: 1740-1747.

Lesseps, F. de 1876. Deuxiéme note sur les lacs amers de l'Isthme de Suez. Comptes Rendus Hebd. des Séances de l'Académie des Sciences Paris 82: 1133-1137.

Lesseps, F. de 1878. Souvenirs de quarante ans dédiés à mes enfants. 2 vols. Nouvelle Revue, Paris.

Lesseps, F. de, 1890. Origines du Canal de Suez. C. Marpon & E. Flammarion, Paris. 220 pp.

Lewinsohn, Ch. Galil B.S. 1982. Notes on species of *Alpheus* (Crustacea, Decapoda) from the Mediterranean coast of Israel. Quaderni del Laboratorio di Tecnologia della Pesca, Ancona 3(2-5): 207-210.

Lewinsohn, Ch., & L.B. Holthuis 1964. New records of Decapod Crustacea from the Mediterranean coast of Israel and the Eastern Mediterranean. Zoologische Mededelingen, Leiden, 40 (8): 45-63.

Lewinsohn Ch., & B. Manning 1980. Stomatopod Crustacea from the Eastern Mediterranean. Smithsonian Contributions to Zoology 305: 1-22.

Liebman, E. 1934. Contributions to the knowledge of Palestine Sea Fishes, Rapports et Procès-Verbaux des Réunions Commission Internationale pour l'Exploration Scientifique de la Mer Méditerranée, 8: 317-327.

Lindner, G. 1987. Interessante Schneckenfunde an der südtürkischen Mittelmeerküste (Reisemitbbringsel der Familie Schmidt, Feldkirchen). Club Conchylia Informationen, 19(3-4): 32-43.

Lotan, R. 1982. The killifish *Aphanius dispar*. Israel Land and Nature 8: 28-30.

Lotan, R., & A. Ben Tuvia 1996. Distribution and reproduction of killifish *Aphanius dispar* and *A. fasciatus* and their hybrids in the Bardawil Lagoon on the Mediterranean coast of Sinai, Egypt. Israel Journal of Zoology 42: 203-213.

Lundberg B., Ogorek R., Galil, B.S., & M. Goren 2004. Dietary choices of siganid fish at Shiqmona reef, Israel. Israel Journal of Zoology 50: 39-53.

MacAndrew, R. 1870. Report on the Testaceous Mollusca obtained during a Dredging-Excursion in the Gulf of Suez in the months of February and March 1869. Annals and Magazine of Natural History, London (4th ser.) 6 (36): 429-450.

Macdonald, R. 1933. An examination of plankton hauls made in the Suez Canal during the year 1928. Notes and Memoirs, Fisheries Research Directorate 3: 1-11.

Maillard, C., & A. Raibaut 1990. Human activities and modifications of ichtyofauna of the Mediterranean sea: effect on parasitosis. In: E. di Castri, A.J. Hansen, M. Debussche (eds.). Biological Invasions in Europe and the Mediterranean Basin. 297-305. Kluwer Academic Publishers, Dordrecht.

Maltagliati, F. 1999. Genetic divergence in natural populations of the Mediterranean brackish-water killifish *Aphanius fasciatus*. Marine Ecology Progress Series 179: 155-166.

Mantelatto, F.L.M., & L.L. Dias 1999. Extension of the known distribution of *Charybdis hellerii* (A. Milne-Edwards, 1867) (Decapoda, Portunidae) along the western tropical South Atlantic. Crustaceanan 72(6): 617-620.

Martens, E. von 1866. Verzeichniss der von Dr. E. Schweinfurth in Sommer 1864 auf seiner reise am rotten Meere gesammelten und nach Berlin engesenden Zoologischen Gegenstande. Verhandlungen der k k zoologisch-botanischen Gesellschaft in Wien.

Martens, E. 1887. Conchylien aus dem Suezkanal. Sitzungs-Bericht der Gesellschaft naturforschender Freunde zu Berlin, 6: 89-96.

Mayerson, P. 1996. The port of Clysma (Suez) in transition from Roman to Arab rule. Journal of Near Eastern Studies 55(2): 119-126.

Mendelssohn, H. 1947. A new locality for *Cyprinodon dispar* Rueppel. Nature 160: 123.

Meshal, A.H. 1975. Brine at the bottom of the Great Bitter Lake as a result of closing the Suez Canal. Nature, 256: 297-298.

Mienis, H.K. 1995. Further records of the Erythrean gastropod *Syrnola fasciata* from the Mediterranean coast of Israel. *Levantina* 82: 7-8.

Mienis, H.K. 1999. On the presence of the Indo-Pacific bivalve *Gafrarium pectinatum* (Linnaeus, 1758) along the Mediterranean coast of Israel. Club Conchylia Informationen, 31(1/2): 37.

Mienis, H. 2002. Is the Lessepsian migrant *Cellana rota* replacing native limpets along the Mediterranean coast of Israel? The Conchologists' Newsletter 163: 275-276.

Mienis, H. 2003. Native marine molluscs replaced by Lessepsian migrants. Tentacle 11: 15-16.

Mienis, H. 2004. New data concerning the presence of Lessepsian and other Indo-Pacific migrants among the molluscs in the Mediterranean Sea with emphasize on the situation in Israel. *In*: B. Öztürk & A. Salman (Eds.): Proceedings 1st National Malacology congress, 1-3 September 2004, Izmir. Turkish Journal of Aquatic Life, 2(2): 117-131.

Mienis, H. K., E. Galili, & J. Rapoport 1993a. On the presence of the Indo-Pacific bivalve *Chama Pacifica* in the eastern Mediterranean (Mollusca, Bivalvia, Chamidae). Gloria Maris 32(2): 13-18.

Mienis, H.K., E. Galili, & J. Rapoport 1993b. The Spiny Oyster, *Spondylus spinosus*, a well-established Indo-Pacific bivalve in the Eastern Mediterranean off Israel (Mollusca, Bivalvia, Spondylidae). Zoology in the Middle East 9: 83-91.

Miller, A.R. & R.G. Munns 1974. The Bitter Lake salt barrier. In: L'Oceanographie physique de la Mer Rouge. Publications du Centre national pour l'exploitation des oceans (CNEXO), Actes de colloques 2: 295-309.

Moazzo, P. G. 1939. Mollusques testacés marins du Canal de Suez. Mémoires de l'Institut d'Égypte, 38: 1-283.

Mohammed, S.Z. 1997. Influence of age structure of *Brachiodontes variabilis* on the community structure of its associated fauna in the Great Bitter Lake, Suez Canal. Journal of the Egyptian-German Society of Zoology, 24: 51-67.

Mohammed, S.Z., H.R. Gabr, A.F.A. Ghobashy, & A.R. Brand 1992. Species composition and distribution of benthic molluscs in Lake Timsah (Suez Canal). Journal of the Egyptian-German Society of Zoology, 7: 161-174.

Monod, Th. 1930. Über einige indo-pazifische Decapoden der Meeresfauna Syriens. Zoologischer Anzeiger, Leipzig, 92(5/6): 135-141.

Monod, Th. 1931. Crustacés de Syrie. In: A. Gruvel (ed.), Les États de Syrie. Richesses marines et fluviales, exportation actuelle – Avenir. Société d'Edition Geographiques, Maritimes et Coloniales, Paris. Pp. 397-435.

Monod, Th. 1932. Crustacés exotiques en Méditerranée. La Terre et la Vie – Revue d'Ecologie, 2: 65-73.

Monterosato, T. A. di 1878. Enumerazione e sinonimia delle conchiglie Mediterranee. Giornale di scienze naturali ed economiche di Palermo 13: 61-113.

Monterosato, T. A. di 1899. Coquilles marines de Chypre. Journal de conchyliologie, 47(4): 392-401.

Morcos, S.A. 1960. Die Verteilung des Salzgehaltes im Suez-Kanal. Kieler Meeresforschungen 16(2): 133-154.

Morcos, S.A. 1967. Effect of the Aswan High Dam on the current regime in the Suez Canal. Nature, 214: 901-902.

Morcos, S.A. 1972. Early investigations of the Suez Canal waters during and after its opening in 1869. Proceedings of the second International Congress on the History of Oceanography. Proc. Roy. Soc. Edinburgh, Section B, 72: 449-458.

Morcos, S.A. 1975. A transitional stage in the current regime in the Suez Canal. Limnology and Oceanography, 20: 672-679.

Morcos, S.A. 1980. Seasonal changes in the Suez Canal following its opening in 1869: newly discovered hydrographic records of 1870-1872. in: Oceanography, the past, eds. M. Sears & D. Merriman. Springer – Verlag, New York. Pp. 290-305.
Morcos, S.A., & M.A. Gerges 1974. Circulation and mean sea level in the Suez Canal. In: L'Oceanographie physique de la Mer Rouge. Publications du Centre national pour l'exploitation des oceans (CNEXO), Actes de colloques 2: 295-309.
Morcos, S.A., & S.N. Messieh 1973a. Circulation and salinity distribution in the southern part of the Suez Canal. Limnology and Oceanography, 18: 267-287.
Morcos, S.A., & S.N. Messieh 1973b. Change in the current regime in the Suez Canal after construction of Aswan High Dam. Nature, 242: 38-39.
Nares, G.S. 1871. Report on the Suez Canal. H.M.S. *Newport*, April 1871. Admiralty, London, 4 pp.
Nittis, K. & A. Lascaratos 1999. Intermediate water formation in the Levantine Sea: the response to interannual variability of atmospheric forcing. In: P. Malanotte-Rizzoli & V.N. Eremeev (eds.), The eastern Mediterranean as a laboratory basin for the assessment of contrasting ecosystems: 441-446. (Kluwer Academic Publishers, Dordrecht).
Norman, J.R. 1927. XXV. Report on the Fishes. In: Cambridge Expedition to the Suez Canal, 1924. Transactions of the Zoological Society of London 3(12): 375-390.
Nourse, J.E. 1884. The maritime canal of Suez from its inauguration, November 17, 1869, to the year 1884. Government Printing Office, Washington. 164 pp.
O'Brien, J., & P. Van Wyk 1985. Effects of crustacean parasitic castrators (epicaridean isopods and rhizocephalan barnacles) on the growth of their crustacean hosts. Crustacean Issues 3: 191-218.
Occhipinti-Ambrogi, A., & B.S. Galil 2004. A uniform terminology on bioinvasions: a chimera or an operative tool? Marine Pollution Bulletin 49: 688-694.
Oksnebjerg, B., R. Enzenross, & L. Enzenross 1997. First record of Rhizocephala (Crustacea: cirripedia) from Turkish waters, with notes on Lessepsian migration. Stuttgarter Beiträge zur Naturkunde, Serie (Biologie), 557: 1-7.
Oldfather, C.H. 1968. Diodorus of Sicily. In: E.H. Warmington (ed.) The Loeb Classical Library. Harvard University Press, William Heinemann. I: 1-470.
Oliverio, M., G. Gerosa, & M. Cocco 1992. First record of *Pinctada radiata* (Bivalvia, Pteriidae) epibiont on the loggerhead sea turtle *Caretta caretta* (Chelonia, Cheloniidae). Bollettino Malacologico 28(5-12): 149-152.
Oren, O.H. 1957a. Changes in the temperature of the eastern Mediterranean Sea in relation to the catch of the Israel trawl fishery during the years 1954-1956. Fishermen's Bulletin, Haifa 12: 4-6. [Hebrew].
Oren, O.H. 1957b. Changes in the temperature of the eastern Mediterranean Sea in relation to the catch of the Israel trawl fishery during the years 1954-55 and 1955-56. Bulletin de l'Institut Océanographique, 1102: 1-13.
Oren, O.H. 1969. Oceanographic and biological influence of the Suez Canal, the Nile and the Aswan Dam on the Levant Basin. Progress in Oceanography 5: 161-167
Oren, O.H. 1970. The Suez Canal and the Aswan High Dam, their effect on the Mediterranean. Underwater Science and Technology Journal, 2: 222-229.
Oren, O.H., & H. Steinitz 1959. Regional bibliography of the Mediterranean coast of Israel and the adjacent Levant countries. Bulletin Sea Fisheries Research Station, Haifa 22: 1-32.
Ozcan, T., T. Katagan, & A. Kocatas 2005. Brachyuran crabs from Iskenderun Bay (Southeastern Turkey). Crustaceana 78(2): 237-243.
Ozturk, B., M. Onen, A. Dogan, & D. Urkmez 2002. The Lessepsian molluscs of the Mediterranean and their distribution along the Turkish coast. In: Ozturk, B., Basusta, N. (eds.) Workshop on Lessepsian Migration. Turkish Marine Research foundation, Istanbul, 9: 1-115.
Pallary, P. 1912a. Catalogue des mollusques du littoral méditerranéen de l'Egypte. Mémoires de l'Institut d'Egypte 7: 69-207, pls. 15-18.
Pallary, P. 1912b. Liste des mollusques marins des côtes de la. Syrie. Feuille des Jeunes Naturalistes, Paris, 42(504): 171-174.
Pallary, P. 1938. Les mollusques marins de la Syrie. Journal de Conchyliologie, 82: 5-57.
Pancucci-Papadopoulou, M.A., Kevrekidis, K., Corsini-Foka, M., & N. Simboura 2005. Changes in species: invasion of exotic species. In: Papathanassiou. E., Zenetos, A. (eds.) State of the Hellenic Marine Environment. HCMR Publ. Athens. Pp. 336-342.
Papaconstantinou, C. 1987. Distribution of the Lessepsian fish migrants in the Aegean Sea. Biologia Gallo-hellenica 13: 15-20.
Paperna, I. 1972. Parasitiological implications of fish migration through interoceanic canals. 17e Congrès International de Zoologie (Monte-Carlo 25-27 Septembre 1972). Theme N° 3: Les consequences biologiques des canaux inter-océans. 9 pp.
Pashtan, A., & U. Ritte 1977. Genetic changes in *Cerithium scabridum* (Mollusca: Gastropoda) following colonization in the Mediterranean. Israel Journal of Zoology 26: 258-259.
Pérès, J.M. 1958. Ascidies recoltées sur les côtes Méditerranéennes d'Israel. Bulletin of the Research Council of Israel, 7B(3/4): 143-150.
Perlmutter, A. 1956. Sea fisheries. Report to the government of Israel. Technical Assistance program USOM Agricultural Report No. 13, 58 pp.

Philippi, R.A. 1836. Enumeratio Molluscorum Siciliae, cum viventium in tellure tertiaria fossilium, quae in itinere suo observavit. S. Schroppii, Berlin. Vol. 1 xiv + 267 pp.
Phillips, W.J., & L.R.G. Cannon 1978. Ecological observations on the commercial sand crab, *Portunus pelagicus* (L.), and its parasite, *Sacculina granifera* Boschma, 1973 (Cirripedia: Rhizocephala). Journal of Fish Diseases, 1: 137-149.
Pisanty, S., & E. Grofit 1991. Limiting effort in the Israeli trawl fishery. Fisheries and Fishbreeding in Israel 24: 100-133. [English Abstract].
Por, F.D. 1978. Lessepsian Migration – the influx of Red Sea Biota into the Mediterranean by way of the Suez Canal. Ecological Studies vol. 23, Springer-Verlag, Berlin-Heidelberg-New York, 228 pp.
Por, F.D. 1990. Lessepsian migration. An appraisal and new data. Bulletin de l'Institut oceanographique, Monaco, 7: 1-10.
Quéguiner, J. 1978. The Mediterranean as a Maritime Trade Route. Ocean Management 3: 179-189.
Rackham, H. 1969. Pliny Natural History. In: E. H. Warmington (ed.) The Loeb Classical Library. Harvard University Press, William Heinemann. II: 1-664.
Reinhardt, E.G., D.J. Stanley, & H.P. Schwarcz 2001. Human-induced desalinization of Manzala Lagoon, Nile delta, Egypt: evidence from isotopic analysis of benthic inverterbrates. Journal of Coastal Research, 17(2): 431-442.
Richards, G.H., & A. Clarke 1870. The Suez Canal. Proceedings of the Royal Geographical Society, xiv: 270.
Rilov, G., Benayahu, Y., A. Gasith 2004. Prolonged lag in population outbreak of an invasive mussel: a shifting-habitat model. Biological Invasions 6: 347-364.
Rogers, R.W. 1929. A History of Ancient Persia from its earliest beginnings to the death of Alexander the Great. Charles Scribner's sons, New York, London. 394 pp.
Roux, J.C. 1901. L'Isthme et le Canal de Suez, historique – état actuel. Hachette, Paris. 2 vols. 516 + 550 pp.
Ruiz, G.M., J.T. Carlton, E.D. Grosholtz, & A.H. Hines 1997. Global invasions of marine and estuarine habitats by non-indigenous species: mechanisms, extent and consequences. American Zoologist 37: 621-632.
Ruiz, G.M., P.W. Fofonoff, J.T. Carlton, M.J. Wonham, & A.H. Hines 2000. Invasions of coastal communities in North America: apparent patterns, processes, and biases. Annual Review of Ecology and Systematics, 31: 481-531.
Saad, A. 2000. Fishes. National Country Study of Biological Diversity in Syrian Arab Republic. United Nation Environment Programme, Syrian Arab Republic, Ministry of Environment, National Biodiversity Unit. Damascus. Pp. 135-138.
Safriel, U.N., Felsenburg, T., & A. Gilboa 1980a. Distribution of rocky intertidal mussels in the Red Sea coasts of Sinai, the Suez Canal and the Mediterranean coasst of Israel, with special reference to recent colonizers. Journal of Biogeography 7: 39-62.
Safriel, U.N., Felsenburg, T., & A. Gilboa 1980b. The distribution of *Brachidontes variabilis* (Krauss) along the Red Sea coast of Sinai. Argamon 7(3): 31-43.
Safriel, U.N., & Z. Sasson-Frostig 1988. Can colonizing mussel outcompete indigenous mussel? Journal of the Experimental Marine Biology and Ecology 117: 211-226.
Sarà, M. 1985. Ecological factors and their biogeographic consequences in the Mediterranean ecosystems. In: Moraitou-Apostolopoulou and M & Kiortsis V (eds) Mediterranean Marine Ecosystems. Plenum Press, New York, 1-17
Sarioglu, F. 2002. Opening of the Suez canal. In: Öztürk, B., & Başusta A., (eds) Workshop on Lessepsian Migration. Turkish Marine Research Foundation, Istanbul 9: 22- 26.
Savary, J. 1679. Le parfait négociant ou Instruction générale pour ce qui regarde le commerce des merchandises de France et des pays étrangers.Second Edition, Jean Guignard, Paris. 398 + 560 pp.
Savigny, J.C. 1817. Description de l'Egypte, ou recueil des observations et des recherches qui ont éte faites en Egypte pendant l'Expédition de l'armée francaise, publiée par ordre du gouvernement. Histoire naturelle. Planches, vol. II. Imprimerie royale, Paris.
Schonfield, H.J. 1969. The Suez Canal in peace and war 1869-1969. University of Miami Press, Florida. 214 pp.
Selim, S.A. 1997. Description and remarks on Suez Canal serpulids (Polychaeta). Journal of the Egyptian-German Society of Zoology, 22: 87-110.
Shafei, A. Bey 1946. Historical notes on the Pelusiac branch, the Red Sea Canal and the route of the Exodus. Bulletin de la Société royale de géoraphie d'Égypte 1946: 231-287.
Shaheen, A.H., & S.F. Yousef 1978. The effect of the cessation of Nile flood on the hydrographic features of Lake Manzala, Egypt. Archiv für Hydrobiologie, 84: 339-367.
Shalla, S.H., & T.J. Holt 1999. The Lessepsian migrant *Pomatoleios kraussii* (Annelida, Polychaeta) – recent formation of dense aggregations in Lake Timsah and the Bitter Lakes (Suez Canal, Egypt). Egyptian Journal of Biology, 1: 133-137.
Sharabati, D. 1984. Red Sea Shells. 128 pp. KPI, London, Boston, Melbourne, Henley.
Sharaf El Din, S.H. 1974. Further studies on tides and the hydrography of the Suez Canal and its lakes. In: L'Oceanographie physique de la Mer Rouge. Publications du Centre national pour l'exploitation des oceans (CNEXO), Actes de colloques 2: 289-294.
Sidebotham, S.E. 1986. Roman economic policy in the Erythra Thalassa 30 B.C. – A.D. 217. Mnemosyne, 91:1-226.

Silvera, A. 1975. Bonaparte and Talleyrand, The origin of the French Expedition to Egypt in 1798. The American journal of Arabian Studies 3: 1-13.

Sneh, A., T. Weissbrod, & I. Perath 1975. Evidence for an ancient Egyptian frontier canal. American Scientist, 63: 542-548.

Snovsky. Z., & J. Shapiro 1999. The fisheries and Aquaculture of Israel 1997 in figures. Ministry of Agriculture, Department of Fisheries, xii+42 pp.

Soliman, G.F. 1995a. Hydrographic features in the Gulf of Suez and the estimation of its water and salt budgets. Bulletin of the National Institute of Oceanography and Fisheries, 21(1): 1-33.

Soliman, G.F. 1995b. The influence of the last widening and deepening of the Suez Canal on its tidal motion. Bulletin of the National Institute of Oceanography and Fisheries, 21(1): 35-66.

Stanley, D.J., G.L. Freeland, & H. Sheng 1982. Dispersal of Mediterranean and Suez Bay sediments in the Suez Canal. Marine Geology, 49: 61-79.

Steinitz, H. 1970. A critical List of Immigrants via the Suez Canal. Biota of the Red Sea and Eastern Mediterranean, pp. 59-63. [mimeo.].

Steinitz, W. 1919. Memorial on the founding of a sea-station on the coast of Palestine for zoological investigations. Manuscript. 16 pp

Steinitz W. 1927. Beiträge zur kenntnis der Küstenfauna Palästinas. I. Pubblicazioni della Stazione zoologica di Napoli 8(3-4): 311-353.

Steinitz, W. 1929. Die Wanderung indopazifischer Arten ins Mittelmeer seit Beginn der Quartärperiode. Internationale Revue der gesamten Hydrobiologie und Hydrographie, 22: 1-90.

Steinitz, W. 1936. Die Herkunft der Fauna in den Küstengewässern Palästinas. Society for Natural Sciences, Tel Aviv. Privarely printed, 8 pp.

Steuer, A. 1936. Cumacea and stomatopoda von Alexandrien in Ägypten. Notizen des Deutsch-Italienischen Institutes für Meeresbiologie in Rovigno d'Istria 21: 1-19.

Steuer, A. 1939. Die Fischereigründe vor Alexandrien. XIX. Mollusca. Thalassia, 3: 3-143.

Tebble, N. 1959. On a collection of Polychaetes from the Mediterranean coast of Israel. Bulletin of the Research Council of Israel, 8B: 9-30.

Theocharis, A., P. Georgopoulos, A. Karagevrekis, L. Iona, L. Perivoliotis, & N. Charalambidis 1992. Aegean influence in the deep layers of the eastern Ionian Sea. Rapports et Procès-verbaux des Réunions, Commission Internationale pour l'Exploration Scientifique de la Mer Méditerranée, 3: 235.

Theocharis, A., & A. Lascaratos 2000. Possible causes, origin, evolution and some consequences of the Eastern Mediterranean Transient during the period 1987- 1999. In: F. Briand (ed.), The Eastern Mediterranean climatic Transient, its origin, evolution and impact on the ecosystem. CIESM Workshop Series, Monaco, 10: 1-86. www.CIESM.org/publications/Trieste

Theocharis, A., K. Nittis, H. Kontoyiannis, E. Papageorgiou, & E. Balopoulos 1999. Climatic changes in the Aegean Sea influence the eastern Mediterranean thermohaline circulation (1986-1997). Geophysical Research Letters, 26 (11): 1617-1620.

Thompson, I.C., & A. Scott 1903. Report on the Copepoda collected by Professor Herdman, at Ceylon, in 1902. Ceylon Pearl Oyster Fisheries – 1903 – Supplementary Reports 7: 227-307.

Thompson, J.M. 1988. Napoleon Bonaparte. Basil Blackwell, New York. 418 pp.

Thorson, G. 1968. Animal migrations through the Suez Canal in the past, recent years and the future. III Symposium Européen de Biologie marine. II. Biologie des eaux à salinité variable. Arcachon. pp. 1-3. (mimeo.).

Thorson, G. 1971. Animal migrations through the Suez Canal in the past, recent years and the future (A preliminary report). Vie et Milieu, suppl. 22: 841-846.

Thresher R.E., M. Werner, J.T. Høeg, I. Svane, H. Glenner, N.E. Murphy, & C. Wittwer 2000. Developing the options for managing marine pests: specifity trials on the parasitic castrator, *Sacculina carcini*, against the European crab, *Carcinus maenas*, and related species. Journal of Experimental Marine Biology and Ecology 254: 37-51

Tillier, J.B. 1902. Le Canal de Suez et sa faune ichthyologique. Mémoires de la Société zoologique de France 15(3-4): 279-318.

Tillier, J.B., & A. Bavay 1905. Les mollusques testacés du Canal de Suez. Bulletin de la Société Zoologique de France, 30: 170-181.

Tillier, J.B., & A. Bavay 1906. Au sujet des mollusques testacés du Canal de Suez. Bulletin de la Société Zoologique de France, 31: 129-131.

Tom, M. 1979. Survey of the commercial shrimp population off the Mediterranean coast of Sinai. I. Fisheries and Fishbreeding in Israel 14 (1): 8-19. [English Abstract].

Torchin, M.E., K.D. Lafferty, A.P. Dobson, J. McKenzie, &, A.M. Kuris 2003. Introduced species and their missing parasites. Nature 421: 628-630.

Torchin, M.E., K.D. Lafferty, & A.M. Kuris 2001. Release from parasites as natural enemies: increased performance of a globally introduced marine crab. Biological Invasions 3: 333-345.

Tortonese, E. 1947. Biologia del Canale di Suez. Historia Naturalis, Roma 2: 1-6.

Tortonese, E. 1948. Ricerche Zoologiche nel Canale di Suez e dintorni. II. Pesci. Archivio Zoologico Italiano 33: 275-292.

Tortonese, E. 1952. Some field-notes on the fauna of the Suez Canal (Timsah and Bitter lakes). Publications of the Hydrobiological Research Institute, Faculty of Science, University of Istanbul, 1: 1-6.

Tortonese, E. 1966. Echinoderms from the coast of Lebanon. Miscellaneous Papers in the Natural Sciences, The American University in Beirut 5: 2-5.

Udekem d'Acoz, C. d' 1999. Inventaire et distribution des crustacés décapodes de l'Atlantique nord-oriental, de la Méditerranée et des eaux continentales adjacentes au nord de 25°N. Patrimoines naturels (M.N.H.N./S.P.N.) 40: i-x, 1-383

Vaillant, L. 1865. Recherches sur la faune malacologique de la baie de Suez. Journal de Conchyliologie, 13: 97-127.

Van der Linden, J., & J.C.A. Eikenboom 1992. On the taxonomy of the recent species of the genus *Chrysallida* Carpenter from Europe, the Canary Islands and the Azores (Gastropoda : Pyramidellidae). Basteria, 56: 3-63.

Vassel, E. 1896. Sur la Pintadine du Golfe de Gabès. Association Français pour l'Avancement des Sciences, Compte-rendu de la 25 session, Cartage, 1896. Masson, Paris. Pp. 458-466.

Villwock, W. 1985. Contribution on natural hybrids between two valid species of *Aphanius* (Pisces: Cyprinodontidae) from the Bardawil-Lagoon, North Sinai, Egypt. Rapports et procès-verbaux des réunions, Commission internationale pour l'Exploration scientifique de la mer Méditerranée 29: 57-63.

Villwock, W. 1987. Further contributions on natural hybrids between two valid species of *Aphanius*, *Aphanius dispar dispar* (Rűppell) and *Aphanius fasciatus* (Valenciennes), (Pisces: Cyprinodontidae) from the Bardawil-Lagoon, North Sinai, and al-Qanatir, west of the Suez Canal, Egypt. In: Krupp, F., Schneider, W., and Kinzelbach, R., eds. Proc. Symp. Fauna Zoogeogr. Middle East. Beih.

Volney, M.C. 1787. Voyage en Syrie et en Égypte, pendant les anneés 1783, 1784, & 1785, avec deux cartes géographiques & deux planches gravées, représentant les ruines du temple du Soleil à Balbek, & celles de la ville de Palmyre dans le desert de Syrie. Volland & Desenne, Paris.

Williamson, M., & A. Fitter 1996. The Varying success of invaders. Ecology 77(6): 1661-1666.

Wilson, A.T. 1933. The Suez Canal, its past, present and future. Oxford University Press, London. 224 pp.

Wimpenny, R.S. 1934. The fisheries of Egypt. Science progress 114: 210-227.

Wirszubski, A. 1953. On the biology and biotope of the red mullet, *Mullus barbatus* L., Bulletin Sea Fisheries Research Station, Caesarea 7: 1-20.

Wolfe, L.M. 2002. Why alien invaders succeed: support for the Escape-from-Enemy Hypothesis. The American Naturalist 160(6): 705-711.

Woodward, S.P. 1866. A manual of the recent Mollusca, a treatise on the recent and fossil Shells. 2nd edition. With an appendix of recent and fossil conchological discoveries to the present time, by Ralph Tate. Virtue & Co. xiv + 542 pp. London.

Wu, P., K. Haines, & N. Pinardi 2000. Toward an understanding of deep-water renewal in the eastern Mediterranean. Journal of Physical Oceanography, 30: 443-458.

Wüst, G. von 1934. Salzgehalt und wasserbewegung im Suezkanal. Naturwissenschaften 26: 447-450.

Wüst, G. von 1935. Fortschreitende Salzgehalt-abnahme im Suezkanal. Ann. Hydrogr. U. Maritim. Meteorol. 63: 391.

Wüst, G. von 1951. Über die Abnahme des Salzgehaltes im Suezkanal von 1869-1937. Erdkunde 5: 241.

Yacobi, Y.Z., T. Zohary, N. Kress, A. Hecht, R.D. Robarts, M. Waiser, A.M. Wood, & W.K.W. Li 1995. Chlorophyll distribution throughout the southeastern Mediterranean in relation to the physical structure of the water mass. Journal of Marine Systems 6: 179-190

Yaron, I., T. Schiøtte, & G. Wium-Andersen 1986. A review of molluscan taxa described by P. Forsskål and C. Niebuhr with citation of original descriptions, discussions of type-material available and selection of some lectotypes. Steeenstrupia 12 (10): 157-203.

Yokes, B., & B.S. Galil 2004. New Records of alien decapods from the southwestern coast of Turkey. Rapport du 37 Congrés de la Commission internationale pour l'Exploration scientifique de la mer Méditerranée. CIESM Publishers, Monaco. P. 556.

Zibrowius, H. 1979. Serpulidae (Annelida Polychaeta) de l'océan Indien arrives sur des coques de bateaux à Toulon (France). Rapports et procès-verbaux des réunions, Commission internationale pour l'Exploration scientifique de la mer Méditerranée, 25/26 (4): 133-134.

Zibrowius, H. 1992. Ongoing modification of the Mediterranean marine fauna and flora by the establishment of exotic species. Mésogée 51: 83-107.

Zore-Armanda, M., B. Grbec, M. Morović 1999. Oceanographic properties of the Adriatic Sea – A point of view. Acta Adriatica, 40 (suppl.): 39-:54.

Epilogue

Canals, Invasion Corridors and Introductions

CHAD HEWITT[1], DAN MINCHIN[2], SERGEJ OLENIN[3] & STEPHAN GOLLASCH[4]

[1] *National Centre for Marine and Coastal Conservation, Private Mail Bag 10, Rosebud, Victoria 3939, Australia*
[2] *Marine Organism Investigations, 3 Marina Village, Ballina, Killaloe, Co Clare, Ireland*
[3] *Coastal Research and Planning Institute, Klaipeda University, H. Manto 84, Klaipeda, Lithuania*
[4] *GoConsult, Bahrenfelder Str. 73a, 22765 Hamburg, Germany*

Aquatic species spread *via* canals as a result of unaided, gradual and often unperceived movements through a canal system, with human activities or as combinations of these. It may be tempting to presume that unaided movements of species through a canal system are attributable to natural spread. However, without the presence of a canal, their spread would not have been possible. For this reason, we even consider these unaided movements to result from human activities and they do not represent a natural spread. Indeed, many species may become distributed with passing traffic as well as by their own abilities to disperse.

Canals act in similar fashion to other recognised vectors of aquatic biological invasions: they eliminate biogeographic barriers and create links at different scales between lakes, watersheds and oceans. This in turn provides opportunities for populations of species that have evolved in isolation over long periods of geological time to expand and interact with other populations.

The three grand engineering projects discussed in this book, the Kiel, Panama and Suez canals, each demonstrably acts as a corridor of invasions. These canals created significant economic, social and political stress during development, yet the economic benefits, and safer shipping conditions they provided were considered to justify the costs. When these canals were built however, no thought was given to the impacts arising from species invasions.

Lessons from existing canals

Despite the demonstrated risks of biological invasions associated with the construction and use of canals, numerous projects have been proposed linking previously isolated waterways, lake systems, and biogeographic regions. To facilitate consideration of these proposals and provide clarity to the management decisions, we have summarised the biological invasion lessons from previous canal projects: canals with environmental barriers to species' migration (e.g. salinity, temperature, pH, DO) reduce the likelihood of biological invasions. The corollary of this statement of course is that canals with no discernable environmental difference *at some point along their length, or between the two endpoints* will lead to an unhindered passage of species.

A salinity barrier reduces the likelihood of species migrations through canals. Much of the length of the Panama Canal consists of Gatun Lake, which provides a freshwater barrier between the Caribbean Sea and the Pacific Ocean. This salinity barrier accounts for the relatively small numbers of alien species that have migrated through Panama Canal and established populations at either end. The Suez Canal also had, for nearly a century, a natural salinity barrier in the form of the high salinity Bitter Lakes. The magnitude of Erythrean invaders is directly related to the decline in salinity in the Bitter Lakes, and to the rise in salinity in its northern half since the damming of the Nile River.

For freshwater and brackish biota too, canals connecting watersheds, lake systems and inland seas serve as invasion corridors. The great majority of canals connect rivers and lakes in the Americas, Asia and Europe. In Eastern and Central Europe, a web of canals connects the Baltic Sea, the White Sea and the Ponto-Caspian region (including the Black, Azov and Caspian Seas). These connections have been well established since the end of the 18th century, joining previously isolated brackish water bodies (Slynko et al. 2002, Olenin 2002). The most important European inland invasion corridor is the Ponto-Caspian – Volga – Baltic waterway; more than 300 million tons of cargo per year was been transported with shipping during the turn of the 21st century (Slynko et al. 2002, Panov et al. in prep., Galil et al. in prep.). Species have spread through these waterways either by gradual expansion from the source region, or in a series of sudden and isolated appearances. This "jump dispersal" mode (Olenin

2002) suggests that shipping is an important vector facilitating the spread of alien species within artificial inland waterways.

A similar and more recent suite of expansions in the ranges of alien species has taken place following the creation of the Main-Danube Canal in 1992 linking the Ponto-Caspian region to northwestern Europe (Bij de Vaate et al. 2002). The existance of the "jump dispersal" mode implies that shipping is an important vector facilitating spread of alien species within artificial inland waterway systems.

Future scenarios

Maritime transport is expected to increase as a consequence of globalization, with a concomitant rise in the number and size of vessels. In order to accommodate these changes in the world fleet, as well as changes in market demands, both the Panama and Suez Canals are being enlarged, thereby producing conditions which may enhance their role as invasion corridors. The construction of new docks, new berthing facilities and the modernisation of port facilities may also enhance opportunities for spread (Connell & Glasby 1999, Bacchiocchi & Airoldi 2003, Glasby et al. submitted). Port urbanisation and riverine canalization (e.g. hardening of banks, establishment of wharves) has been shown to alter native biodiversity and create new habitats (Bacchiocchi & Airoldi 2003, Glasby et al. submitted).

The numbers of small craft are likely to continue to expand within inland navigations and coastal areas (Minchin et al. 2006). Some national plans exist to expand inland canals, usually over small distances or by restoring disused routes. The European Union's Water Framework Directive will almost certainly see changes to improve water quality, including canals and this is likely to provide more suitable conditions for invasion.

Global warming is likely to result in shifts in both native and alien species distributions, i.e. change the composition of native communities and allow temperature-limited alien species to expand beyond their present distributions. It has already been considered responsible for expansions of some aquatic invading species in Europe (Clark & Frid 2001, Franke & Gutow 2004), and predicted for the North American Great Lakes (Mandrak 1989). Further, climate warming is expected to result in a contraction of the Arctic ice sheet. This would open summer shipping routes *via* the waters north of Russia and Canada to enter the Pacific Ocean through the Bering Strait in the near future. This would provide the first cold water shipping route between the Atlantic and

Pacific oceans with likely movements of temperate and boreal species between these oceans; whereas before such species, carried with ships, would have been exposed to tropical and/or semi-tropical sea temperatures and have expired. These northern routes may also have consequences for trade through the Panama and Suez canals with fewer vessels passing during the northern summer months (Minchin 2006). Any changes to climate have consequences for invasive species spread and canals are a mode of distributing them.

Prevention and management measures

Canals are just one of the known vectors of marine alien species. Whilst many alien species cause little discernable impact, our experience suggests that any alien species can cause negative environmental, economical, and societal impacts (e.g. Carlton 2001, Hewitt 2003). Management of marine alien species spread is a global responsibility; however, national governments need to lead such management approaches.

Fig. 1. Electric fish barrier in the Chicago Sanitary and Ship Canal near Lemont, Illinois, USA. Photo by Sergej Olenin.

Some technologies may reduce the movement of biota through canals, including electric barriers and air-bubble curtains. Electric barriers, such as micro-pulsed DC electric barriers, are used in some freshwater systems (Fig. 1), but are unlikely to be of value in brackish or marine canals. Air curtains are used, with or without sound emissions, to reduce the entrainment of fish at some industrial water intakes (McKinley et al. 1989, Moy 2000), and larger scale systems may prove suitable for canals. However, these structures may impede the movement of native as well as alien species, with consequent impacts on commercial and recreational fisheries as well as other leisure activities, and they are unlikely to be effective against species adhering to hull surfaces. The most efficient method may be the insertion of an environmental barrier such as a salinity lock to mimic the most successful obstacle to aquatic translocation in canal systems known to us – the freshwater Lake Gatun in the Panama Canal! In an era of heightened environmental concern, engineering solutions will be found eventually, if awareness of the problems caused are recognized.

Several international and regional agreements oblige signatory nations to limit, minimize and/or eliminate harm to native biodiversity resulting from anthropogenically aided dispersal of non-native species. Such issues are often of concern to neighbouring states. The United Nation Convention on the Law of the Seas (UNCLOS 1982) provides a general obligation to Parties in Article 196 stating that Parties should take measures "to prevent, reduce and control pollution of the marine environment resulting from... the intentional or accidental introduction of species alien or new, to a particular part of the marine environment, which may cause significant and harmful changes thereto."

The Convention on Biological Diversity (CBD 1992) is one of the few instruments that explicitly addresses the obligations on Parties to manage biological invasions (Meliane & Hewitt 2005, Doelle et al. in press, Hewitt et al. in press). Currently, 179 nations are party to the Convention. This Convention includes three specific Articles concerning biological invasions or their consequences that apply to the consequences of creating and maintaining a canal, which require signatory nations to take action to (as paraphrased below):

- ensure that activities within their jurisdiction or control do not cause damage to the environment of other States or of areas beyond the limits of national jurisdiction (Article 3);
- prevent the introduction of, control or eradicate those alien species which threaten ecosystems, habitats or species (Article 8(h)); and,
- ensure that the environmental consequences of their programmes and policies that are likely to have significant adverse impacts on biological diversity are duly taken into account (Article 14.1).

Governments should adopt and implement a biosecurity approach that recognises the significant values that we put at risk. This approach should adopt similar procedures and protocols to quarantine activities such as those associated with animal, plant and food safety (Office International des Épizooties, IPPC 1951, Codex Alimentarius). However, when implementing internationally consistent regulations they should not be used as a legal barrier to, or an argument against, other jusrisdictions taking stronger action to protect their waters.

Acknowledgements

Dan Minchin and Sergej Olenin were provided with support for this work under the EU 6th Framework Programme project ALARM (GOCE-CT 2003 506675).

4. References

Bacchiocchi, F., & Airoldi, L. 2003. Distribution and dynamics of epibiota on hard structures for coastal protection. *Estuarine, Coastal & Shelf Science* 56:1157-1166.

Bij de Vaate, A., Jazdzewski, K., Ketelaars, H.A.M., Gollasch, S. & Van der Velde, G. 2002. Geographical patterns in range extension of Ponto-Caspian macroinvertebrate species in Europe. *Canadian Journal of Fisheries and Aquatic Sciences* 59:1159-1174.

Carlton, J.T. 2001. *Introduced Species in U.S. Coastal Waters: Environmental Impacts and Management Priorities*. Pew Oceans Commission, Virginia, U.S.A. pp 28.

Clark, R.A., & Frid, C.L.J. 2001. Long-term changes in the North Sea ecosystem. Environm. Res. 9: 131-187.

CBD. 1992. Convention on Biological Diversity, Rio de Janeiro, Brazil. www.cbd.org, accessed 30 Jan. 2006.

Connell, S.D., & Glasby, T.M. 1999. Do urban structures influence local abundance and diversity of subtidal epibiota? A case study from Sydney Harbour, Australia. *Marine Environmental Research* 47: 373-387

Franke, H.-D. & Gutow, L. 2004. Long-term changes in the macrozoobenthos around the rocky island of Helgoland (German Bight, North Sea). Helgol. Mar. Res. 58: 303-310.

Hewitt, C.L. 2003. The diversity of likely impacts of introduced marine species in Australian waters. *Records of the South Australian Museum Monographs Series* 7: 3-10.

IPPC. 1951. International Plant Protection Convention. Food and Agriculture Organization, Rome, Italy. www.fao.org/Legal/treaties/004s-e.htm, accessed 30 January 2006.

Mandrak, N.E. 1989. Potential invasion of the Great Lakes by fish associated with climate warming. Journal of the Great Lakes Research 15(2): 306-316.

McKinley, R.S., Patrick, P.H., Matousek, J.A. & Wells, A.W. 1989. Field testing of behavioural barriers for fish exclusion at cooling-water intake systems: Ontario Hydro Pickering Nuclear Generating Station: Final report. Technical Report EPRI-GS-6246, 130 pp. www.osti.gov/energycitations/product.biblio.jsp?osti_id=6240492, visited February 15th 2005

Meliane, I., & Hewitt, C.L. 2005. Gaps and priorities in addressing Marine Invasive Species. IUCN Information Document, The World Conservation Union, Gland, Switzerland. www.iucn.org/themes/marine/pubs/pubs.htm, accessed 20 December 2005.

Minchin D. 2006. The transport and the spread of living aquatic species. In: Davenport J & Davenport JL (eds) The Ecology of Transportation: Managing Mobility for the Environment. Kluwer Acad. Publishers.

Moy, P.B. 2000. Development of an aquatic nuisance species barrier in a commercial waterway. In: Perderson, J. (ed.) *First National Conference on Bioinvasions, USA*, Massachusetts Institute of Technology (MIT), MIT Sea Grant Program, Center for Coastal Resources, Cambridge, USA, 24.-27. Januar, 1999. Conference proceedings, 357-362.

Olenin, S. 2002. Black Sea - Baltic Sea invasion corridors. In: *Alien marine organisms introduced by ships in the Mediterranean and Black Seas*. CIESM Workshops Monograph. F Briand (ed.) Comission Internationale pour l'Exploration Scientifique de la mer Mediterranee, Monaco: 29-33

Slynko, Yu., Korneva, L., Rivier, I., Papchenkov, V., Scherbina, G., Orlova, M. & Therriault, T. 2002. The Caspian-Volga-Baltic Invasion Corridor. In: Invasive aquatic species of Europe - distribution, impact and management. Leppäkoski, E., Gollasch, S. & Olenin, S. (eds). Dordrecht, Boston, London. Kluwer Academic Publishers: 399-411.

UNCLOS. 1982. United Nations Convention on the Law of the Sea. United Nations General Assembly, Montego Bay, Jamaica. www.un.org/Depts/los/index.htm, accessed 30 January 2006.

Index

MONOGRAPHIAE BIOLOGICAE

Series Editor: **Prof. Henri J. Dumont**
State University of Ghent, Institute of Animal Ecology, Ghent, Belgium,
henri.dumont@rug.ac.be

Published titles within the series:

Volume 83 **Bridging Divides - Maritime Canals as Invasion Corridors**
Gollasch, Stephan, Galil, Bella S., and Andrew N. Cohen
2006, 315 p., Hardcover, ISBN: 1-4020-5046-1

Volume 82 **Biogeography and Ecology of Bulgaria**
Fet, V.
Forthcoming, Hardcover, ISBN: 1-4020-4417-8

Volume 81 **Monograph of the Spathidiidae**
Foissner, W., and Xu, K.
2006, 486 p., Hardcover, ISBN: 1-4020-4210-8

Volume 80 **Nouragues**
Dynamics and Plant-Animal Interactions in a Neotropical Rainforest
Bongers, F.; Charles-Dominique, P.; Forget, P.-M.; Théry, M. (Eds.)
2002, 456 p., Hardcover, ISBN: 1-4020-0123-1

Volume 79 **Environmental Change and Response in East African Lakes**
Lehman, J.T. (Ed.),
1998, 250 p., Hardcover, ISBN: 0-7923-5118-5

Volume 78 **Monograph of the Oxytrichidae (Ciliophora, Hypotrichia)**
Berger, H.
1999, 1092 p., Hardcover, ISBN: 0-7923-5795-7

Volume 77 **The Pantanal of Poconé**
Biota and Ecology in the Northern Section of the World's Largest Pristine Wetland
Heckman, C.W.
1998, 592 p., Hardcover, ISBN: 0-7923-4863-X

Volume 76 **An Introduction to Saline Lakes on the Qinghai-Tibet Plateau**
Zheng Mianping
1997, 328 p., Hardcover, ISBN: 0-7923-4098-1

Volume 75 **Rocas Alijos**
Schmieder, Robert W.
1996, 520 p., Hardcover, ISBN: 0-7923-4056-6

For more information on our ongoing series, please visit
www.springer.com